W9-CCP-435

The best of ROY PARKER JR.

Reliving Fayetteville's storied military history

including Fort Bragg and Pope Air Force Base

as published in

The Fayetteville Observer

Copyright© 2007 • ISBN: 1-59725-088-0

Published by Pediment Publishing, a division of The Pediment Group, Inc. www.pediment.com Printed in Canada

Contents

CHAPTER FOUR:
World War II

CHAPTER FIVE:
Korea & Vietnam

CHAPTER SIX:
Modern Military

Foreword

Roy Parker Jr. was born Feb. 12, 1930, in Ahoskie, N.C., and was involved in his family's newspaper business from boyhood. He graduated from the University of North Carolina with a journalism degree in 1952 and has worked for his family's weekly newspapers as well as The News & Observer in Raleigh, The Fayetteville Times and The Fayetteville Observer. He was the founding editor of the Times and retired as a contributing editor of the Observer.

A member of the UNC Journalism Hall of Fame, Roy has received many awards throughout his career. Most notably, Roy received The North Carolina Award, the state's highest civilian honor, in 2006. The award is bestowed for contributions to the arts, public service and science.

In addition to this book, Roy is also the author of "Cumberland County: A Brief History," an official publication of the state Division of Archives and History.

Roy's column on military history begin in the Observer in January 1993, and he has since written more than 700 of these columns. We've included 135 of Roy's best in this book, covering the region's rich history from the 18th century and the dawn of the American Revolution to the 21st century and operations in Iraq and Afghanistan.

In these columns, Roy recounts events from the perspective of generals and fresh-faced young soldiers, often using their own words as preserved in letters and other historical documents.

Although the names of the battles and wars have changed over the years, the story remains the same. Fayetteville and the Cape Fear region have been home to fighting men (and now women) who have played key roles in every major military action in our country's history. These are but a few of their stories.

CHAPTER ONE

American Revolution

Planter Played Both Sides Of The Fence

Originally published September 14, 1995

Planter-politician Farquhard Campbell (circa 1730-1808) changed sides so many times during the American Revolution, it was said he "had several flags in his bag."

But in North Carolina's notorious "Regulator War" of 1768-1771, Campbell was firmly on the side of the royal governor, William Tryon.

His loyalty led him to recruit and command one of the first military outfits to take the field from Cumberland County.

Apparently, the Cumberland Detachment did not arrive in time to be on the field at the Battle of Alamance on May 17, 1771, when a force of 1,400 militia commanded by Tryon routed 2,000 enraged farmers who had taken the law into their own hands protesting high taxes and fees, arrogant and corrupt courts and the building of Tryon's Palace in New Bern.

Campbell may have been there, however, serving as an "assistant adjutant general" to Gov. Tryon, an official of the governor's headquarters unit.

Campbell had no previous military experience, but he quickly answered the call when Tryon set quotas of volunteer militia companies for an army to suppress the Regulators.

His zeal for one side in 1771 stood in sharp contrast to Campbell's alleged behavior a few years later in the Revolutionary War.

A longtime county official and legislator in the Colonial government, he played both sides of the fence for several months, at times mystifying the Loyalists and Patriots as to his true allegiance.

During that time, the Scottish-born Campbell gave Royal Gov. James Martin early reason to count him an ally.

But a few weeks later, Martin complained that Campbell was "an ignorant man who has settled from his Childhood in this country, and has imbibed all the American popular principles and prejudices."

On the other hand, Patriots accused Campbell of backing the Loyalists before their defeat at the Battle of Moore's Creek Bridge in February 1776.

They arrested him and even sent him as a paroled prisoner all the way to Philadelphia.

Before independence was won, however, he had regained property and much of his reputation. He went on to be a postwar legislator in the state assembly.

In April 1771, Campbell had no qualms. As soon as Tryon called for militia, he pledged to raise a company of 50 men "to March against the insurgents."

"Council of war"

Campbell's appetite for military service may have been whetted as far back as 1768, when he and several other nearby members of the Colonial assembly attended a "Council of War" that brought two dozen of North Carolina's top militia officers together at Hillsborough to discuss strategy against the rambunctious Regulators.

In 1771, with Tryon now determined to put the Regulators down by force, he quickly accepted Campbell's service.

A captain's commission was forwarded to Campbell, along with two blank commissions for other officers. He was authorized 150 pounds from the Colonial treasury for pay and equipment. Each recruit was issued a bounty of 40 shillings.

There is no record of how large a unit the Cumberland Detachment was, although it probably numbered something more than 30 members.

A daily report of one such company, which may have been the Cumberland Detachment, was more elaborate and gave a good idea of a typical 18th-century military organization.

It listed four officers, two sergeants, one musician (no doubt a fifer or bagpiper), 54 rank-and-file, one servant, one wagoneer, 12 horses and one wagon.

The names of some of the rank-and-file are known.

According to Colonial records, the following men were among 17 enrolled for Campbell's command by officials at the county's largest village, Cross Creek, now Fayetteville: Neil Buie, Daniel Campbell, Dugald Campbell, James Gray, Andrew Ingram, Alexander McDonald, Thomas McDonald, Augustine McDuffie, Neill McGachy, Archibald McMullin and Thomas Nuten.

As the names indicate and as Tryon later noted, most of the Cumberland volunteers were "highlanders," recent immigrants from Scotland who at that time were flooding into the Cape Fear River valley.

These newly arrived Scotchmen were intensely loyal to the British crown, which at the time was encouraging their immigration with promises of land bounties in the colonies.

When the Revolution came four years later, they would be the core of Loyalist opposition to the Patriot cause.

Militia gathers

In early May 1771, the militia began gathering in the Orange-Alamance area.

At the time, Gov. Tryon's records list a "Lt. Campbel" as an assistant adjutant general on his staff. Some sources think this refers to Farquhard Campbell.

If that is true, Campbell could have been on the field on May 16, 1771, when Tryon's army broke up the Regulator force.

Whether or not "Lt. Campbel" was Farquhard, there is no doubt that the Cumberland Detachment joined Tryon's army five days after the Battle of Alamance, on May 21, 1771, at its camp near the site of the fight.

Gov. Tryon's official day-by-day war diary reported:

"The company from Cumberland County arrived. The Cumberland Detachment, mostly highlanders, were formed into a Corps of Light Infantry of the Line, under command of Capt Farquar Campbell who is to receive orders from C-in-C."

For another several days, at least, this unique "special force" attached to Tryon's immediate entourage marched and camped with the militia army as it roamed through

Regulator territory, breaking up gatherings, questioning suspected Regulators and generally mopping up.

The day after it arrived, the Cumberland unit teamed with a detachment from Wake County and "the Light Horse" company for a march to a nearby gristmill, with orders to "escort flour" back to the main body.

Records of payments

There is no record of when the Cumberland unit completed its military service.

But after every Colonial military campaign, there was always a lot of paperwork, mostly the submission and payment of bills for pay, equipment, forage, rum, rations and incidentals.

Among the latter was the five pounds paid to Neil Buie (referred to as "young Bewes") of the Cumberland Detachment for "cornfield destroyed by Horses."

The final account of Capt. Farquhard Campbell listed total payment not of the originally agreed-upon 150 pounds, but rather of 270 pounds.

Local Patriots Wrote, Signed Defiant Document

Originally published July 13, 1995

The spring and summer of 1775 were fateful days in American history.

Every schoolchild knows what happened 220 years ago:

The famous ride of Paul Revere, and the clashes at Lexington Green and Concord Bridge where Patriot minutemen fired "the shot heard 'round the world" at the British Redcoats. It was the beginning of the War of Independence.

Within a few weeks of the fighting on April 19, 1775, the news of what had happened was known in every colony. It reached North Carolina in early May.

By June, Patriots were stirring, vowing support for their Massachusetts brethren, promising even to go to war against Great Britain.

The village of Cross Creek, today's Fayetteville, was part of that warlike stirring.

There was no shootout on the green or at the bridge.

But there was the "Cumberland Association," a defiant document signed in late June 1775 by 55 Cumberland County men.

The document declared that the colony would be "thoroughly justified in resisting force by force."

They vowed to "associate as a band in her defense against every foe," and promised that if the call came they would "go forth and be ready to sacrifice our lives and fortunes to secure her freedom and safety."

Many of those who gathered in a village tavern on that June day 220 years ago made good on their promise.

Several of Cumberland County's most significant Revolutionary War military leaders would emerge from the group.

The most notable of all was Robert Rowan (circa 1738-1798), who drew up the document (copying a form in general circulation) and was the first to sign it.

Rowan was already a leading citizen, a colonial sheriff and legislator. He was also named colonel of the county militia but gave up any idea of being a military commander in deference to others he felt were more qualified.

Workhorse of war

Nonetheless, Rowan was the workhorse of the war effort in Cumberland County.

Throughout the Revolutionary War years from 1775 to 1782, he served in a series of commissions to procure supplies for Continental regiments and for the militia.

With Rowan as the central figure, the settlement at Cross Creek became a major logistical center for equipping and feeding Patriot troops.

Among Rowan's associates in the supply business were James Gee, Lewis Barge and Peter Mallett, all signers of the Association.

As late as 1790, Rowan was presenting claims for his services as "superintending commissioner" for the Wilmington (militia) District.

When he died in 1798, Rowan was already counted as one of the county's most indefatigable "Sons of Liberty," an organization he apparently joined as early as 1770.

The honors continue to this day. The local chapter of the Daughters of the American Revolution is named for him.

Among those who signed the defiant document and worked in the supply capacity was tavern keeper George Fletcher, a militia captain, whose enthusiasm was such that the North Carolina assembly told him "it gives us hope you will be able to do something clever."

Fletcher was among five Cross Creek tavern keepers who signed the Association document. Others were Lewis Bowel, Aaron Verdie and Martin Leonard.

"Liberty Point"

Long tradition holds that the Patriots who signed the document gathered in a tavern at what is now the intersection of Person and Bow streets in Fayetteville. Since early in the 19th century, the spot has been called "Liberty Point."

Fletcher was probably the host taverner.

Only one of the 55 signers is known to have died in the uniform of a Continental soldier.

Young Arthur Council died in 1777 while serving as a lieutenant in the 6th North Carolina Regiment.

Young Joshua Hadley, son of wartime Sheriff Thomas Hadley, signed the Association and became a lieutenant and then captain in the North Carolina Continental Line.

Among others who played important military roles were James Emmet and Walter Murray.

Emmet was colonel of the Cumberland militia in the last two years of the war when forces loyal to Great Britain, "Loyalists" to the British, "Tories" to the Patriots, often had Patriots on the run.

He doggedly defended Cross Creek from bands of Tories and kept up a steady drumfire of alarming reports to state military authorities.

In the summer of 1781, several other Patriot leaders, including Robert Rowan, were briefly held prisoner when a Tory band seized the village for a day. An embarrassed Emmet had to report the incident to the state authorities.

After the war, Emmet was himself a tavern keeper on Green Street in Fayetteville.

Walter Murray, who held a succession of public offices before and after the Revolution, was "first major" under Emmet during the tough times of 1780-1781.

The Evans-Carver family, big planters on upper Rockfish Creek, supplied no less than seven signers of the 1775 document.

David Evans (1753-1812) was briefly commander of a militia company raised in late 1776.

William Carver (1754-1836), who signed the Association with his brothers Robert and Samuel, was the last of the 55 signers to pass away. His obituary described him as "the last of the gallant band, who in the face of discouragements of a numerous and implacable Tory foe, united in the Association known as 'the Cumberland Association.' "

Like his kinsmen, Carver was in the Revolutionary militia, "at different times for several years a volunteer in the service of his country."

Sunshine Patriots

In the roll call of the signers at Liberty Point, at least two turned out to have been sunshine Patriots. That is, they switched their allegiance.

Aaron Verdie was almost surely the same man captured less than a year later as a Loyalist at the Battle of Moore's Creek Bridge in February 1776. The sometime Cross Creek taverner was listed as a "Waggon Master" for the Loyalist forces smashed in that battle.

So, too, was the man who stepped up first after Robert Rowan to sign the Association. Maurice Nowlan was apparently "a captain of 20 men who fought for the Loyalists at Moore's Creek."

The stories of many of the signers of the 220-year-old document that launched the Revolution in Cumberland County have yet to be uncovered.

If you want to see their names, they are carved in a memorial stone located in a small park at the Person-Bow intersection.

Cumberland Had Key Role In Revolutionary Battle

Originally published February 25, 1993

The last week in February in the Cape Fear River region is always associated in military history with the Battle of Moore's Creek Bridge, an early clash of arms in the American War of Independence.

It happened on Feb. 27, 1776, at a little bridge spanning a black water stream named for the Widow Moore, whose plantation home was nearby.

The early morning shootout at the isolated spot 15 miles north of Wilmington pitted Americans against Americans as Patriots battled Scottish Highlander immigrants only recently arrived in North America. The latter were known also as "Loyalists" because of their loyalty to the British crown. In derision, Loyalists called Patriots "Whigs" and Patriots called Loyalists "Tories."

In the predawn hours, while a fog still hung over the low marshes, a column of Highlanders moved into position along the banks of the stream near the bridge. Across the slow-moving water, more than a thousand Patriot soldiers, mostly militiamen, waited.

As the sun rose, the Loyalists stormed the bridge, only to meet disaster.

The Patriot forces had removed the planking and greased the runners of the Widow Moore's bridge.

As Loyalists slipped and slid along the timbers or attempted to wade the river, Patriot riflemen and a pair of small cannon (named "Mother Covington and Her Daughter") took a heavy toll. In a few minutes, the battle was over.

More than 30 Loyalists were dead and dozens were wounded.

The others raced away, some riding "three upon a horse," according to a battle report. Thomas Rutherford, a well-to-do Cumberland County planter who had joined the Loyalists, "ran like a lusty fellow," but was not able to escape. The bag of prisoners included most of the officers who had spent several months recruiting the incipient "Royal Emigrant Regiment" and who were leading it toward Wilmington in hopes of rescuing the royal governor of North Carolina, Josiah Martin, from his enforced exile on a ship off that port town.

In the shootout at the bridge, only a single Patriot militiaman was killed. The Patriot commanders, Cols. James Moore, Alexander Lillington and Richard Caswell, were hailed as Revolutionary heroes. British hopes of rallying Loyalist forces all over the Carolinas were dashed for years and the victory spurred sentiment for independence from Great Britain. On April 5, 1776, the North Carolina Provincial Congress formally instructed delegates to the Continental Congress "to concur ... in declaring independency."

The village of Cross Creek, later named Fayetteville, was the scene of key events leading up to the battle at the Widow Moore's creek.

As early as Christmas 1775, small groups of Loyalists from the far reaches of Cumberland County and from present-day Moore and Harnett counties drifted into the village, checking on reports that Loyalist officers had put out a call for them to rally to the royal standard there.

When that call did go out in early February, Loyalists, mostly Scotch Highlanders who had settled in the Cape Fear area only in the past three or four years, responded in good numbers.

By mid-February, more than 1,600 Loyalists were said to be gathered in the vicinity of Cross Creek. They had been cobbled together as an "army" commanded by Gen. Donald McDonald, a professional soldier who had actually fought with the Highlander army at the fateful Battle of Culledon Moor in 1746 when the British had suppressed a Highlander rebellion. They were answering the call of the crown to whom they had sworn an oath of allegiance, and who had earlier encouraged them to take up lands in the royal colonies.

Since their orders were to advance to Wilmington, the Loyalist forces camped south of the Cross Creek village on the road along the west bank of the Cape Fear River that led toward the port town 100 miles away.

Tradition places the main camp at about where present-day West Mountain Drive connects with N.C. 87. That was in the vicinity of the big riverside plantation and landing known as Springhill. Other camps were closer to the village, on the ridgeline west of town that would later be known as Haymount.

Meanwhile, Patriot forces had also gathered, determined to block the Loyalist move.

Col. Moore commanded the 1st North Carolina Continental Regiment and was thus in overall command. He was joined by contingents of militiamen. Col. Lillington headed a force of militia from Wilmington, and Col. Caswell the New Bern militia. Lt. Col. Ebenezer Folsom of Cumberland County directed a screening force of mounted militia. Other militia units were marching for Cross Creek from the west.

The Patriot force arrived at Rockfish Creek, two or three miles south of the Loyalist camp, at about the same time that McDonald finished gathering his force.

For two days beginning Feb. 18, the leaders of the two forces parleyed and negotiated.

Gen. McDonald tried to bluff Col. Moore out of his path.

Instead, Moore set in motion several columns that effectively blocked all the land routes to Wilmington. Hard marching by the Patriots enabled them to get a day's march on the Loyalists, time to mount the defenses at the Widow Moore's Creek bridge.

In the days following the defeat at Moore's Creek, many of the Loyalists who escaped capture found their way back to Cross Creek, the seat of the county where so many lived and where Gen. McDonald had made his camp.

It could be that some of them were searching for something they knew to be there — a chest containing gold pieces worth 15,000 English pounds, cash provided by Gov. Martin for expenses of the Highlanders, which Gen. McDonald had sequestered there.

Instead, well-placed tradition says the money was recovered by Patriot forces who were told by a black man that he had seen the Highlanders conceal the chest in a local barn.

Another tradition with less authenticity, but which has become firmly imbedded in North Carolina folklore, is that the noted Highlander heroine, Flora MacDonald, who had immigrated to North Carolina in 1774, was on hand at the Cross Creek camp when the Loyalist force marched for Wilmington and even made a speech exhorting them to loyalty to the crown.

Allan MacDonald, Flora's husband, was indeed an officer in the Loyalist force and was captured at Moore's Creek. However, there is no mention of Flora MacDonald in contemporary records. Her brief but enduring moment of fame had come 30 years earlier, in 1746. Then, as a teenage Scottish lass, she had helped Charles Stuart, the "Young Pretender" to the British crown, elude British forces after his Highlander army had been defeated at Culledon Moor.

Many of the Loyalists who made their way back to Cumberland County after Moore's Creek or were released as prisoners later in the year did eschew their old oath and take an oath of loyalty to the independent state of North Carolina.

Many others, however, never joined the Patriot cause. Some immigrated to Canada. Others, including Flora and Allan MacDonald, returned to Scotland.

Meanwhile, as the War for Independence continued, Cross Creek was generally controlled by Patriot officials. In some cases, the property of absent Loyalists was confiscated. The settlement became a supply point for Revolutionary commissary provisions, with several merchants engaged in buying food, clothing and salt for Continental troops.

After Moore's Creek, often glorified as "the Lexington and Concord of the South," there would be little military activity in the area for four years.

Patriots Take The Spoils Of Moore's Creek Triumph

Originally published May 8, 1997

The Battle of Moore's Creek was over. The little Loyalist army that had marched from the Cape Fear River village of Cross Creek toward Wilmington was beaten and scattered after the battle on Feb. 27, 1776.

Patriots were triumphant in one of the first military encounters of the American Revolution in the South.

As spring of 1776 came to Cumberland County, captured leaders of the defeated Loyalist band were marched off to the Patriot capital at the tiny Roanoke River town of

Halifax.

More than 800 "common soldiers" of the band, many of them recently arrived Scotch Highlander farmers who lived in the county, "were made prisoners, disarmed and discharged."

Now came the work of cleaning up after a battle and then preparing Cumberland County for the next round of a fight that looked more and more like a War of Independence from the British mother country.

First, there were the prisoners and the loot.

A report from Halifax jail listed the top officers of the ill-fated Loyalist army:

"Kingsborough McDonald, Mr. Rutherford, Hector McNeil, and Alexander McDonald, Captains Morrison, McKenzie, Ure, Legate, Cross, Parsons, McCoy, Mase, Mickeson, McCarter, and Adjutant Frazer; Lieutenants McIver, and Hewes, Cameron, Donald Hewes, Donald Cameron, and Sundry other Lieutenants and Ensigns. Kenneth McDonald, Aide-de-Camp; James Hepburn, Secretary; Parson Beatty, and Dr. Morrison, Commissary."

The report concluded: "General Donald McDonald and Brigadier-General McLeod (the latter of whom was killed) set out at the head of this banditti with the avowed intention of carrying (Royal) Governor Martin into the interior of the Province."

Two months later, 26 Loyalist prisoners were listed in Halifax jail, "destined for Philadelphia."

The report included the first names of the men listed above as well as others.

Halifax prisoners who gave their address as the village of Cross Creek included Neill MacArthur, "captain of 55," and Maurice Nowlan, "capt. of 20," as well as Aaron Verdie, "waggon-master to MacDonald," and "Hepburn, secretary to General McDonald."

Then there was merchant George Milne, who had been "entrusted with gunpowder which he gave to McDonald."

And Malcom McNeill, who before the march to Moore's Creek had "encouraged Daniel Treadaway to reveal secrets to Capt. (Ebenezer) Folsom, (and) afterward bore the (Loyalist) colors afterward erected as a standard at Cross Creek."

Many of these men would never see Cumberland County again. But others would return, either on parole during the war or after the war.

A lot of loot

Then the loot. It was considerable, a veritable war chest now available for the Patriot cause. Two weeks after the rout at Moore's Creek, a report listed it:

"The Conquerors have already taken 350 guns and shot-bags, about 150 swords and dirks; 1,500 excellent rifles; two medicine chests fresh from England, one of them valued at 300 pounds sterling, a box containing half Joaneses and Guineas (gold coins), secreted in a stable at Cross Creek, discovered by a negro and reported to be worth 15,000 pounds sterling; also 13 wagons with complete set of horses."

To get some control over what the Patriots described as "Tory property," the Revolutionary assembly of North Carolina named three Cumberland men to inventory the goods and especially the lands of Loyalists who were either in Halifax jail or who had fled elsewhere.

Named to the task were Cross Creek merchant Peter Mallett, planter William

Rand of the upper part of the county (near present-day Lillington in Harnett County), and Robert Cobb.

County militia leaders

In April, the assembly, known as the Provincial Congress, put the Cumberland military house in order by naming officers of the county militia regiment and picking "arms procurers" officials.

The militia command consisted of much-respected Alexander McAllister as colonel and fiery Ebenezer Folsom as lieutenant colonel.

A few weeks later the older McAllister gave up the post. Folsom, who had raised and commanded a busy cavalry troop during the Moore's Creek campaign, was made colonel. David Smith was lieutenant colonel; Phillip Alston, first major, and John Armstrong, second major.

Folsom and John Blocker were at the same time named arms procurers. Alexander Hostler was given a special assignment to procure military supplies for a company of Anson County militia stationed at Deep Inlet near the mouth of the Cape Fear.

The militia members were not the only Cumberland military men.

Capt. Joshua Gist of Cumberland marched off with a small company to join a state regiment of the newly forming Continental Line.

The county had been given a quota of 63 men for the unit.

Headquarters

Then in early May 1776, Cumberland became a military headquarters.

Worried that Loyalists might try again to get to Wilmington or that British troops might land there, the Provincial Congress called Folsom to active duty and authorized him to raise "100 light cavalry and 200 infantry" to be stationed in the village, with orders to scout widely for trouble as far west as the Pee Dee River.

Folsom set up his command, with Henry Giffard as his commissary officer, or logistical chief.

Salt and bills

Then there was the matter of salt, a commodity that would occupy much attention in Cross Creek for all the years of the Revolution.

Merchants Robert Rowan and Peter Mallett were named to disburse the remainder of nearly 2 tons of salt gathered at various stores in the village. Patriot Committees of Safety in counties surrounding Cumberland were eligible to dip into the supply, for a price.

Meanwhile, the Provincial Congress paid some logistical bills for the Moore's Creek campaign.

Alexander McCortle of Rowan County was allowed 19 pounds, tenpence for his "waggon, team, and driver" hired for the "late expedition to Cross Creek against the tories."

George Davidson got 39 pounds for his wagon for the "expedition against the Highlanders" and William Knox recieved 32 pounds, fivepence for "waggon hire, Rowan County to Cross Creek."

The campaign had caused problems for Cumberland County's government, however.

For one thing, somebody had stolen the government records, the minute book of the county Court of Pleas and Quarter Sessions.

Patriots blamed Loyalists. Col. Folsom was informed that the records were "being concealed by disaffected persons," and he was authorized to search them out and turn them over to William Rand, the acting clerk of the court.

Apparently the search was not immediately successful. Most of Cumberland's records for 1776 have never been found.

For another, several of the county's colonial leaders were among the Loyalists. That included even some members of the county Committee of Safety, the networking organization formed in 1775 to further the Patriot cause.

On June 21, 1776, the Provincial Congress ordered a new committee to be elected for Cumberland, in view of the fact that many "original members, both in the county and in Campbellton, are removed out of the province and some remaining decline to act."

Cumberland County was going on a war footing, and Patriots were firmly in control.

Moore's Creek Victory Puts Patriots In Charge

Originally published October 31, 2002

Patriots took quick military control of North Carolina in the early months of the American Revolution by pulling together a scratch force of militiamen good enough to whip a Loyalist "army" at the Battle of Moore's Creek Bridge in February 1776.

It took another year to sort out the new military establishment of the independent state, often having to find replacements for Loyalists who held commands in the prewar colonial militia.

In Cumberland County, where Loyalists were as much as a third of the population, the sorting out was severe.

But in 1777, when Cumberland's new organization was complete, Patriots were firmly in the saddle.

The prewar commander of county militia, Col. Thomas Rutherford, a well-to-do planter, actually marched with the Loyalists to Moore's Creek Bridge.

When the Patriots routed his command, a Patriot account reported that Rutherford "ran like a lusty fellow."

He was not able to escape, however, and died four years later on board a British warship as a recently freed prisoner of war.

On the Patriot side, men not earlier identified as military figures rose to the occasion.

Robert Rowan, the county's leading Patriot, recruited a company from the vicinity of the village of Cross Creek and would occasionally call it to arms when rumors arose of British invasions or Loyalist uprisings.

While he apparently never took the field as a commander, Rowan ranked as a colonel in the designation of his wartime friend, Gov. Richard Caswell. Throughout the war, Rowan was an important military supply officer for the state militia and the Continental Army.

Ebenezer Folsom, a planter who lived near present-day Lillington, then part of Cumberland County, took it on himself to recruit a company of horsemen in time for the Moore's Creek campaign.

His command was credited with blocking wagon roads leading from Cross Creek, capturing Loyalist supplies and arms, and rounding up many fleeing Loyalists.

In April of 1776, Folsom became colonel of the county militia, named by the Patriot state government to replace Rutherford. He was also given a cash award of 100 pounds "for the vigorous manner" of his service in the campaign.

Folsom goes down in local history for presiding over what must have been Cross Creek's first formal celebration of the Fourth of July, albeit a little late.

In August, the state government commanded the new militia colonel to call a meeting in the village and read the Declaration of Independence of July 4, 1776, to the assembled company.

Folsom's career as head of the county military establishment ended in late 1776 after complaints that a volunteer company he commanded during the summer didn't perform well. It was also alleged that its colonel doctored financial accounts and even skimmed the rum ration.

Later in the war, Folsom again took the field and served well against the Loyalist uprisings of 1780 and 1781.

By the summer of 1777, the Provincial Congress, the Patriot government of the state, passed a new Militia Act that at least on paper organized North Carolina into six districts, each commanded by a brigadier general, with each county assigned one regiment.

Expanding on a similar structure in colonial days, a county was generally divided into geographical "captains' districts."

In October 225 years ago, Cumberland County was able to report its new structure, comprised of 12 districts.

The list of districts, the captains, and the tax officers (the districts also served as local government tax jurisdictions) appeared in the minutes of the Cumberland County Court of Pleas and Quarter Sessions, the county governing body. The list is the earliest record of the county's military establishment after the Declaration of Independence.

The dozen captains were all staunch Patriots, and several would see active service in the war.

Jacob Duckworth, for instance, would go on active duty in the summer of 1781 when he assembled a band of Patriot horsemen from his district along the Deep River in today's upper Moore County.

He would make local military history by getting the best of the notorious and much feared Loyalist commander, Col. David Fanning.

Fanning tells the story in his Journal:

"The day Lord Cornwallis defeated General Greene at Guilford (the Battle of Guilford Courthouse), I was surprised by a Capt. Duck with a company of Rebels where I sustained the loss of all our horses and arms. We had one Man killed on each side. The day following, myself and three more of the Company furnished ourselves with arms and pursued the Rebels, who we discovered had parted and gone to their Respective homes with their plunder. We visited one of the Houses and found 14 horses."

Not all the 1777 captains were newcomers to the roster of Cumberland militia officers.

John Armstrong, whose district was along the Lower Little River, was a captain as far back as 1770 in the colonial militia list of that year.

Thomas Dobbin, from the "Barbecue District" in present-day Harnett County, was a lieutenant in the 1770 list.

Hugh Gilmore was a 1770 lieutenant in the district of "the forks of the little rivers."

Two of the captains, Gilmore and William Seale, would move up in county government to exchange their military role for 1779 appointments as justices of the peace, members of the county government body. Alexander Avera would serve as militia captain and justice at the same time.

Other 1777 captains were John Matthews, Robert Cobb, Walter Murray, John Cox, Charles Campbell and John Stevens.

Most of the 1777 captains would be around when the Revolution was won.

Several had notable postwar careers, especially Alexander Avera.

In 1791, the General Assembly provided for a town to be laid out on Avera's land near the present-day county line between Harnett and Cumberland.

A crossroads village named Averasboro thrived there for a while in the early 19th century.

The place gave its name to another chapter in local military history, the 1865 Civil War fight in nearby pinewoods that came to be known as the Battle of Averasboro.

Today, the area that was home to the Revolutionary War captain is preserved as a Civil War historic site.

"Free Man Of Color" Made Mark On Fayetteville

Originally published January 30, 1997

He was barely 15 when the young fifer-soldier spent that unforgettable winter of 1777-78 with the ragged Continental Army in its winter quarters near the little Pennsylvania village of Valley Forge.

And like all the others who lived through that winter, Isaac Hammond, "free man of color," undoubtedly remembered it as a defining moment in his life.

The light-skinned youngster, who probably grew up in the Roanoke River valley of North Carolina, served only a year in the Continental Line. But that year and his subsequent service as a militia fifer earned him a unique place in the military history of Fayetteville.

To this day, he is the only individual from Fayetteville to be honored with a monument dedicated to his military service.

When the Revolution was won, young Isaac Hammond became a citizen of the Cape Fear town that recently had changed its name from Cross Creek to Fayetteville.

In 1787, he married a local free black woman named Dicey. When the first federal census was taken three years later in 1790, Hammond was listed as a free black householder with four "other free" individuals in his household.

Hammond was among 32 free black men listed as "head of household" in the 1790 census of Fayetteville.

In those years, free black people constituted as much as 10 percent of the village's 2,000 to 3,000 inhabitants.

Although hedged about by restrictions that did not apply to whites, free black men were free to order their own families and vocations, and they could vote for presidential electors and members of the state legislature.

They could also serve as citizen soldiers in militia companies.

So in 1793, when Fayetteville's white gentry organized the unit known as the Fayetteville Independent Light Infantry Company, Hammond stepped forward and offered to serve as the company fifer.

For nearly 30 years thereafter, Hammond's music would sound during drills, balls, Fourth of July parades and at funerals. When the company's first captain, Robert Adam, died in 1801, the company gathered "on six successive Sundays, with music playing, consisting of drum and flute." Hammond was undoubtedly the fifer for this elaborate memorial.

His example would be taken up by other young black men.

Among them was Nelson Henderson (1791-1874). He was a slave barber, bought his own freedom in 1813, and played horns and beat the drum for other militia units right up until the Civil War. The Marquis de Lafayette probably heard Henderson's music when the famous Frenchman visited Fayetteville in 1825.

When Henderson died in 1874 at 83, he was buried "with full military honors" at a funeral attended by black Civil War veterans wearing the Union blue and white Confederates wearing gray.

While census reports and other records are silent on Hammond's vocation, tradition holds that he, too, was a barber, a skill that was wholly dominated by black men until long after the Civil War.

Hammond's status as a Revolutionary War veteran and popular member of the FILI also made him something of a political power in the little town.

In closely fought yearly elections for the "town member" of the North Carolina legislature, the votes of the few dozen free black men could be critical.

In an 1849 affidavit seeking a Revolutionary War pension for Hammond's descendents, a white petitioner said that in early politics, Hammond's influence among free black voters could "frequently shape the result" of town elections in Fayetteville.

A vivid picture of how Hammond operated is painted in affidavits describing a disputed election in 1810.

To woo his voters, Henderson employed a campaign tactic already well established. He tossed a barbecue. It must have been a great time. An affidavit said:

"There was plenty of provender and spirits. They appeared merry, and there were frequent Huzzas. ... The fare included two shoats roasted and boiled victuals."

Hammond's high spirits were not the only side of his character. The only mention of him in Cumberland County public records lists his conviction on March 16, 1809, for "assault and battery on Lucretia Bass." He was made to post a bond of 50 pounds to keep the peace with her for the coming year. Bass was a free black woman who lived near the Hammond home, probably on today's Old Wilmington Road.

Hammond's legend was well established when he died in 1822, and it grew.

He apparently requested to be buried near the Cool Spring, on a muster ground (and Fourth of July partying spot) of the FILI, next to Cross Creek where it flows under a present-day bridge on Cool Spring Street.

Twenty years later in the 1840s, a poem celebrated his deathbed wish and burial in an unmarked grave.

Written by Louola Miller, a young lady of the town, it contained such lines as: "And when ye rest beside the spring, At morning's dawn or evening's gloom, Discharge a volley o'er the spot, and cheer the silence of the tomb."

More than a century later, after World War II, the FILI honored its old fifer by putting down a small engraved stone at the traditional spot of his grave. It stands today near a newer and much more elaborate monument honoring the FILI itself.

Hammond's wife, Dicey, outlived him by many years. Much of what we know of this early military hero comes from her 1849 application for a Revolutionary War pension for their descendants.

In the petition, she said that he was the son of a barber, and that both his parents were "Mulattoes or Mustees having no African blood in them."

His Revolutionary War service, she said, was in the 10th North Carolina Regiment, and he served for 12 months.

The 10th Regiment was recruited in the northeastern part of the state in the summer of 1777. Plagued by bad leadership, disease and desertion, it mustered only a few score soldiers when it arrived in Philadelphia that winter to join the Main Army of the Continental forces.

During the Valley Forge winter, its survivors were absorbed into other North Carolina regiments. It was disbanded on June 1, 1778.

Dicey Hammond died at 80 in 1852, a well-known figure in the town. She left her estate to her last surviving offspring, her daughter Rachel Lomack (b. 1794), whose father-in-law, William Lomack of Robeson County, was also a Revolutionary War veteran.

Other free black men from the Cape Fear region who were soldiers with Isaac Hammond in the War of Independence and who later drew pensions for their service were Philip Pettiford (1754-1825), who raised a Fayetteville family that later included many artisans, and Louie Revels, whose descendant, Hiram Revels, would be a U.S. senator from Louisiana after the Civil War. John Lomax, Thomas Bell and Thomas Hood were others.

City Owes Its Existence To Revolutionary Patriots

Originally published March 6, 2003

It goes without saying these days that Fayetteville and the military are peas in the same pod, city and Fort Bragg all one community.

From the point of view of history, you might stretch it to say that Fayetteville is actually a creation of the military.

At least it took war, the American War of Independence, to transform the informal colonial settlement known as Cross Creek into Cumberland County's official town and set the stage for a new name once the war was won.

Fayetteville's transformation began 225 years ago in the winter of 1778.

By then, the War of Independence was nearly three years old, and Cumberland County was firmly controlled by Patriots who had seized the local government soon after royal sympathizers were routed at the Battle of Moore's Creek Bridge in February 1776.

At the time, the village of Cross Creek was a prospering commercial place of merchant store owners, taverners, millers and traders crowded beside an intersection of colonial trading roads.

In the 20 years since its first water wheel mill was built and the first tavern opened, the settlement a mile from the Cape Fear River had far surpassed the official town, known as Campbellton, established in 1762 on a 100-acre riverside tract where Cross Creek flowed into the Cape Fear.

By 1778, Campbellton was little more than a boat landing, a few warehouses and a decaying one-room courthouse where the local government, the Court of Pleas and Quarter Sessions of Cumberland County, convened every four months.

As early as 1767, several dozen of the residents and neighboring planters of Cross

Creek petitioned the royal government to allow the county court to "move up" to Cross Creek.

The petition was ignored, although the royal governor of North Carolina did acquiesce in the construction of a jail at Cross Creek, because no one could be found to act as jailer at Campbellton.

The War of Independence changed the political situation.

In the winter of 1778, many of the prewar petitioners favoring Cross Creek had become Patriot leaders of the settlement and of the county.

The Patriot government of North Carolina, the Provincial Congress, was sympathetic to these local leaders, many of whom were active military men or who were engaged in military supply activities supporting the Continental Army and the state militia.

And so 92 names were quickly gathered for a new petition.

The list was a roll call of ardent Patriots in the War of Independence.

They included such military activists and local government officials as Robert Rowan, the Armstrong brothers, James Dick, Thomas Hadley, Stephen Gilmore, George Fletcher, Robert Cochran, James Patterson, Lewis Barge, James Gee and Peter Mallett.

Rowan was the settlement's leading Patriot. In 1775, he circulated the petition known as the Cumberland Association, a defiant anti-British document, and had recruited a volunteer Patriot militia company from the Cross Creek community in the fall of 1775.

In 1778, Rowan was a member of the senate of the Provincial Congress and was a key figure in the state's military supply organization.

In the same assembly session that took up the new Cross Creek petition, Rowan was on a committee that compiled a list of "goods to be furnished by each county" to the military establishment.

Cumberland was listed as the third largest potential supplier, with a quota of 53 hats, 216 linens, 105 woolen or cloth outfits, 105 shoes and 105 stockings.

Early leaders

The assembly also named recruiting officers for the Continental Army for each district of the state. Rowan, Philip Alston, David Smith and William Rand were the Cumberland recruiters.

The county was assigned a goal of 73 recruits for a new Continental battalion.

Among other Cross Creek petitioners, the Armstrongs and Hadley would become active militia commanders. Fletcher and Dick were junior militia officers and active commissary agents. Mallett was a key civilian supply officer.

With such backing, the Cross Creek petition quickly became law.

It officially established the town of "Upper Campbellton" and "Lower Campbellton," and decreed that the courthouse would be erected in the former.

It named a commission to lay off streets in "Upper Campbellton" in a grid pattern like the on-paper model of the 1762 act that established Campbellton.

The commission, which ranks as Cross Creek's first municipal government, was composed of Peter Mallett, Robert Cochran, Lewis Barge, Daniel Sutherland, James Patterson, George Fletcher and Robert Rowan.

The task of the commission would not be accomplished during the years of the War of Independence.

It would be another five years before the legislature of the postwar state would name another committee for the job and change the name of the old settlement to Fayetteville.

By midsummer of 1778, however, Cross Creek was firmly established as the county's seat of government as well as a leading military supply base.

A small new courthouse was quickly erected on Maiden Lane, where the Cumberland County Headquarters Library stands today.

And before the War of Independence was won, the new official county seat that owed its standing to war would also briefly be the scene of wartime events.

Three years after the 1778 petition went to the assembly, the British army under Lord Charles Cornwallis marched through the village in April of 1781 on its way to Wilmington from the Battle of Guilford Courthouse.

Courthouse taken

And four months later, the little courthouse itself was seized by a band of hard-riding Loyalist partisans who swept into town while the Court of Pleas and Quarter Sessions was meeting.

The raid was so swift that resistance was impossible, and among the embarrassed officials was Rowan himself, briefly held prisoner by the invaders.

Fayetteville and war have been synonymous at least since then, wouldn't you say?

War Seemed Far Away During 1778

Originally published June 25, 1998

In 1778, Patriots were in such firm control of Cumberland County you might think the War of Independence had been won already.

Two years after the Declaration of Independence, Patriots held all county offices, with William Rand as clerk of the Court of Pleas and Quarter Sessions, the governing body of county government, and Thomas Hadley as sheriff.

Former Col. Ebenezer Folsom, who in 1776 had commanded militia forces stationed at the village of Cross Creek guarding against a rumored British attack up the Cape Fear, was now serving in the senate of the North Carolina assembly.

He recommended that unusable muskets captured by Patriots from Loyalists two years earlier at the Battle of Moore's Creek Bridge be sold to the highest bidder.

Things seemed so calm in the spring of 1778 that Folsom's successor, Col. David Smith, and his deputy, Maj. John Armstrong, handed in their resignations to Gov. Richard Caswell.

The governor promptly recommended the county's most respected Patriot leader, Robert Rowan, again become colonel of militia.

Caswell used glowing terms to describe Rowan, whose service in the cause of independence ran back to the summer of 1775, when he had circulated the Cumberland County petition known as "the association," which pledged signers to fight if necessary to redress British oppression. Caswell wrote to the Provincial Congress, the wartime General Assembly:

"He (Rowan) is a person well acquainted with the different manners and customs of the people in that country, and one whose past conduct in military matters will give him weight with these people."

Business as usual

Emphasizing the almost peacetime calm, county government was busy naming overseers to maintain new roads or bridges, setting ferry rates, naming executors for estates, registering more than 200 land grants made by the Provincial Congress, levying and collecting taxes, issuing tavern licenses, and setting prices for the rum, wine, beer and cider that they sold.

In April, the residents of Cross Creek, present-day Fayetteville, achieved a 10-year-old goal.

Congress, meeting in New Bern, heeded the petition of 92 leading residents and approved an act allowing the village to absorb the official colonial town of Campbellton, a mile away on the Cape Fear River.

The act decreed that the county courthouse could be "moved up to Cross Creek."

That was especially desired because earlier petitions described low-lying Campbellton as "a swamp and morass" and bemoaned that it was bereft of even minimal comforts.

It was so dismal that attendees at the quarterly meetings of county government had to hike to taverns in Cross Creek.

The act also called for a committee of local leaders to lay out the village "in a regular manner" of straight streets and squares to replace the haphazard development along a web of old colonial roads that converged on Cross Creek at present-day Green Street.

A tiny new courthouse was quickly built on the site of the present-day Cumberland County Headquarters Library on Maiden Lane in downtown Fayetteville.

However, the plan for laying out the village would languish until after the War of Independence.

While there were no guns firing in the region in 1778, there was wartime logistical activity as the state government strove to supply clothing, rations and arms to the state's soldiers on duty with Gen. George Washington's army in New Jersey and New York.

The army had emerged from its bitter winter quarters at Valley Forge in stronger shape than the year before. Hopes were high that the new alliance with France would bring a summer victory over British troops in Philadelphia and Rhode Island.

But the British evacuated Philadelphia and a storm scattered the French fleet off Newport.

After a running fight with the retreating redcoats at Monmouth in New Jersey, Washington's army drew a defensive arc around New York, and the Revolutionary War in the northern colonies became mostly a waiting game.

Change in tactics

The British, in fact, had decided on a southern strategy that would be launched in December of the year.

Meanwhile, in Cumberland County, Rowan as sometime militia colonel and a commissary official for the state's Provincial Congress learned that Cumberland was expected to supply 53 hats, 216 yards of linen, 105 yards of wool, 105 pairs of shoes and 105 pairs of stockings for Continental uniforms.

Militia captains were ordered to "make up a list of the clothing" owned by each militiaman that could be collected in their districts (now townships).

Recruiting for the Continental regiments was also a responsibility of local officials. Rowan, Phillip Alston, former militia Col. David Smith, and Rand were detailed by the Congress as recruiters.

The county's quota for a newly authorized regiment of Continentals was 73 men in a total of 2,648.

At least one Cumberland County Continental soldier was not with Gen. Washington's army.

Jesse Hix was "on furlough from indisposition."

But Justice of the Peace Alexander Avera was suspicious. He charged Jesse's father, Drury Hix, with "harboring" a deserter and fined him 50 pounds of Continental money.

Hix told the county court his son "had a dangerous disorder" that made it impossible for him to report to his regiment.

The court sympathized, and ordered Avera to return the fine.

In both supplying and recruiting, Cumberland officials, like their compatriots elsewhere, found a lessening interest in a war that seemed so remote from the Cape Fear valley.

But while the War of Independence seemed far removed from Cumberland County 220 years ago, there nonetheless were tensions boiling below the seemingly calm surface.

Patriots Mobilize For Defense

Originally published February 25, 1999

In the three years after the Patriot victory over Loyalists at the Battle of Moore's Creek Bridge in 1776, an uneasy truce prevailed in the Cape Fear region.

But as 1779 began, the war flames smoldered again.

Since Moore's Creek, the major theater of the War of Independence was far off in New York and New Jersey.

In the Cape Fear, Patriots controlled state and local governments. Their main contribution to the war effort was in raising supplies for the North Carolina Continentals serving with Gen. George Washington's army.

The losers at Moore's Creek and their sympathizers, while numbering in the hundreds, were obliged to bide their time, waiting for a turn in the fortunes of war that might encourage them again to take up arms in the king's defense.

Meanwhile, many endured harsh confiscation laws that often meant loss of their property and which forced several score into exile from the state.

Then, as the third anniversary of Moore's Creek arrived in early 1779, the scene changed abruptly.

High-level strategic decisions far from the Cape Fear set the stage for the civil war between Patriot and Loyalist to erupt again.

The village of Cross Creek, now Fayetteville, which had been a rallying place for both sides in 1776, was once again feeling the war fever.

In late December 1778, a British fleet landed at the little port of Savannah in Georgia, seized the town, defeated a small Patriot army under North Carolina's Gen. Robert Howe, and set out on the new strategy of seizing control of the Southern colonies, especially the Carolinas.

The British issued proclamations urging Loyalists to take up arms, to join such units as the Royal North Carolina Regiment, commanded by Lt. Col. John Hamliton, a former merchant in the North Carolina town of Halifax. Hamilton's unit landed with the British invasion force at Savannah.

A few of Hamilton's men were from the Cape Fear, and he hoped to recruit many more as the mixed army of redcoats and Loyalists invaded the Carolinas.

Archibald McDugald of Cumberland County made his way to Savannah hoping to join Hamilton but was captured by the Patriots and held prisoner. He later escaped from Charleston and became an ensign in the Royal Regiment.

In its first fight, however, Hamilton's Loyalist force was badly beaten at Kettle Creek, upriver from Savannah, on Feb. 17, 1779.

Mustering militia

By then, North Carolina Patriots were again hastily mobilizing militia to both guard against a Loyalist uprising in the state and to march for South Carolina, where the British army was moving northward toward Charleston, arousing the strong Loyalist sentiment of that state along its way.

Attention quickly turned to Cumberland County, where Patriots had been in firm but uneasy control since 1776, but where even they admitted that several hundred of their neighbors were "disaffected" from the cause of independence.

In early January 1779, the Provincial Congress, North Carolina's wartime legislature, ordered the recruiting of five companies of militia, 250 foot and 25 horse, to be dispatched to Cumberland, Anson and other counties along the South Carolina border.

The units were ordered "to apprehend those Disaffected to the American cause and believed to be the Ringleaders of the people called Highlanders."

The Congress hoped the new militia could strike before the Loyalists got wind of the move to disarm them. Militia commanders were urged to "observe as much secrecy as you can."

Meanwhile, the village of Cross Creek became an assembly and supply point for other militia companies marching to reinforce Patriot forces in South Carolina.

The village of only a few hundred residents was the largest settlement between North Carolina and the Patriot forces gathered near Charleston, and as such was the jumping-off point and supply base for the reinforcements.

Militiamen from Bertie County and the village of Kingston (present-day Kinston) were expected in late January.

Gov. Richard Caswell apparently came to the village to spur the reinforcements and check on the supply situation.

Peter Mallett, a local merchant who purchased rations for the military, reported about the same time that he had "purchased 400 hogs from David Smith and Mr. Turner," spending all the appropriation available to him.

Peaceful business

Amid all these new scenes of war, the county government of Cumberland held its first meeting of 1779 to handle the business of peace.

The county Court of Pleas and Quarter Sessions, a gathering of justices of the peace, met in the unfinished new courthouse on Maiden Lane in Cross Creek.

The Provincial Congress only a few months earlier had heeded a long-standing petition of merchants and justices asking that the seat of county government be located in Cross Creek, rather than in the swampy riverside place known as Campbellton a mile away, where it had been for a decade.

The first order of business was to instruct the county sheriff to complete the courthouse by furnishing a door and windows.

Next, the justices chose Patriot taverner George Fletcher and Edward Winslow to be constables in the district of Upper Campbellton, which was supposed to be the new name of Cross Creek.

Another court order paid James Dick six pounds of "proclamation money" for rendering inventories of taxable property of "inhabitants of Cross Creek."

The court named James Patterson overseer of the road leading from Cross Creek to the riverside at Campbellton.

Passive resistance

Despite the routine business, however, there were hints of the divisions below the surface calm.

The Patriot justices assigned to collect county taxes in the dozen tax districts reported nearly 120 delinquents from 1777 and 1778, largely in districts in present-day Harnett and Moore counties where Loyalist sentiment was strongest.

The "disaffected" were showing their defiance.

County Dodged British Invasion

Originally published October 7, 1999

Along with the flowering dogwood, the early spring of 1779 brought sighs of relief to the Patriots of Cumberland County.

It appeared that they, and all of North Carolina, had dodged a bullet, at least for the time being.

The threat of a British invasion, which had so alarmed the previous six months of the fifth year of the American War of Independence, abated.

The month of May 220 years ago brought welcome news that the British and their Loyalist allies were falling back from a short-lived siege of Charleston, returning to their base at Savannah, which they had seized in December 1778 as part of Britain's new "southern strategy" for victory over the Patriots.

But June brought a sobering footnote promising that the war was far from over in the South.

Gen. Benjamin Lincoln's little army of Continentals and militiamen, marching on the heels of the British force, was badly bloodied by the redcoat rearguard in a sharp fight at a river crossing known as Stono's Ferry.

In a quirk of history, the British heroes at Stono's Ferry would become a household word in Cumberland County.

They were men of the 71st Regiment of Foot (Fraser's Highlanders), a unit that in the next 30 months would fight across the Carolinas, march through Cross Creek with Lord Cornwallis in 1781, and surrender at Yorktown.

Forty years later, a township in Cumberland would be named in honor of the kilted redcoats.

The British presence in the South was doubly threatening to Patriots of Cape Fear, who rightly surmised that simmering Loyalist sentiment would flare into open defiance as it had three years earlier, only to be thoroughly suppressed at the Battle of Moore's Creek.

Some Loyalists from the area were already in the fight.

The British forces in Savannah included units of Loyalists, among them the North Carolina Royal Regiment under Lt. Col. John Hamilton.

Some men from the Cape Fear region were serving under Hamilton's royal flag and may have taken part in the shootout at Stono's Ferry.

The threat of British invasion turned Cross Creek into a busy staging point for North Carolina militia units called out by Gov. Richard Caswell and ordered south to join Benjamin Lincoln's army.

Cross Creek merchants acting as commissary officers did a brisk business provisioning units marching from as far away as the Chowan River valley.

The operations of the provisioning system were evident in a letter that Caswell wrote in June to Cross Creek merchant Peter Mallett: "Find enclosed a warrant from the Treasury for $5,000 which I request you immediately apply to purchasing flour, grain, for the Continental Troops and militia who may make your way to the southward.

"You will also be pleased to provide them flesh provisions from the public stock.

"I should have wrote you on this subject sooner, but it was til very lately altogether uncertain if these articles would be necessary at Cross Creek.

"Let me to entreat you to give every assistance in your power to the troops on their march.

"Col. Lamb, with a party of men hired by the militia will, I suspect, be with you in a few days."

The summer of 1779 saw a return to less hectic times for the village's few hundred residents.

The county government, holding its quarterly session in July, attended to an agenda of mostly peaceful pursuits, such as laying off roads and approving tax collectors for the various "captains' districts."

There was even time to levy a fine of five pounds on one John Murphy "for being drunk on jury duty."

Until he paid up, Murphy was to be "taken into custody."

But there was also war-related business. The court proceeded to prove that Patriots were still firmly in control of local affairs in Cumberland.

Even with the Loyalist rumblings under way across the Carolinas, the Patriot county government proceeded to enforce a "confiscation act" seizing the property of Loyalists who had already left the county.

Perhaps meeting in a new courthouse on Maiden Lane in "Upper Campbellton" (the new official name for Cross Creek), the justices of the peace who comprised the local county governing body appointed three notable Patriots as "commissioners of confiscated property."

They were taverner George Fletcher, merchant Pat Travers and planter Thomas Armstrong. The latter, who did not live in Cross Creek, failed to make it to meetings and was replaced in October by John Matthews.

By August, the alarms from South Carolina were so muted that a body of mounted militiamen stationed at a "camp near Cross Creek" since earlier in the year was disbanded.

Its commander, Lt. Col. James Thackston of Orange County, reported to his superior, Brig. Gen. Jethro Sumner, on his collateral duties organizing militia units for the southward march.

He wrote: "I have just finished discharging all the soldiers whose time has expired on the first and fifth of this month.

"Those whose times does not expire until December next, and the old soldiers, we by order of Gen. Lincoln, formed into companies, officered and sent to Charleston, a return of which, together with a return of the Brigade for the last month, you will receive withall.

"All arms and accoutrements in possession of the men now discharged have been sent to the assistant quartermaster general in South Carolina, except for about 30 stand that were went with the prisoners (apparently deserters) to Salisbury.

"By order of Gen. Caswell, the whole brigade was mustered out and paid off, up to the last day of this month. I will send (muster rolls) to you. In the hurry of business, Lt. Col. Lytle, who was appointed muster of the regiments (when he left me), carried them off with him.

"I was under the necessity when before I left headquarters of applying to Gen. Lincoln for money, on the account of this state, to purchase necessaries for the officers on the march home, which was cheerfully granted, to the amount of $4,180, which sum I put in the hand of Col. Madearis, to enable him to furnish the requisites.

"I have had a very troublesome and tiresome time of it since you left, but I have at length got through with it, I hope to your satisfaction.

"I have given orders to all the officers to come in with the men now discharged to be diligent in apprehending deserters who may be lurking in the counties adjacent to the places where they are until they receive further orders from you."

With that, the 1779 Cross Creek military establishment of the War of Independence shut down for the season.

71st Named For The Redcoats

Originally published June 15, 2000

The Southern pine woods were shimmering in the heat of a typical early June on the day in 1780 when the column of redcoats marched into their new bivouac on a South Carolina plateau known as the "High Hill of Cheraw."

The 71st Regiment of Foot of Lord Charles Cornwallis's command, informally known as Fraser's Highlanders, was taking up its position as the vanguard of a planned British invasion meant to smash the American Revolution in his Brittanic Majesty's colony of North Carolina.

The British soldiers experiencing the Southern heat that day were volunteers from the windswept Highlands of Scotland, where a summer temperature of 60 was considered a heat wave.

The new camp of the 71st was barely a day's march from the border between North and South Carolina.

Two days beyond that, at the end of a curling wagon road through the seemingly endless pine-covered Sandhills, was the village of Cross Creek on the Cape Fear River.

In that crossroads of a few hundred inhabitants, the Patriot government of North Carolina had a major military supply depot.

Since the first of the year, food, clothing, rum and camp supplies had piled up there, meant for Patriot forces, Continentals and militiamen, ordered to the defense of Charleston, S.C.

But now as Lt. Col. John Maitland's Scots took up their position just west of the Pee Dee River (the Yadkin in North Carolina), Charleston was in British hands, surrendered along with the entire defending force of Americans on May 12, 1780.

British redcoat detachments and battle groups had since fanned out into upcountry South Carolina, establishing a major base at Camden on the Santee River, two days march from Maitland's post, at the other end of the wagon road to Cross Creek in North Carolina.

By now, too, John Maitland's Highlanders were veterans of the American Revolutionary War, with a reputation for tenacity and bravery in action, admired and feared on both sides.

They had arrived in Boston four years earlier just in time to see their colleagues of Lord Howe's command evacuate the town and leave it to George Washington's Continentals.

For two years, they were in redcoat forces pursuing Washington's army across New York and New Jersey.

Washington became familiar with Maitland's battalion because it was especially active in harassing Patriot picket lines.

A war story says that in an exchange of notes, Maitland informed the American commander in chief that his troops could recognize the Highlanders by the red feather that they wore in their caps as a regimental emblem.

In late 1778, the Highlanders came south as the point unit of the new British strategy to put down the American War of Independence by conquering the southern colonies.

The strategy expected that the invading British forces would trigger a massive uprising of Americans still loyal to the Crown. These Loyalists were numerous in the Carolinas.

The 71st was the main body of Regulars in the 3,000-man British force that captured Savannah, Ga., in October 1779.

The Highlanders experienced their first southern summer in 1779, when they marched with an unsuccessful attempt to capture Charleston.

Standing as a rear guard when the British force turned back for Savannah, a detachment of the regiment fought one of the bloodiest small-unit actions of the Revolution at Stono River south of Charleston on June 20, 1779.

There, American soldiers learned the hard way the deadly tenacity of Maitland's firing line.

Now a year later, the Highlanders were savoring the victory at Charleston and looking forward to the next mission in the Southern Strategy — the invasion of North Carolina.

Cornwallis himself showed up to inspect the Cheraw Hill outpost a few days after it was established, and only a few days after he had been named commander in chief of the British forces in the South.

He wrote to his superior that the outpost included Maitland's soldiers, commanded at the moment by Maj. Archibald McArthur, "a troop of dragoons, and a six-pounder" artillery piece.

The bulk of the British forces, including a gathering of hundreds of Loyalist militiamen lately raised in South Carolina, were stationed at Camden.

His plans called for firmly securing the South Carolina backcountry before marshaling his forces for the invasion of North Carolina.

Meanwhile, he urged Loyalists in North Carolina to lay low, because bodies of Patriot militia and some Continental units were on alert in the state awaiting the British move.

Among the forces, he reported, were 1,000 militiamen camped near the supply base at Cross Creek, commanded by none other than Gen. Richard Caswell, the state's wartime governor and now full-time commander in chief of Patriot militia forces.

In the year to come, the 71st Regiment of Foot would become all to familiar to North Carolina, to Cross Creek and to the Patriot militia.

And, as you may have guessed, many years after the Revolution a township of the Sandhills over which the road from Camden to Cross Creek (later Fayetteville) passed would take the name of the military unit of Scots volunteers from the craggy hills above the Loch Ness.

"Harrington's" One Of Area's Historic Hills

Originally published July 6, 1995

There is "Smoke Bomb Hill," a noted area of Fort Bragg. And there is "Haymount Hill," where the ruins of the Fayetteville Arsenal look down on the old part of town.

And there is another bit of high ground with a military heritage.

It is "Harrington's Hill."

To get there, head north out of downtown Fayetteville on Ramsey Street, pass the intersection with the Central Business District Loop, and then notice as you climb up to a ridge several hundred yards ahead.

When you get up there, you are on Harrington's Hill. About a mile out from the Market House.

If you had been there 215 years ago, you would have observed a Harrington's Hill bustling with Revolutionary War soldierly activity.

The summer of 1780 was a time of crisis for the American cause in North Carolina. After four years of relative peace and of Patriot ascendancy, a redcoat invasion and a Loyalist uprising loomed.

A British army had seized Charleston in May, capturing thousands of veteran soldiers and experienced officers manning the defenses of that port.

As June came, a redcoat army under Lord Cornwallis was marching north through South Carolina, aiming for North Carolina, where the British commander expected Loyalist sympathizers would rise up in large numbers.

To meet the threat, militia drums were beating assembly all over North Carolina.

The state's tiny government, located at Hillsborough, drew a line along the Pee Dee and Cape Fear rivers. It ordered bodies of Patriot militia to gather at points on those lines with orders to keep a sharp eye out for the approach of Cornwallis or of Loyalist (they called them "Tories") bands.

The settlement a mile from the Cape Fear known as Cross Creek, now downtown Fayetteville, was a key position, an important depot for military supplies.

Goods stockpiled

For several years, energetic Patriot military officers and public officials at Cross Creek had been acquiring, storing and shipping salt, leather, beef, pork, wheat and forage for North Carolina troops of the Continental Line and militia.

So, when the plans for defense were issued to brand-new Brig. Gen. Henry William

Harrington (1747-1809), commanding officer of the Salisbury District, he was ordered to keep a vigilant watch along the river lines and expressly enjoined to "guard the stores at Cross Creek."

With about 450 militiamen under his command, Patriots from Duplin, Onslow, Bladen and Cumberland counties, the 33-year-old Harrington immediately set up his posts.

A large camp was established on the ridgeline that looked southward to the Cross Creek settlement and the river landing a mile away known as Campbellton.

For the next five months, the Cross Creek camp served as a command post, observation post and occasional bivouac for the small groups of horsemen and infantry of the Salisbury District.

Gen. Harrington was frequently at the post. He would come over from his home a few miles south of present-day Rockingham in Richmond County to receive the latest intelligence or to issue orders to roving units of militia cavalry.

Command was short

Perhaps even by then, the camp's site was being called Harrington's Hill. It would be called that for a century and a half.

However, Harrington's tenure in his military command was as short as the name he left on the land in Cumberland County was long.

His had been a temporary appointment while a more experienced militia general had been attending to defenses farther west.

Thus, the state assembly officially relieved him from the command on Sept. 3, 1780.

Despite that, Harrington continued active direction of the Salisbury District until he resigned his commission in November.

By then, Cornwallis' army had defeated the Patriots at the Battle of Camden in South Carolina, while over-mountain Patriots had smashed a Tory force at the Battle of King's Mountain in North Carolina.

By then, too, the uprising of Loyalist forces had started. Much of the energy of Harrington's command was spent rushing to places threatened by newly emboldened Loyalist bands.

Just south of Cross Creek, Loyalist partisans were especially active in Bladen County.

There, Patriot partisan leader Col. Francis Marion of South Carolina, the famed "Swamp Fox," commanded forces battling to suppress the Loyalists.

A rare contemporary letter from an aide of Marion described the action. Writing on Sept. 20, 1780, to the "Board of War" established by the North Carolina assembly, he reported:

"Tories have embodied in Bladen County; they have been twice routed by the Colonel; in one Action he killed and took Thirty. A Detachment of Four Hundred from Lord Cornwallis are sent to join them; This Movement of the enemy makes it highly necessary that your Brigade should move, or as many as are equipped, without delay to Cross Creek to join General Harrington. They burn all the Houses of our friends that are absent."

Importance stressed

On that same day, Cross Creek's significance as a military supply point was emphasized in a letter from the board chairman to Gen. Harrington:

"We shall be in great want of Salt if any Accident happens to what is at Cross Creek. Should you think it necessary to remove the Stores there you order Waggons to be impressed and send them on. As I am very desirous to procure Shoes for the Regular Troops here, I shall be glad to know if Leather can be got at Cross Creek, one Hide tanned for two raw; if so, do give directions for sending on as much as can be got, and I will order the Raw Hides from this place by the Waggons that return or go for the Stores."

For London-born Harrington, who had been colonel of Richmond County militia and a member of the Patriot assembly, the command at Cross Creek was his last military service.

But he had a long and distinguished career, otherwise.

Returning to Richmond County during the last year of the Revolutionary War, this "very intelligent gentleman" saw his farm ravaged by Tory partisans and his wife even held as a hostage.

When the war ended, however, his fortunes were restored.

State's "first farmer"

In the early 19th century, he was styled both "the first farmer of the state" and "the father of export cotton in North Carolina!"

He was an original trustee of the University of North Carolina, a legislator, and the "leading citizen of Richmond County after the Revolution."

A son and namesake served in the Navy during the War of 1812 and was later a militia colonel.

The eminence that bore Gen. Harrington's name was mentioned frequently as a landmark in early 19th-century Cumberland County deeds.

Harrington's Hill in 1858 was described by Fayetteville lawyer-historian James Banks. Narrating an excursion on the railroad line under construction north out of the village of Fayetteville, he wrote:

"We look ahead and the eye rests on 'Harrington's Hill' covered with neat white residences and crowned with lofty towering pines. The Hill was so-called in honor of Col. H.W. Harrington of Richmond County, father of Col. H.W. Harrington of the same county. Here General Harrington was for some time encamped in the Revolutionary War, on the American side, overlooking as it were the Loyalists in and about Cross Creek. But the Mile-branch is crossed, Harrington's Hill fades from our sight."

The name was "still common" in the early 20th century, but was generally out of use by 1950, according to the late John A. Oates, Fayetteville's leading historian of the first half of the century.

Patriot Stayed True Despite War's Hardships

Originally published July 25, 2002

In late summer of 1780, the fifth summer of the War of Independence, James Emmet could make a good argument that he was being chosen as a lamb for sacrifice.

His friends Robert Cochran and Edward Winslow had recommended Emmet to become lieutenant colonel and commander of the Cumberland County militia.

They wrote to North Carolina's wartime governor, Richard Caswell, describing Emmet in glowing terms:

"His general good character, his experience in Military Affairs, and his steady, spirited and uniform Conduct during the whole of the present contest, all conspire to entitle him to such a command."

Emmet could have retorted with something like, "yeah, but look fellas, if it's just as well with you, I decline."

By the late summer of 1780, five years of Patriot control over the Cape Fear River valley was in serious jeopardy.

Emboldened by a British invasion of South Carolina and the capture of the entire army of the Southern Department of the American military establishment at Charleston in May, with British troops standing on the border of North Carolina, suppressed Loyalists were rising up, mustering in hard-riding fighting bands, challenging short-handed Patriot forces and exacting violent revenge on Patriot families.

The struggle in the Cape Fear region would not be Continental regulars versus the redcoat regulars of the British army. Rather, it was neighbor-against-neighbor civil war. It was savage, bloody, unmerciful, Patriot against Loyalist, or as they often referred to each other, "Whig against Tory."

James Emmet, however, was not a sunshine Patriot. His motto could well be that old adage, "when the going gets tough, the tough get going."

A couple of weeks later, the English-born Emmet, who was among the earliest avowed Patriots and Continental officers in the village of Cross Creek, now Fayetteville, was back in military service.

So, in this story of a Revolutionary War hero from Cumberland County, you won't be reading about personal derring-do or about victory. More often, it is a story of defeat and frustration.

But I hope you will agree that Emmet's perseverance and judgment fulfilled the confidence of his wartime colleagues and made him worthy of the postwar honors he would enjoy because of his service.

In the summer of 1775, Emmet was among the 55 signers of the defiant anti-British petition known as the Cumberland Association, circulated in the village of Cross Creek by Robert Rowan.

In the spring of 1776, he put his patriotism to military service, commissioned among eight captains of the 3rd North Carolina Regiment of the Continental Army.

In the wartime years since then, however, Emmet had spent most of his time in Cumberland County as an off-duty officer and recruiter for Continental troops. Apparently, his health had forced him from active campaigning.

In that role, he was also active in the military supply apparatus that centered in the village. As a merchant and an ally of the Mallett family of Wilmington, he acted often as a commissary officer and purchasing agent for supplies destined for the North Carolina militia and for the Southern Department troops in South Carolina.

By the spring of 1781, Col. Emmet's worst forebodings were proving accurate.

British Gen. Lord Charles Cornwallis had marched his regulars into North Carolina, fought a stand-off battle with a Patriot army under Gen. Nathanael Greene at Guilford Courthouse near present-day Greensboro and then retreated along the Cape Fear toward Wilmington, marching through Cross Creek for two days in early April.

Emmet and a hard-riding band of militiamen, under direct orders from Gen. Greene, had been trailing Cornwallis's column for more than a week. When they rode back to Cross Creek on April 4, Cornwallis was gone.

The situation would get worse.

On April 28, for instance, Emmet reported to Greene:

"This town, and indeed many parts of the county, is much infested with the small pox, which makes it out of my power to keep a single man in this place."

As summer went on, Emmet's reports were gloomier and more urgent.

In a letter to the new Patriot governor in the last days of July 1781, he first offered "a sketch of the unhappy situation in this County."

Then he warned:

"I must beg leave to observe that if we do not Garrison X Creek (Cross Creek) very soon, the Enemy will."

Emmet knew that a Tory force commanded by David Fanning had already pulled off a spectacular raid on another courthouse town, nearby Pittsboro, on July 16, 1781.

There, they captured the entire Patriot government and militia high command of Chatham County.

Emmet knew that other Loyalist bands were controlling much of the swampy western end of Bladen Precinct, present-day Robeson County, sweeping out to raid and intimidate the countryside in western Cumberland County.

He was officially required to keep an eye on other Loyalist bands operating along the Cape Fear River as far up as the Bladen courthouse near Elizabethtown, carrying out the orders of British Maj. James Craig in Wilmington to raid Patriots throughout the Cape Fear valley.

Emmet managed to keep a small command of his own on call and sometimes mustered them when Loyalist raids raised alarms.

But luck ran out and the worst happened on Aug. 14, 1781.

On that day 221 years ago, a veritable host of Loyalists rode into Cross Creek from three directions. They captured the men of the Patriot county government even as they held their regular quarterly session.

Emmet wasn't there. But in attempting to scout the attacking bands, he ran into one of them and was captured for a brief time.

It was the lowest point of Patriot fortunes.

In October, the news of the surrender of Lord Cornwallis at Yorktown in Virginia helped Patriots finally put Loyalists on the run.

The war over, Emmet reaped his rewards as a stalwart of the Revolution.

County sheriff

In 1784, he was elected sheriff of Cumberland County and also chosen as a Cumberland member of the lower house of the General Assembly.

The grateful government of North Carolina authorized a grant of 1,600 acres "reserved for officers and soldiers of the Continental line in this state." And he also was entitled to a "moiety of 5,000 acres over the mountains," meaning in the territory that would become the state of Tennessee.

By then, Emmet was a much-admired member of the coterie of Revolutionary War veterans who were dominating civic life in the newly named town of Fayetteville.

He was referred to always as "Col. Emmet." As well as being a public official, he was apparently associated with the Mallett family commercial enterprises and also was a taverner.

But in 1786, Emmet died, apparently suddenly.

My assumption that Emmet was British-born stems from the provision in his will in which he left a yearly stipend of 20 pounds sterling to his "Ancient mother in

London."

It is also possible that Emmet was the son or brother of John Emmit (always spelled with the "i") who was a carpenter and resident in Cross Creek by 1765.

Emmet was in business with and kin to the Mallett family. His widow, Margaret, disputed his will apparently because it devised property to Mary Mallett, eldest daughter of Peter Mallett, who was appointed along with Margaret as executor.

Margaret lived until 1806 and was the hostess of the family tavern, which was on the west side of Green Street, a door or two from Cross Creek.

James Emmet's name is among the 55 engraved on the stone at the intersection of Person and Bow streets in Fayetteville listing the first Patriots of Cumberland who signed the so-called "Liberty Point Resolves" in 1775.

For Emmet, service in the War of Independence turned out to be more than merely signing a petition. In good times and bad, he gave his all to the cause.

British Deserter's Lies Forced Precautions

Originally published November 20, 2003

This is the story of the "British deserter."

But first, some background.

If you know military history, you know that Cumberland County's 71st Township is the only political jurisdiction in the country named for a regiment of the British army.

You knew that, didn't you? OK, which one?

Why the 71st Regiment of Foot, "Fraser's Highlanders," they were called at first.

A proud boast of the official historian of the 71st is that, although it served in North America throughout the American War of Independence, none of its men were known to have gone over to the enemy.

Despite years away from their homeland in Scotland, shot at in dozens of fights, often poorly fed and clothed, they stuck with the mother country.

That is in sharp contrast to the notorious Hessians, the hired soldiers from Germany, who played such a big role in that conflict. Thousands of them found a new life in North America by deserting the ranks.

But there seems to have been at least one redcoat of Fraser's Highlanders who walked away from the fight.

In the archives of the Patriot war effort in North Carolina is the rollicking story of the British deserter who showed up at the village of Cross Creek, now Fayetteville, one evening in 1780.

Before it was over, he was given another name, "the lying Deserter."

The story begins on a hillside north of the village, where a small Patriot garrison camped with orders to keep watch for any British movement toward the valley of the Cape Fear River or the South Carolina border.

Letter to Gen. Gates

From this "camp near Cross Creek," militia Brig. Gen. H.W. Harrison sat down at 10 o'clock on the night of Sept. 7, 1780, and wrote to Gen. Horatio Gates, the overall commander of the forces of Independence in the South: "This evening about sunset I received information that a British Deserter was in Cross Creek: I went immediately and examined him. He says he left Anson courthouse on Monday evening last, that he belonged to Capt. Sutherland's company of the 71st Regiment, that an officer beat him with a sword, which caused him to leave them, in a hour after the enemy left the courthouse."

The deserter then proceeded to a long litany of information about the alleged size, destination, and command structure of the British forces said to be just across the line in South Carolina, where Lord Charles Cornwallis's army a few weeks earlier had smashed a Patriot army under Horatio Gates at the Battle of Camden.

He listed "Lord Rawdon in command, with his own corps of the 23rd, 33rd, and 71st, (as well as) Delancy's, the Prince of Wales, Hamilton's (Loyalist) North Carolinians," and also "troops of horse, wagons and draught animals."

These forces were said to be moving on North Carolina.

Harrison continued that the British soldier "told his story with seeming simplicity."

On the other hand, the boastful accounting of "such great strength makes me doubt his veracity."

However, Harrison couldn't take chances. That very night he reported: "I sent a party of horse towards Pee Dee by Coles Bridge (on Drowning Creek)," and would send yet another if the first found anything.

As a further precaution, Harrison gathered his stores and wagons and moved them more than a mile south of his camp to the northeast edge of the village of Cross Creek.

"I do not want to retreat until I hear something of the matter. Will send an express (message) as soon as I know the truth of what this fellow says."

In closing, Harrison took the opportunity to make an appeal that was often heard from outpost commanders: "I am much in want of Regular horse."

The British deserter's alarm was destined to last less than a day.

At 8 o'clock Sept. 8, 1780, Harrison went to his tent at the "camp near Cross Creek" and wrote another dispatch to Gates: "I was just now informed that some of my soldiers have seen McClean, the Deserter from the 71st Regt, before this.

"On further examination he produced a pass signed, Sir, by you on the 9th of August, yet he insists on the truth of last night's tale, and adds that the Tories carried him back to Camden after he received that pass.

"From the circumstances of the pass I cannot give the least Credit to him; yet is a matter of the highest moment, I shall not lose sight of him or his story till I am convinced in the clearest manner of its truth or falsehood, and of which sir I shall inform you the moment I know the certainty."

A few moments later, before handing the dispatch to a courier, Harrison appended a postscript:

"Since writing I have seen a man in whom I confide who left Pee Dee last Tuesday, and says the Deserter's tale is false, and that the Enemy are encamped between Camden and Lynche's Creek."

Deserter's identity

A few days later, in another dispatch about another matter, the identity of "the lying Deserter" was authentically given as Pvt. James McLean.

With that offhand entry, the story disappears into the mists of history, waiting for more information to emerge.

Did James McLean remain in North America? Or was he sent back to his regiment, which sailed for home in 1783 after surrendering with Cornwallis's army at Yorktown in October of 1781.

Certainly there are still McLeans living in the sprawling sandy reaches of 71st Township in Cumberland County.

That jurisdiction, home to many families with Scots background, got its official name 41 years after that September in 1780.

In 1821, the county government of Cumberland gave new names to tax districts that formerly had been designated by the names of captains of the local militia, so-called "captains' districts."

Under the old designations, 71st formerly had been "Gillis's."

There is no record of the reasoning for the naming.

But the long-held assumption is that it was a salute from a heavily Scottish neighborhood to the brave heritage of a famous fighting unit from the Highlands of Scotland, the erstwhile regiment of James McLean.

Scattered Patriots Rallied At Cross Creek

Originally published October 12, 2000

It had been a near thing.

But as of the middle of October in 1780, Gen. Charles Cornwallis was pulling back from his invasion of North Carolina.

Word arrived at the village of Cross Creek that the redcoat force that had deployed out of the little town of Charlotte was turning back, returning to camps on the border between the Carolinas.

The village on the Cape Fear River that would become postwar Fayetteville could take a breather after a busy season of Revolutionary War military comings and goings as Patriot regulars and North Carolina militiamen maneuvered against the British army. The redcoats had captured Charleston in May and quickly marched north and west along invasion routes.

Cornwallis paused along the border between the Carolinas in June and July, rallying Loyalist volunteers to augment his redcoat regulars. He also fought summer sicknesses that rendered a large part of his army unfit for fighting.

Hardest hit by the fevers of typical Southern "scorcher" weather was the crack 71st Regiment of Foot (Fraser's Highlanders), stationed at Cheraw Hill southwest of Cross Creek.

Meanwhile, the shattered remnants of the Patriot garrison at Charleston congregated at Hillsborough, and Gen. Richard Caswell of North Carolina rallied more drafts of militiamen.

As Cornwallis gathered strength for an invasion, Gen. Horatio Gates came south from the Main Army of Gen. George Washington to take command of the Southern Department of the Patriot cause. He boldly marched against Cornwallis, joined by Caswell at the South Carolina village of Camden.

On the morning of Aug. 16, 1780, the British marched out and crashed into the startled Patriots.

The Battle of Camden became a rout as the few hundred Continentals and thousands of militiamen broke under the charge of such redcoat units as the 71st Highlanders.

September 1780 became a month of desperate maneuvering as the British army marched for Charlotte and Patriots tried to cobble together a defense force.

Cross Creek was designated as a rallying point. On Aug. 27, the governor wrote from Hillsborough to Cross Creek. It said:

"To officer of militia at or near Cross Creek. You are required to halt all the militia in neighborhood of CC and at Campbellton, apply to Colonel Rowan for arms and ammunition.

"These men you are to employ in defending and securing the public stores at that place. And to furnish guards if thought expedient to march them to any part of the state, in that case you may receive orders from the commissary of stores. Make a return to Kingston."

But as October came, fortunes turned.

On Oct. 9, 1780, Cornwallis's western wing of Loyalist militia was roundly trounced by Patriot militia at Kings Mountain west of Charlotte.

Harassed by Patriot militiamen, the British general called off any campaign into North Carolina for the season.

In the few weeks before the Patriot army marched for Camden, Cross Creek continued its role as a commissary depot for North Carolina militiamen.

Since the first of the year, the little village of a few hundred residents, a collection of taverns and stores, of water wheel mills and wagon yards, had seen a constant stream of Continentals and militia first marching for Charleston, and then retreating from it.

A militia assembly point known as the "camp near Cross Creek" saw a constant traffic in part-time soldiers.

A permanent reconnaissance force of militia was stationed on a ridgeline north of the village that for many years would be named for the commander, Brig. Gen. Henry William Harrison.

Local merchants who provided food, fodder and clothing for the soldiers found it harder and harder to find the supplies sought by Caswell and Gates.

As scarcity drove up prices, the economy suffered from the falling value of money. The phrase "not worth a continental" referred to the money issued by the Continental Congress.

North Carolina money was even less valued. Some financial drafts were in dollars; others continued in the British system of pounds, shillings and pence.

In addition to Continental and state money, troops from other states often brought their own paper. "Virginia money" had been common in colonial times.

In an effort to sort out the various currencies and set norms for buyers and sellers, the government of Cumberland County named three men as "inspectors of circulating currency." They were David Smith, Lewis Barge and Daniel Buie.

Meanwhile, Caswell sought to round up North Carolina militiamen for the march south. He met stiff resistance from some county governments. They wanted men for their own defense.

And some individuals used creative means to escape the call.

The Rev. Charles Pettigrew had been sent to the camp near Cross Creek in a small party of militiamen from Chowan County in the far northeastern corner of the state.

On June 27, Caswell got a letter from Cross Creek about Pettigrew. It certified that

Pettigrew, "of draft from Chowan, is discharged, he having produced Zachariah Carter, an able bodied man, in his Room."

Someone, perhaps Pettigrew himself, wrote a note on the back of the certification: "A 7,500 Dollar Touch. The price of a Clergyman's exemption from Military Service in NC."

The note reflected both the troubles Caswell had with recruiting and the wild depreciation of the currency.

The respite afforded by Cornwallis's pause at Charlotte gave time to reorganize Cross Creek's military leadership.

First, Cumberland County got a new colonel of militia. James Emmet, a former major in the Continental Army, took the post in mid-September.

In the same week, George Fletcher, a Cross Creek innkeeper, replaced merchant Peter Mallett as chief commissary for the southeastern district of North Carolina.

Cross Creek was girding for what might come in the next season of the American War of Independence.

As 1780 Waned, War Plans Grew

Originally published December 7, 2000

As December arrived in the Cape Fear 220 years ago, British and Patriot military commanders prepared for campaigns in the American War of Independence.

And the village of Cross Creek, today's Fayetteville, figured in their plans in these final days of 1780.

For the past year, since British forces began executing their southern strategy to crush the American Revolution, the little community of water wheel mills, tanyards and merchant stores had been a busy logistical center for Patriot forces, a supply base and rendezvous for marching units of Continental soldiers and militiamen.

After the year of battles and marches, British and Patriot forces faced each other across the long border between the two Carolinas.

The British, under Lord Charles Cornwallis, controlled south of the border.

Following the capture of Charleston on May 12, 1780, the redcoats and their Loyalist allies swept across central South Carolina and paused to gather strength for an invasion of North Carolina.

In August, a thin army of Patriots commanded by Gen. Horatio Gates marched against the main British camp at Camden, only to be routed by the redcoats.

Cornwallis then marched on North Carolina, seized the village of Charlotte and prepared to move across the state.

But on Oct. 7, 1780, Cornwallis's forces were dealt a stunning defeat when Patriots clashed with his Loyalist left wing in a bloody shootout on Kings Mountain west of Charlotte.

A week later, the British army withdrew from Charlotte and took up posts centering on Winnsboro and Camden in South Carolina.

On Dec. 3, 1780, the Patriot army got a new commander in Gen. Nathanael Greene.

Greene decided to move his tattered battalions into South Carolina, watching warily lest Cornwallis's small legions march out against him in a winter campaign.

Greene only got to South Carolina on the day after Christmas, camping near Cheraw in what he called his "camp of repose" and what later historians would describe as "the

Valley Forge of the South."

The Patriot camp was not far from where redcoats of the 71st Regiment of Foot (Fraser's Highlanders) spent several fever-plagued weeks in the previous summer.

Meanwhile, the Patriot government of North Carolina rallied small bodies of militia to protect supply centers at Cross Creek, Hillsborough and Wilmington.

Both armies prepared for the next round of the Revolution in the Carolinas.

In October, Cross Creek figured in the discussions when British commanders mapped strategy for the next campaign season.

One proposal called for Cornwallis to stand on the defensive in central South Carolina while a British invasion force seized Wilmington and then marched up the Cape Fear valley.

The object would be to trap Nathanael Greene's little Patriot forces between redcoat pinchers.

Gen. Alexander Leslie, who commanded British forces occupying Portsmouth, Va., drew up plans to command such a force, to move inland while Cornwallis moved east toward Cross Creek, where the two would join.

Gen. Henry Clinton was more interested in holding the British foothold at Portsmouth than in any early drive into North Carolina by Cornwallis.

In November, he wrote to Cornwallis:

"My plan would one to establish a post at Hillsborough, feed it from Cross Creek, and be able to keep the camp at Portsmouth."

But logistics ruled out any quick military action in the Cape Fear region.

A British commander wrote:

"The Country from Cape Fear (Wilmington) to Cross Creek (the Highland settlement) produces so little it would be prerequisite, in penetrating through it, to carry your provisions with you."

On the day after Christmas 1780, from his camp on Hicks Creek near Cheraw, Greene published his own plans.

They again involved the village of Cross Creek in its usual role as a military supply point and rendezvous.

Greene wrote to Col. Nicolas Long of his plans to get his troops into action against Cornwallis by striking across the swamps of the Pee Dee River.

For that, he would need boats.

He ordered Long to send nails and necessary tools to Cross Creek and move them if the British invaded the lower Cape Fear.

But Greene's hopes for early action were also hampered by logistics.

If he was counting on the region around Cross Creek for provisioning his skimpy battalions, he would find the situation much as the British commander described it.

A year of war had created scarcity. Commissary officers were short of money and the value of what money they had was constantly falling. Loyalists were withholding supplies, and even those in sympathy with the Patriot cause were wary of the state government to fend off the expected British invasion.

Cross Creek merchant John Ingram, who was assistant commissary officer in the village, summed up the supply picture in a report to the North Carolina government's Board of War.

Ingram said the only provisions at "this Place are two or three hundred barrels of pork."

Ingram said he had not been able to send a full commissary report for the months since the fall of Charleston because much of the material under his inventory had been captured there. Or it had failed to reach there before the Patriots surrendered and had since been used up by marching militia units or by outpost garrisons.

No salt was available at Cross Creek, he said, but there were perhaps 300 bushels at Wilmington that could be boated up the Cape Fear.

However, Ingram said, the Wilmington salt was under the direct command of Greene and could not be brought upriver without Greene's express command.

There were other supplies of salt on board coastal vessels at Wilmington and New Bern.

Ingram ended his report by asking for instructions so he could inform other supply officers of the needs and orders of the Board of War.

With that clouded picture, the war year of 1780 came to an end in the village on the Cape Fear.

Revolutionary Sides Unfold

Originally published February 1, 2001

The Revolutionary War village of Cross Creek seldom heard such good news — or such bad news — as that of the final week of January 1781.

The week 220 years ago began when a rider arrived with word that a Patriot army had won an overwhelming battlefield victory at a place called Cowpens, 120 miles away in upper South Carolina.

The Patriots under Brig. Gen. Daniel Morgan thoroughly defeated a mixed force of Loyalists and British Regulars commanded by the hated British officer known as the "Green Dragoon," Lt. Col. Guy Tarlton.

By smashing the western wing of Gen. Charles Cornwallis's forces, the win at Cowpens might postpone the British commander's planned invasion of North Carolina and give the tiny Patriot army of Gen. Nathanael Greene time to gain strength for the campaign that all knew was coming.

Cross Creek, long a supply depot and rendezvous for Patriot military forces, was already figuring in Greene's plans.

On Jan. 21, 1781, he ordered salt and other supplies hopefully available at Wilmington to be moved up the Cape Fear River to Cross Creek.

The busy little village of merchant stores, taverns, a slaughtering yard, gristmills and sawmills was only two or three days march from the camp of the Patriot army, which since late December was at Hicks Creek near Cheraw, S.C.

But before the order to the military commissary at Wilmington could be delivered, the bad news came.

On Jan. 28, 1781, a small fleet of British ships sailed into the Cape Fear and anchored at the Wilmington waterfront.

Six companies of hard-bitten British soldiers, troops of the 82nd Regiment of Foot, debarked.

Their commander, Maj. James Craig, came ashore to raise the king's colors over the New Hanover County courthouse.

For the first time since late 1777, when local Patriots ejected royal Gov. Josiah

Martin from his official residence, Wilmington was under crown control.

Ever since that time, periodic rumors of a British return to the Cape Fear River valley were common in Cross Creek.

Patriots who firmly controlled local government and directed the military supply activities in the village nonetheless realized that much of the population in the valley was either loyal to the British crown or indifferent to the outcome of the War of Independence that had been going on for so many years.

This was especially true of the scores of families of Highland Scot settlers who were prevalent in those parts of Cumberland County that would later become Moore (1784) and Harnett (1858) counties.

Cornwallis was indeed counting on the long-quiescent Loyalists to rally to the king's standard. He wanted them to gather in militia companies that could augment his redcoat regiments.

The arrival of Craig's troops and the impending invasion by Cornwallis did indeed bring Loyalists into the open.

More than likely, however, the bands of Loyalists who formed throughout the Cape Fear valley remained detached from the British army.

Instead, they operated as guerrillas ("partisan" was the 18th-century word), sallying out from camps in swamps to harass and often overwhelm hastily assembled Patriot militia.

For more than a year after Craig's arrival in Wilmington, the Cape Fear valley was destined to flame with bitter neighbor-against-neighbor fighting, a bloody civil war that Patriots would call the Tory War.

Soon after the news of Wilmington's capture, several notable Loyalists stirred themselves.

In Wilmington, John Slingsby reported to Craig and was immediately commissioned as a lieutenant colonel of Loyal Militia.

Slingsby, a British-born merchant who had operated stores in Cross Creek and Wilmington before the war, was ordered to rally a band of Loyalists and advance up the Cape Fear toward the village.

Meanwhile, at his home near the Presbyterian meeting house known as Bluff on the Cape Fear near the mouth of Lower Little River, Hector McNeill got word of the British arrival.

McNeill had settled in the area before Cumberland County was carved from Bladen Precinct in 1754.

He had been sheriff of Cumberland during prewar years.

But in 1781, his Loyalist leanings were well-known.

Several of his relatives were already serving in Loyalist units with Cornwallis or in other states.

As early as the summer of 1777, the county government had ordered him to come to Cross Creek to answer charges, "being suspected of speaking many words against the state."

McNeill would soon make his way to Wilmington to receive a commission from Craig, and the man known as "Old Hector" McNeill would become Lt. Col. McNeill, commander of the Bladen County Loyal Militia.

The long-expected British arrival at Wilmington came just as the little armies of Cornwallis and Greene made their first moves in what would become a six-week campaign across the central part of North Carolina.

Patriots in the Cape Fear would largely be left to their own devices to defend against anything Craig might attempt.

The Patriots had some assets, especially in the military leadership of the local militia.

Col. James Emmet, lately named new colonel of Cumberland militia, was a veteran officer who had been a captain in the Continental Army in the early months of the war.

Bladen County militia leaders included Col. Thomas Robeson and his brother, Capt. Peter Robeson, and Lt. Col. Thomas Brown, all of whom had long experience in keeping an eye on the large Loyalist population in the area that is today Bladen, Sampson and Robeson counties.

Within a week after Craig's troops landed, Patriot militia under the noted Col. Alexander Lillington of New Hanover County set up blocking posts along roads leading out of Wilmington toward Cross Creek.

For the time being, Craig contented himself with raiding in the immediate vicinity of the port village and encouraging the upriver Loyalists to take up arms.

There was a momentary calm before the violence of 1781 would begin.

War Divides Neighbors

Originally published April 5, 2001

On April 5, the civil war had come to the Cape Fear area.

Not the war in 1861-65 between blue and gray.

But rather the war in 1781-82 between Patriots and Loyalists, "Whig" and "Tory."

It was a savage neighbor-against-neighbor struggle, on one side those fighting for American independence, on the other those swearing loyalty to Great Britain.

On Feb. 19, 1781, the commander of Patriot militia in Bladen County — today's Bladen, Sampson and Robeson counties — wrote to a fellow Patriot military commander:

"The greatest part of the good people in this County is Engaged back against the Tories and Seems very Loth to go against the British And lieve their families exposed to a set of Villains, who threatens their Destruction."

Col. Thomas Brown wrote that assessment to Brig. Gen. Alexander Lillington, a Patriot commander near Wilmington.

Lillington was pleading for reinforcements to help contain a force of British redcoats who had occupied Wilmington in late January.

Loyalists rally

The British commander, Maj. James Craig, was energetically rallying long-suppressed Loyalists to band together in their own militia outfits. He wanted them to seize control of the Cape Fear region.

Lillington's plight grew out of the military situation in North Carolina in these opening weeks of the seventh year of the American War of Independence.

The Cape Fear region was having to fend for itself against Craig's operations.

It could get no help from the small Patriot army commanded by Gen. Horatio Gates.

That army was fighting for its life in a marching race across the central part of the state, staying a few days ahead of the British army commanded by Lord Charles Cornwallis.

Militia commanders from counties outside the Cape Fear were giving little succor.

In his letter to Lillington, written from the tiny Bladen County seat of

Elizabethtown, Brown complained:

"Enclose letter from Col. Emmet to inform how infamously New Bern District has behaved."

Emmet was Col. James Emmet, commander of the Cumberland County militia, who had begged for Patriot militia troops from the New Bern area to come to the aid of the Cape Fear region.

Despite his problem with the internal threat of resurgent Loyalists in his own area, Brown offered some help to Lillington.

He finished his report:

"I will guard the (Cape Fear) river on acct of the Baggage & as far as lies in my power. I intend setting out for Wilmington on Thursday with what few I can raise; at which time you shall hear from me."

After years of relative quiet in the American War of Independence, the Cape Fear was exploding into the bitter partisan warfare that would be called the Tory War.

Loyalists, who made up as much as a third of the sparse population along the Cape Fear and the upper Little Pee Dee rivers (the latter known as Drowning Creek in North Carolina), had been largely quiescent since the defeat of a little Loyalist "army" at the Battle of Moore's Creek Bridge five years earlier in February 1776.

But the presence of British regulars in Wilmington and the appeals of Maj. Craig were producing results.

While Lillington was able to contain raids by the redcoats in the immediate vicinity of the port town, Brown and other Patriot leaders in the upper Cape Fear were finding it increasingly difficult to contain emboldened Loyalists.

The Loyalists were eager to avenge themselves of what they considered years of oppression by local Patriot officials who generally controlled the county governments of Bladen, Cumberland (which included present-day Harnett and Hoke counties), as well as Richmond and Anson west of the Pee Dee River.

These newly energized Loyalists who Brown called a "set of villains" included many of the colonel's erstwhile neighbors from plantations along the Cape Fear.

Such was the case, too, in the lower part of Bladen, present-day Robeson County, where Loyalists were numerous along Drowning Creek (now the Lumber River) and its tangled boggy tributary, known as Raft Swamp.

Patriots prepare

In a few weeks, Brown and his compatriot in the Bladen militia, Lt. Col. Thomas Robeson, would be desperately rallying their Patriot neighbors to join in the looming civil war.

So, too, in Cumberland County, Patriots were stirring in the districts where Scottish Highlander immigrants were numerous, in present-day Moore and Harnett counties, especially along the Deep River and the Upper and Lower Little rivers.

The growing Loyalist insurgency had yet to intrude on the long Patriot domination of the village of Cross Creek, future Fayetteville, the largest settlement in the upper Cape Fear.

The village of merchant stores, gristmills, sawmills and taverns was still regarded as a key supply base and depot for the Patriot army of Gen. Nathanael Greene and for militia from the southeastern part of the state.

On Feb. 20, 1781, however, Cornwallis broke off the chase of Greene's forces and occupied Hillsborough.

That village, on the Haw River north of Cross Creek, until then had been, like Cross Creek, a Patriot stronghold, where the Revolutionary War governor and other state officials were often located.

Cross Creek was increasingly an isolated Patriot island in a sea of surging Loyalists.

From Hillsborough, Cornwallis once more prepared to take up the chase after Greene's army.

He paused long enough to send a dispatch to Maj. Craig at Wilmington.

Cornwallis directed that the Wilmington forces carry out the long-standing plan to ship supplies up the Cape Fear to Cross Creek.

There they would be available for the British regulars who by then had abandoned most of their supply train in order to move more swiftly after Greene.

With that mission in mind, Maj. Craig stepped up his campaign to bring Loyalists into action in larger and larger numbers.

As spring approached in the Cape Fear region 220 years ago, a new and bloody chapter was about to begin in the Tory War.

Cornwallis' Cross Creek Encounters

Originally published May 16, 1996

Col. James Emmet and his hard-marching band of Cumberland County militiamen trailed into the village of Cross Creek late on April 4, 1781, to find the enemy gone.

Emmet knew it two days earlier. On April 2, 1781, from a camp on the Cape Fear River near present-day Wade, he sent a report to Gen. Nathanael Greene, commander of Patriot forces in the Carolinas. He wrote:

"I received information from a person I sent for that purpose that the enemy left CrossCrick yesterday, they took the road to Rockfish. I am now on the march to Crosscrick, where I shall be happy to receive your commands."

For a week, the veteran Patriot soldier and his part-time troops had been shadowing the long column of weary redcoats.

His orders from Greene were to report every move of Lord Charles Cornwallis' British army as it marched painfully away from the battleground near Guilford Courthouse (present-day Greensboro). There, on March 15, 1781, Cornwallis' men had battered Greene's army in a famous turning point of the Revolutionary War.

Although Cornwallis' hard-bitten regulars had won the field, the cost in dead and wounded redcoats was so high that the general gave up chasing Greene's ragtag force of Continentals and militia. He decided instead to fall back toward Wilmington.

Cornwallis was hoping to stop in Cross Creek, the village consisting of a courthouse, mills, taverns, stores and homes that sprawled around the crossing of a creek by that name, in what is now downtown Fayetteville.

He hoped to find rations for his hungry brigades and time to rest and regroup in a friendly place.

After all, the village of 60 or 70 households was in the heart of the Cape Fear River region, where Loyalist sentiment was said to be burning strongly among recent immigrants from Scotland. Those immigrants were significant in the sparsely settled population.

They would find familiar names even among their own troops, men of the 71st Regiment of Foot, raised in Glasgow and known as "Fraser's Highlanders."

But Cornwallis did not reckon with the tenacity of Patriot leaders such as Emmet and former militia Col. Robert Rowan. Emmett and Rowan had labored for years to make the Cross Creek village a stronghold of the Revolutionary cause.

Emmet and Rowan were early to that cause. Nearly five years earlier, in the summer of 1775, Rowan had circulated the defiant petition — later known as the "Liberty Point Resolves" — that was signed at a Cross Creek tavern by more than 50 men, among them James Emmett.

Since the Loyalist defeat at Moore's Creek Bridge in February 1776, Cross Creek had served as a collection point and depot for provisions supplying Patriot forces in North Carolina.

Emmet became a captain and then a major in the Continental Line. Rowan stayed in the village and directed commissary activities.

In the early months of 1781, Emmet put aside his Continental commission and took command of the Cumberland County militia.

That was how it was when Cornwallis' weary brigades, perhaps 2,500 men, arrived in the village, probably on March 31 or April 1, 1781.

Instead of supplies and friends, he found few rations and a sullen population. Few Loyalists revealed their colors. Out beyond his rear guard, Emmet's militiamen were creeping closer, keeping watch on his every step.

(If you have been reading current headlines, you know that this 18th-century invasion contrasts sharply with the invasion 215 years later, when thousands of British paratroopers are jumping at Fort Bragg and experiencing the hospitality of 20th-century Fayetteville.)

Cornwallis later wrote that he had hoped for hospitality: "From all my information, I intended to have halted at Cross Creek as a proper place to rest and refit the troops, and I was disappointed on my arrival there to find it totally impossible."

What happened next in 1781 was reported to Gen. Greene in a dispatch from Emmet dated April 4 from Cross Creek itself. He wrote:

"The Enemy makes rapid marches, seemingly toward Wilmington, & by information reached Elizabethtown in Bladen at 12 o'clock yesterday, where I learn they will halt some time, to settle the business of your Excellency's flag as well as to bury Col. Webster, died since they left this place."

"The English, according to custom, have left small Pox behind, which makes it out of my power to keep my men in town."

Emmet was wrong about Cornwallis' plans for the tiny Bladen courthouse settlement at Elizabethtown, although correct about the sad story of Col. James Webster.

That gallant British officer was a hero of the Battle of Guilford Courthouse, leading his 33rd Regiment of Foot in the final smashing blow against the American lines.

But he was badly wounded at the climax of the charge.

Carried on a stretcher along with many other wounded, he survived the march until the column passed through Cross Creek. Webster died at some point short of Elizabethtown. He was buried beside the road.

For years afterward, "the grave of Webster" was a legendary but ill-defined spot in Bladen County folklore.

Cornwallis didn't linger at Elizabethtown. By April 7, 1781, the redcoat column was in Wilmington, safe under the guns of the Royal Navy.

Six months later, most of the men who survived the march along Cross Creek's sullen streets that April would lay down their arms at Yorktown in the climax of the American War for Independence.

At least one member of the British column did stay in Cross Creek.

An infant girl, offspring of a camp follower, was left on a bed in the home of widow Betsy Brown.

Nearly 70 years later, the 1850 census listed "Betsy Allen" in the household of Dr. Ben Robinson, Fayetteville's most notable physician-politician. The census-taker noted that she had been "left by Cornwallis."

A 19th-century historian said the abandoned baby was later married to "a man named Allen," and still later as a widow herself "lived for a number of years in the family of Dr. Robinson as housekeeper."

And 40 years after the march in 1821, the 71st Regiment of Foot would leave its name on the land of Cumberland County. The descendants of Scotch settlers who lived on rich farms west of Fayetteville chose "71st" as the name of their township.

Col. Emmet, whose Patriot militiamen shadowed the redcoats, would go on to be postwar sheriff of Cumberland County and keeper of a noted tavern where toasts were raised in 1783 when the name of Cross Creek was changed to Fayetteville. He died in 1787.

Revolutionary War Hero Voluntarily Gave His Life

Originally published June 27, 2002

If the Medal of Honor had existed during the War of Independence, Capt. Denny Porterfield of old Cross Creek, now Fayetteville, would surely have deserved it.

See if you think so.

It happened on Sept. 8, 1781, at the Battle of Eutaw Springs in South Carolina.

On a battleground little larger than today's Pike Field on Fort Bragg, two armies of about 2,000 each, Patriot and British, clashed in a bloody four-hour scramble that left a quarter of the combatants dead, wounded or captured.

The Patriots, commanded by Gen. Nathanael Greene, opened the battle by charging the hastily assembled British line and were at first successful in pressing them back into their encampment on the grounds of a stout two-story brick house.

British rally

But the British rallied and in wild combat on the flanks pressed the Patriots backward.

At the end, the little Patriot force retired from the field.

But on the other side, the battered force of British redcoats and Loyalist militia was so depleted that Lt. Col. Alexander Stuart soon gave up the position and retired toward the port town of Charleston. There, remaining British forces in the South were bottled up. To the north, at Yorktown in Virginia, Gen. George Washington forced the surrender of Lord Charles Cornwallis's army, the band played "The World Turned Upside Down," and American independence was born.

Denny Porterfield never left the field of Eutaw Springs. The young officer of the North Carolina Continental Line was killed in action.

Here is a story of his death written 76 years after the battle, apparently based on interviews with descendants.

It appeared in The Fayetteville Observer in 1857:

"Foremost in this intrepid charge was the high-souled and valorous Denny Porterfield, who seemed to have a charmed life, as he exposed himself upon his mottled charger, with epaulet and red and buff vest on, to the murderous fire of the enemy.

"Lt Col. Campbell (commander of the Virginia brigade) received a mortal wound while leading the successful charge. Porterfield and his brave companions rushed on to avenge his death, and took upwards of five hundred prisoners.

"In their retreat, the British took post in a strong brick house and ploquested (fenced) garden, and from this advantageous position, under cover, commenced firing.

"At this crisis in the battle, Gen. Greene desired to bring forward reinforcements to storm the house. To save time, it became important that someone should ride within range of the British cannon. It was in reality a forlorn hope.

"The American general would detail no one for the enterprise, but asked if anyone would volunteer.

"Instantly Denny Porterfield mounted his charger and rode into his presence. Gen. Greene inquired if he was aware of the peril, if he knew that his path lay between converging fires, and in full sight of the British army.

"Porterfield replied that when he had entered the American army he had subjected his powers of mind and body to the glorious cause, and if need be he was prepared to die in its behalf.

"Greene communicated the command, which was to order into service a reserve corps that lay in ambuscade, ready to advance upon receiving the signal agreed upon.

"With a brave and undaunted bearing, Porterfield dashed off upon his fleet courser, and so sudden was his appearance among the British, and so heroic the deed, that they paused to admire his bravery, and omitted to fire until he was beyond reach of their guns, but on his return, they fired and shot took effect in his breast, and the brave Denny Porterfield fell, and sealed his devotion to the cause with his blood on the plain of Eutaw.

"His horse escaped unhurt, galloped into the American ranks, and never halted until he reached his accustomed place in the ranks.

"General Greene, who witnessed the instinct of the animal, shed tears, and ordered David Twiggs, father of Miss Winny Twiggs now of Fayetteville, to take charge of the horse and carry him to Mrs. Porterfield in Cross Creek. (Twiggs) brought with him the red buff vest that Porterfield wore."

Porterfield's body was apparently buried a few miles from the battlefield.

The article concluded:

"His horse lived for several years, a pensioner, roaming at pleasure on the banks of Cross Creek, known and beloved by all who venerated the valor and chivalry of Denny Porterfield."

Despite its flowery language, the 1857 account of Porterfield's service at Eutaw

Springs fits with the general story of the battle. He was one of three North Carolina captains from the Continental Line killed on the field.

Throughout the years, special military honors have often been awarded to those such as Porterfield who voluntarily take on deadly missions.

The first North Carolinian to receive the modern Medal of Honor, Pvt. Robert L. Blackwell of Hurdle Mills, won this highest of military honors in World War I when he volunteered to go for reinforcements for his hard-pressed platoon and was killed in the attempt.

Porterfield was one of two Continental captains from Cross Creek who fought at Eutaw Springs.

The other I have already written about, Joshua Hadley, who went on to a long life as a pioneer of the new state of Tennessee.

Porterfield and Hadley were young men together in the little colonial village. Both were from merchant families who became early and ardent Patriots.

Porterfield's Revolutionary War career was in some ways a carbon copy of Hadley's.

Both were in the local company raised in the fall of 1775 and later incorporated into the 6th North Carolina Continental Regiment. The regiment joined Gen. George Washington's Main Army near Philadelphia and fought at the Battle of Germantown on Oct. 11, 1777. By then, Porterfield had been a lieutenant for six months.

Porterfield apparently was with the Main Army in the winter of 1777-78 when it suffered in the snow at Valley Forge in Pennsylvania.

Units consolidated

By spring of 1778, the North Carolina brigade was so reduced that the tiny units were consolidated, and Porterfield was transferred to the 1st North Carolina Regiment.

He apparently served in the Main Army throughout 1778 and was promoted to captain on Feb. 1, 1779.

Then it seems Porterfield was among officers either sent home to recruit or put on half-pay. He was not among the prisoners when the entire North Carolina Continental Line was captured at Charleston in May 1780.

However, by the summer of 1781, Porterfield had returned to duty in the small North Carolina "brigade" of 350 men that Gen. Jethro Sumner commanded in the army of Gen. Greene. Porterfield left no direct descendants.

But his name did live on.

His fighting companion from Cross Creek, Joshua Hadley, named his own son Denny Porterfield Hadley.

Loyalists Fought From Swamps

Originally published October 16, 2003

The military climax of the American War of Independence occurred in October 1781 at Yorktown, Va., when Lord Charles Cornwallis and his redcoats surrendered to Gen. George Washington and the French.

But in the region between the Cape Fear River and Drowning Creek in North

Carolina, the chaotic civil war between Patriots and Loyalists was far from over.

As late as the spring of 1782, local Patriot leaders were still trying to bring the area under control, to flush out bands of Loyalists who were camped in the swamps of what is now Robeson County, to sort out the local government activities after a year of confusion and bloody neighbor-against-neighbor warfare.

By January 1782, British and Loyalist military activity was dying down.

Since November 1781, when the British occupying force at Wilmington pulled up stakes and sailed for Charleston, S.C., accompanied by several hundred Loyalist fighters, organized military operations were at an end.

British regular troops posted in upper South Carolina also were falling back to Charleston.

In their wake, bands of Loyalists who had been so dominant at the height of the Patriot-Loyalist civil war in 1781 also were making their way to the protection of the British there.

Pay claims

Among the most noted of such bands was Lt. Col. Hector McNeill's "Regiment of Loyalists."

That unit, recruited in the Cape Fear and Drowning Creek area, had joined with the forces of the most famous Loyalist leader, Lt. Col. David Fanning, to terrorize the area in the summer of 1781.

Now, in Charleston, McNeill's command presented its claims for pay. It listed companies from Shoe Heel, Gum Swamp, Long Swamp and Great Swamp.

Many of the names on the rolls of pay claims are familiar in the area to this day.

Despite their departure, the swamps were apparently still thick with Loyalists.

Especially troubling was a large zone along the South Carolina border that the famed Patriot leader Francis Marion had declared a "sanctuary" or truce area, where Loyalists could go for safety on their promise to stop fighting.

Lt. Col. Thomas Robeson of the Bladen County militia wasn't so confident of their peacefulness.

"Tories are still active"

In January, he wrote to the governor of North Carolina that "the Tories (Loyalists) are still active," and that he was thinking of "stationing a regiment there."

The alleged trouble with "swamp Loyalists" continued during much of 1782.

As late as October, Robeson and other Bladen Patriot leaders were complaining that Loyalists "were doing mischief in Raft Swamp."

By then, however, there was dispute about who was harassing whom. Some observers were accusing Robeson's Patriot militia of acting in the role of "rioters."

In December 1782, the War of Independence in the region could be said to have definitely ended when the British forces evacuated Charleston.

The confused state of local government and the lack of strong law enforcement in North Carolina gave much room for plain criminal activity.

In January 1782, Cumberland County government issued warrants charging two men with "plundering and stealing confiscated property."

Jeremiah Matthews and Reuben Johnston were charged with stealing from the plantations of absent Loyalists whose property had been seized — "confiscated" was the official word — by the Patriot government.

Francis Danielly, who was also accused of "plundering," apparently didn't take the charge lying down.

In January, he was accused by John Robinson of threatening two Patriot leaders, William Rand and Ebenezer Folsom, because they had "informed the commanding officer of Cumberland County militia" that Danielly was stealing confiscated property.

Loose person

Rand and Folsom told the county government, meeting in the village of Cross Creek, that they believed Danielly was "a loose and disorderly person," and that the confiscated property "stands in danger" of being stolen.

The Court of Pleas and Quarter Sessions, the local government of Cumberland County, ordered a peace warrant to be served on Danielly.

The task of settling the fate of confiscated Loyalist property was taken over by a new team of "commissioners of confiscated property" named by the county government.

The commissioners were Robert Rowan, the county's most important Patriot leader, Thomas Armstrong and McNeill Cranie.

The reorganization of the commission took place in the aftermath of one of the war's saddest incidents, a duel to the death between Armstrong and an earlier colleague on the commission, George Fletcher.

The loser in the "affair of honor" was Armstrong's subordinate in the Patriot militia, and while accounts differ, the duel apparently grew out of complaints by Fletcher that Armstrong, serving as head of the militia in the autumn of 1781, was not tough enough on suspected Loyalists.

Peacetime normalcy

By the summer of 1782, there were definite signs that the area was getting back to the normalcy of peacetime.

The county government of Cumberland busily reconstituted local government, naming various civil officers, picking tax collectors in each militia district and appointing no fewer than 24 road overseers responsible for enlisting residents (and their slaves) to maintain various roads and segments of roads in the far-flung county that in those days also included the present-day counties of Moore, Harnett and Hoke.

A striking sign of returning normalcy was the return to Cumberland County of Jane Rutherford, widow of Thomas Rutherford.

Her husband, a prewar county official and ardent Loyalist leader, was captured in 1776 at the Battle of Moore's Creek Bridge and died in 1780 in Charleston after being freed as a prisoner of war.

Jane Rutherford came back to lay claim to her husband's confiscated property, including the country plantation known as "Tweedside."

She would be successful in her quest, marry James Simpson and live for many years as an honored citizen of the new state.

A Revolutionary Patriot

Originally published October 5, 2000

Fayetteville streets named for soldiers begin with Bragg Boulevard and include streets such as Rowan (for Robert of the Revolution), Lee (for Robert E.), Pershing (for John J. "Blackjack" of World War I) and Devers (for Jake of World War II).

And there is Dick Street, named for James Dick.

The street is among the city's oldest. It appears as such first in a deed of 1804, when James Dick sold Joshua Carman a lot on the east side of the street and next to Dick's corner. Before then, the street was referred to simply as the "50-foot street" south off Russell Street.

By then, the street was dotted with several structures, most notably the big angular mansion today known as the Woman's Club, but in 1804 a residence later to house a branch of the United States Bank.

In 1804, James Dick was one of Fayetteville's best-known tavern hosts and captain of a colorful militia organization known as Capt. James Dick's Company of Riflemen.

His place in military history isn't one of uniforms or the smoke of battle.

But his diligence in a variety of tasks for the Patriot cause in the Revolutionary War made him worthy of his postwar esteem. James Dick's story as a Patriot, and his first appearance in the record of Cumberland County, appears appropriately in the summer of 1775 when he was among the 59 signers of the defiant document known as the Cumberland Association, later glorified as the Liberty Point Resolves.

In 1777, Dick, who was a carpenter, was named one of four patrollers in the Campbellton District. Patrollers were part-time officers who enforced curfew regulations for slaves and kept a watch for slaves attempting to escape bondage.

The Patriot county government also chose Dick as a constable in the local militia district.

That office made Dick an important cog in the wheel of the Patriot government of the county, because it was the constable's task to collect taxes from both Patriots and from the sullen and often defiant Loyalists who wanted no part in a rebellion against Great Britain.

By then, the young carpenter was associated with tavern owners who would be his model for his postwar fame as a militia captain and tavern host.

In 1778, for example, he signed the performance bond on the tavern license of George Fletcher, host at the public house in the village of Cross Creek where the association document was circulated in 1775.

Dick was also on the bond of taverner Lewis Bowel, another signer of the 1775 document.

Dick's enthusiastic loyalty for the Patriot cause won him the contract in 1779 to work on a tiny new courthouse for Cumberland County, erected on the site of today's Headquarters Library building on Maiden Lane in downtown Fayetteville.

The new structure was in a sense thumbing a nose at the royal crown.

Before the Revolution, the royal government had turned aside petitions from Cross Creek to locate the courthouse in the village, replacing a structure in the swampy official riverside location of Campbellton.

The Revolutionary General Assembly of North Carolina quickly granted a new petition, and the county government was meeting in the Maiden Lane building by 1779, when carpenter Dick was paid 104 pounds, four shillings and eight pence for work on the courthouse.

Dick's next Revolutionary War service apparently grew out of his connection with James Emmet, a future taverner and clerk of county court who had served in the Continental Army in the early days of the war. By 1779, Emmet was back in the village as a sometime commissary officer for North Carolina militia.

When British forces captured Savannah and threatened an invasion of the Carolinas, Gov. Richard Caswell rallied the militia and at one point in the early spring of 1779 came to Cross Creek to personally supervise supply preparations.

Young James Dick was called on as a dispatch rider for the governor.

He rode from Cross Creek to lower South Carolina with letters for Gen. Benjamin Lincoln, the commander of Patriot forces gathering across the path of the British invaders.

The arrangements for Dick's vital mission were detailed in a letter from James Porterfield, a merchant who was supplying feed, forage and animals for the militia forces.

He wrote to Caswell:

"You request I send Mr. Dick forward with letters from South Carolina.

"I have drawn 162 pounds for the express southward. He is to have 8 dollars a day and 7 and a half for horse.

"It is great wages, but I can not get it down on better terms.

"I have advanced from my own pocket because we have not one shilling of public money on hand.

"The horse he rode south must be left behind, which must be paid for.

"That is 300 pounds for the horse, which I must advance.

"If it is convenient, I would be glad to have $4,000 by Mr. Dick."

Home again

With the war over, Dick returned to his role as a town constable and his work as a carpenter.

A colorful incident in Dick's postwar story came in 1786 when he and four others were convicted of "making a riot" after they publicly horsewhipped tanner Thomas Cabeen and tavern keeper Mary Brown.

The General Assembly remitted the fines of the rioters when word came from the village that the victims were imminently deserving of their bruises because they had "rendered themselves obnoxious by their many immoralities."

Dick's standing in postwar Fayetteville grew considerably in 1788 when he was named a town commissioner, one of five officers who constituted the government of the bustling community. That year, the town raised a brick courthouse and a brick Town House, played host to a session of the General Assembly of North Carolina and hoped to be named permanent capital of the state.

Dick continued in the town government role for several more years, serving with both younger men and with the old Revolutionary War leader of Patriots in Cumberland County, Robert Rowan.

By the first U.S. census in 1790, James Dick's household included him and his wife and children — six boys and four girls.

By then also, Dick may have changed from carpentry to innkeeping.

His household also listed a slave and two "free men of color," including perhaps Isaac Hammond, a Revolutionary War veteran and fifer for local militia organizations.

Dick's role as a postwar militia captain was under way by 1804, when the now middle-aged Revolutionary War dispatch rider was well known as a tavern host.

That year, it was reported that Fourth of July was celebrated at a banquet of the "Rifle Company of Capt. Dick," assisted by Robert Cochran.

A year later, the report was even more elaborate:

It said that the local militia outfit known as Capt. James Dick's Company of Riflemen, "on the Fourth of July, following a discharge of cannon from Liberty Point at noon, met on the Forest Ground where they were joined by a large number of Republicans of the town and its vicinity.

"At 2 p.m. the crowd sat down to a sumptuous meal prepared by Mr. David Sheppard, with Captain Dick as president, and Capt. William McKerral as vice president."

With those festivities, the historical record of James Dick begins to recede.

He was still living when the census of 1810 was taken.

In that year, his daughter Elizabeth married merchant James Neate.

In Wilmington, a William Dick, who presumably was James Dick's son, was operating a tavern.

A commodious motel

In 1809, Dick's was advertised as a large handsome commodious marine hotel, located at the Sign of the Spread Eagle.

William Dick was still in Wilmington when the 1820 census was taken.

And in Fayetteville, the name of militia captain James Dick, ardent Patriot of the Revolution, was perpetuated on the official map of the town published in 1825.

CHAPTER TWO
Civil War

First Months After Secession Were Busy

Originally published February 23, 2006

North Carolina was late joining the Confederacy, but things began to pop soon after it did. The first 120 days were crowded with events in Fayetteville and elsewhere.

North Carolina seceded in early May 1861 after the governor in April refused President Lincoln's call for troops to quell the rebellion.

"You will get no troops from North Carolina," Gov. John Willis Ellis wired.

Even before the formal secession, Fayetteville witnessed a wartime scene when a scratch force of several hundred local militiamen marched to the U.S. Arsenal on Haymount and demanded its surrender.

No blood was shed, but the few federal soldiers at the garrison were soon on their way north.

Within a few months, more than 26,000 North Carolinians donned Confederate uniforms and formed companies that soon gathered in regiments.

In early June, the 1st North Carolina Volunteer Regiment, with its two Fayetteville companies, saw action in a confused skirmish at Big Bethel Church in Virginia.

The North Carolinians drew first blood in what would be the terrible and disastrous conflict that ultimately wrecked the South and preserved the Union.

The regiment, which became known as the Bethel Regiment, had its regimental flag cut and stitched by "the ladies of Fayetteville" and delivered to its bivouac at the grandly named Camp Fayetteville on the Virginia peninsula near Yorktown on Sept. 9, 1861.

A few days later, the regiment changed camps, setting up its tents at Ship Point on Chesapeake Bay about 10 miles south of Yorktown.

The Observer reported: "A letter has been received here from one of our Fayetteville boys who say that he is better pleased with the place than with Yorktown, though the water is not so good. He thinks the change will improve the health of the regiment."

Another account in the Observer mentioned shortages and sickness in the Virginia camps.

"At Yorktown, the best accommodations for the sick are in tents. These tents, at least as so far as the Fayetteville companies are concerned, were furnished here, and not by the state, as were the clothing, blankets and much of the food consumed.

"With great difficulty, some of our volunteers have secured lumber, paying for it themselves, to floor their tents and sheds for cooking and eating under, and we have heard of one mess that paid $50 for cooking apparatus procured recently from Richmond.

"And none of our men have received any pay, except the $10 (enlistment) bounty. We say that it is a moderate estimate that this town has spent $20,000 upon its two companies at Yorktown."

Meanwhile, back home a new regiment was forming, at least on paper.

The unnumbered regiment was being raised by James Sinclair, a sometime Presbyterian minister in Fayetteville. Styling himself "Colonel, Commanding," Sinclair ran a notice in the Observer:

"VOLUNTEERS WANTED. For Young Men of North Carolina. Your state is invaded and calls you to action.

"I have established my Headquarters for the present in Lumberton, Robeson County, and invited the young men of Columbus, Brunswick, Bladen, Robeson, Cumberland, Richmond, Anson, Moore and Montgomery to rally to the flag of their country, hitherto victorious in all pitched battles, and help drive the enemy from our borders.

"Your services will be accepted for 12 months or during the war. In the twofold capacity of Captain in the army of the Lord and of my country, I propose to lead the men under my command to victory over their spiritual as well as their temporal foe.

"God and country! Fellow citizens to the rescue."

In Fayetteville, the new Confederate commander of the captured Fayetteville Arsenal was preparing a major expansion of his command.

He placed advertisements in The Observer inviting contractors to supply 1.7 million "good hard brick" and more than 3,000 separate pieces of cut timber in a variety of lengths and sizes.

In Wilmington, the Civic and Military House, a men's tailoring firm, advertised to "officers of Companies" that it could supply "1,000 yards of gray cashmere and 1,000 yards of gray clothes," as well as "The Original and Elegant North Carolina State Arms Button" for their uniforms.

The book publishing firm of E.J. Hale and Sons, which published The Fayetteville Observer, offered for sale a new collection of "Military Books. For Cash Only."

Among them was a new "colored map of Virginia" and a new map of the Confederate States of America, which was the first to include North Carolina.

For the fighting soldiers the titles included "Roberts Handbook of Artillery," "Instructions on Outpost Duty," "Cooking by Troops" — which was authored by the famous nurse Florence Nightingale — "Infantry Camp Duty," "Science of War," and "The Volunteers Handbook."

The first cloud to appear in these sunny and enthusiastic first months came in the last days of August 1861.

Federal troops waded ashore at North Carolina's Hatteras Island and captured sand forts that were defended by a few Confederate soldiers. The Confederates laid down their weapons and became some of the earliest prisoners of war in the conflict.

Among those captured was Capt. J.A. DeLagnel, who the previous April was wearing the uniform of the U.S. Army in the garrison of the U.S. Arsenal in Fayetteville when it surrendered to the local militia.

DeLagnel had taken the arsenal detachment back to New York but had then resigned his U.S. commission and joined the South.

Four months later, DeLagnel was wounded in a brief skirmish at Hatteras between the fort's garrison and the Union attackers.

The young officer was carried off to the federal POW camp in New York, which was named, of all things, Fort Lafayette.

U.S. Arsenal Early Prize In Civil War

Originally published April 11, 1996

In April 1861, the U.S. Arsenal in Fayetteville was like Fort Bragg in April 1996.

It was by far the most important military installation in North Carolina.

So, when the state prepared to leave the Union and join the Confederacy, the handsome collection of tile-roofed brick buildings on Haymount became an early prize of what would soon be the Civil War.

It happened the afternoon of April 22, 1861. It was contemplated in a short notice by the editor of The Fayetteville Observer published that morning:

"We regret we can not delay the paper this afternoon, but must, to supply the mails, anticipate our usual hour of publication.

"The Editors and all other adults employed about the establishment will be engaged with their respective companies in executing an order from Gov. Ellis to take the U.S. Arsenal at this place.

"It is hoped that the Officers of the U.S. Army in charge of the Arsenal will not require that any blood shall be shed in executing this purpose, for the conflict would be a very unequal and bootless one, the force now being mustered against it being about 900 determined men (against 60), as follows:

Five Town Companies 400
Capt. Bulla's company 200
Capt. Phillips Rockfish 60
Capt. McDonald's (Uniformed) 40
Capt. Marsh, Grays Creek 60
Capt. Nixon, Carvers Creek 60
Capt. Ray, Pine Foresters 40
Company at Cedar Creek 50
"Total of 910.

"The whole 33rd Regiment, commanded by Col. John H. Cook, under the immediate command of Brig. General Walter Droughon.

"It may be well to correct an erroneous impression in regard to the relative position of the U.S. Officers at the Arsenal.

"The Commandant is Capt. J.A.J. Bradford of the Ordnance Corps, and not Brevet Major Anderson. The latter commands the company of Artillery stationed here at the request of the citizens, for the protection of the Arsenal, and is subject to the call, for that purpose, of Capt. Bradford."

Arsenal surrendered

A week later, the Observer reported:

"In a postscript to the larger part of our edition Monday, we simply mentioned the fact that the U.S. Arsenal in this place had been surrendered to the state of North Carolina.

"Instead of 900 men we had been led to expect would obey the call of the colonel of the 33rd Regiment, there were actually mustered 2,050, all of this county except a small volunteer company of about 20 from Robeson County.

"They were all armed with muskets, mostly drawn from the State Arsenal, and provided with ball cartridges.

"After the necessary formalities, the articles of capitulation were signed, the U.S. flag lowered and saluted, and a Southern Confederacy flag was hoisted and saluted.

"This occurred at about 3 o'clock P.M.

"Agreeably to the stipulations, the Southern flag was also taken down after the salute, and neither flag will be raised until after the departure of the U.S. troops.

"The officers and soldiers heretofore stationed at the Arsenal are still here. The latter will soon leave for New York, we suppose, a safe conduct having of course been guaranteed to them, and it being a point of honor with them to rejoin their regiment.

"As to the officers, Capt. Bradford, Brevet Major Anderson and Lt. (J.A.) DeLagnel, none of them will ever draw sword against the South. Of that we are sure. They have many warm personal friends here, who will part from them with great regret.

"The Arsenal Buildings and machinery cost the United States more than a quarter of a million of dollars.

"The machinery is especially very perfect for the manufacture of many implements of war.

"There are 4 brass 6-pounders and 2 brass 12-pounder howitzers, forming a complete 'battery,' in military phrase, with all the horse trappings therefore, and two old-make 6-pounders, 37,000 muskets and rifles, with other military stores, and a large quantity of Powder.

"Lt. John A. Pemberton of the Lafayette Light Infantry is temporarily in charge of the Arsenal."

The men who took over the Arsenal were members of militia companies.

There were volunteer outfits such as the Lafayette Light Infantry and the Fayetteville Independent Light Infantry, and part of the regular state militia establishment.

The 33rd Regiment was the state militia unit covering Cumberland County.

Within a few weeks, several of these companies, the Lafayette Light Infantry first, and many of the other men, would be marching off as soldiers of the Confederacy.

The garrison that surrendered included a company of Regular Army artillerymen of the 2nd Regiment of Artillery. It had two officers (Anderson and de Lagnel), and 56 enlisted men.

The artillery unit had been at the Arsenal only since November 1860.

It was sent down from 2nd Artillery Regiment headquarters in New York at the request of Fayetteville authorities worried that rebellious slaves might attempt a raid such as the ill-fated attack by abolitionist John Brown and his little band against the U.S. Arsenal in Harper's Ferry, Va., a year earlier.

Muskets moved

After that raid, the U.S. Army had moved 37,000 flint muskets from Harper's Ferry for storage in the Fayetteville Arsenal.

A few days later, with flag furled and in silence, men of the company marched down Hay Street to the Cape Fear River boat landing where they boarded steamers and sailed back to New York.

Fifty-eight Aprils would pass before another U.S. Army unit would be stationed at a Cumberland County installation.

In the spring of 1919, soldiers of the 19th and 21st Field Artillery Regiments would arrive at brand-new Camp Bragg.

The editor of the Observer was right about one officer of the artillery company. Lt. Julius Adolph DeLagnel, a 34-year-old native of New Jersey but an adopted Virginian, would resign his U.S. commission and become a lieutenant colonel in the Confederate army.

For nearly three years, he would be second in command of the Confederate States Ordnance Department as that agency strove to supply the guns and bullets for Rebel armies. He died in Washington in 1912. Incidentally, DeLagnel's father of the same name had been an ordnance captain who in the 1830s had drawn general blueprints for a series of arsenals. The Fayetteville Arsenal was built to a modified de Lagnel plan.

The editor was also right about the "Commandant," Capt. James Andrew Jackson Bradford, a Kentucky native then 57 years old.

Bradford, for whom Bradford Street in Fayetteville is named, was the Arsenal's most famous personage, associated with it since construction began in 1838.

He was the first officer on the site, raised the first Stars and Stripes, issued the first contracts, superintended construction, fought for appropriations and whenever posted to some other duty managed to return to his baby on Haymount. He considered Fayetteville his home and was involved in elaborate real estate schemes in the vicinity of the arsenal.

After the flag came down in 1861, Bradford would also don a Confederate uniform. But bad health kept him from duty, and he died in 1863.

The Fayetteville Arsenal would live up to its importance.

Much expanded and with gun-making machinery captured at Harper's Ferry, it would be for three years a vital Confederate arms and munitions factory.

In March 1865, Union Army engineers from Michigan with Gen. William Tecumseh Sherman's bluecoat forces demolished its handsome buildings and set fire to the ruins.

Today, ruins of the arsenal are a state historic site.

On Saturday, April 20, almost 135 years to the day after the 1861 event there, reenactors wearing both Confederate and Union uniforms will stage an "Arsenal Encampment" with displays of drills, campsites and even the firing of artillery.

The encampment is being held in conjunction with the second annual Civil War Conference conducted by the N.C. Civil War Tourism Council Inc.

Fayetteville Represented In First Civil War Scrap

Originally published June 27, 1996

It was only a skirmish, but for the young Confederate soldiers from Fayetteville it was Civil War immortality.

The Battle of Big Bethel Church, just outside Hampton Roads, Va., on June 10, 1861, is considered the first real fight of the North-South conflict.

Most of the guns fired from the Confederate side on that June day 135 years ago were in the hands of men from the 1st North Carolina Regiment.

And two of the regiment's 10 companies were former volunteer militia outfits from Fayetteville, the Lafayette Light Infantry (Company F) and the Fayetteville Independent Light Infantry (Company H).

Only a single Confederate died in the shootout: young Henry Lawson Wyatt, 18, of Edgecombe County. Eighteen Union soldiers were killed or mortally wounded.

The news of the 2 1/2-hour skirmish near a Virginia barn made big headlines North and South. The accomplishment of the "Bethel Regiment" was hailed as proof that one Confederate was equal to 10 Yankees.

In Civil War history, North Carolina used Big Bethel for its justifiable boast that the state's Confederate soldiers were "first at Bethel, farthest at Gettysburg, and last at Appomattox."

The 1st North Carolina, which was enlisted for only 10 months, was soon disbanded, although many of its men would sign up for other units and wear Confederate gray until the surrender at Appomattox four years later.

That the regiment could even be on the field at this first clash of the war was a tribute to the swift mobilization of North Carolina troops once the state reluctantly joined the Confederacy following the firing on Fort Sumter and President Lincoln's call for troops on April 15, 1861.

A week later, on April 22, the U.S. Arsenal at Fayetteville was peacefully surrendered to the local militia.

By May 1, a "camp of instruction" was established in Raleigh where volunteer companies were to be prepared for war. It was "Camp Ellis," named for Gov. John Ellis.

On that same day, the 104 officers and enlisted men of the Lafayette Light Infantry were on their way to the camp.

The Fayetteville Observer reported it:

"The Lafayette Light Infantry took its departure yesterday on the steamer Hurt for Wilmington to go thence by rail to Raleigh.

"An immense crowd attended them to the boat, where a brief address was delivered by the Hon. J.G. Shepherd, with a response by Lt. Pemberton, and a fervent and touching prayer from the Rev. James McDaniel.

"The officers and men of the company were covered with bouquets which were showered on them by the ladies, and departed with the blessings and prayers of wives, mothers, sisters, and friends."

The Lafayette Light Infantry had been chartered in 1856. It joined the 68-year-old Fayetteville Independent Light Infantry as a town volunteer militia outfit.

Before Fort Sumter, it consisted of fewer than 50 officers and men.

In barely two weeks, recruits doubled that number, and the Observer reported that "18 or 20 more" were expected to join the company at Camp Ellis in a few days.

Starr leads company

The "remarkably fine company" was commanded by 31-year-old Capt. Joseph B. Starr, a notable Hay Street merchant.

Starr's father had come to Fayetteville from Connecticut in 1819 and the young Starr was a graduate of Middleton College there.

Soon after returning from his schooling, Starr was instrumental in chartering the militia company and was its founding captain.

Among the company's officers, Lt. John A. Pemberton was also a merchant. Lt. Benjamin Rush Jr. was son of a Cape Fear River steamboat captain.

Many of the men who volunteered after Sumter were from the Rev. McDaniels' congregation at Fayetteville Baptist Church. Others were young farmers from the countryside.

A week later, on May 7, the volunteers of the Fayetteville Independent Light Infantry, "with knapsacks on their backs and all other equipments ready for starting," were about to embark on the Hurt when word came to wait a few days. Camp Ellis was overflowing. The volunteers set up camp about two miles out of town.

On May 8, the orders came. The Observer reported:

"Cheered by an immense crowd of citizens who accompanied them to the wharf," the 108 officers and men of the unit already known as "the second oldest in the country" boarded the steamboat Hurt.

By 3 o'clock the next afternoon they were in Raleigh. They marched from the railroad station to the governor's mansion to be greeted by Gov. Ellis and by former Congressman Warren Winslow, a Fayetteville resident who was Ellis' chief aide for mobilizing the state's men and supplies for the Confederacy.

At Camp Ellis, the two Fayetteville companies were among the first assigned to the 1st Regiment.

They may have mustered in a hurry, and many did not yet have muskets.

But Fayetteville's new Confederates quickly became veterans at something as old as soldiering.

They griped about the rations.

A letter in the Observer only five days after the Lafayette Light Infantry arrived at Camp Ellis prompted an inquiry and a reply in which another correspondent rebutted "complaints about the fare of troops stationed here."

Under a headline, "Glad to Hear It," the second writer assured readers that, "these complaints are unfounded, we have plenty of good and substantial plain food."

The first Fayetteville company to arrive was praised by the same correspondent: "The Lafayette Light Infantry are in good health and spirits, and are improving rapidly in drill."

On May 20, 1861, the Fayetteville companies and a company from Richmond County, nicknamed the "Southern Stars" and designated as Company K, were the first North Carolina troops to "take the cars" for Virginia.

The other seven followed within a few days, and by early June the regiment was in the small Confederate force standing guard on the road from Norfolk to Yorktown.

They waited barely a week before they carved their niche in history by being there when the first shots of the Civil War rang out at Big Bethel.

Newspaper Provides Window To Civil War

Originally published April 18, 1996

In World War II, we got our news over the radio from Edward R. Murrow, H.V. Kaltenborn and Gabriel Heater.

In the Persian Gulf war, it was on television from CNN.

In the Civil War, it was in the columns of newspapers.

When the Civil War started 135 years ago this month, The Fayetteville Observer became a veritable official gazette of military affairs, a weekly compendium of war news from all over.

In the spring of 1861, columns once filled with political news and stories of the county fair were filled with reports of local men flocking to Confederate colors, listing the names of every recruit.

At least one of its four pages was devoted to war news from looming battlegrounds in the west and in Virginia.

Editorially, the newspaper encouraged civilians to donate money for "comforts" for new soldiers. It also sternly warned against wasting gunpowder on civilian pursuits.

As new Confederates marched off to war, there was practical advice to the troops from an 1861 equivalent of the armchair military expert, usually a retired general or admiral like you see today on television.

In April 1861, departing soldiers could read how to survive the rigors of military life from a short article addressed to "young soldiers" and signed simply: "An Old Soldier."

Under a headline, "Advice to Volunteers — How to Prepare for the Campaign," the expert wrote:

"1. Remember that in a campaign, more men die of sickness than by the bullet.

"2. Line your blanket with one thickness of brown drilling. This adds but four ounces in weight, and doubles the warmth.

"3. Buy a small India Rubber Blanket (only $1.50) to lay on the ground, or to throw over your shoulders, when on duty during a rainstorm. Most of the Eastern troops are provided with these. Straw to be on is not always at hand.

"4. The best military hat in use is the light-colored soft felt; the crown being sufficiently high to allow space for air over the brain. You can fasten it up as a continental in fair weather, or turn it down when it gets wet or very sunny.

"5. Let your beard grow, so as to protect the throat and lungs.

"6. Keep your entire person clean; this prevents fevers and bowel complaints in warm climates. Wash your body each day, if possible. Avoid strong coffee and oily meat. General Scott said that the too free use of these two (together with neglect in keeping the skin clean) cost many a soldier his life in Mexico.

"7. A sudden check of perspiration by chilly or night air often causes fever and death. When thus exposed, do not forget your blanket."

From his reference to Gen. (Winfield) Scott and his focus on Mexico, it is clear the "Old Soldier" picked up his campaign wisdom in the Mexican War of 1847-48.

Quaint as it may seem, his advice was sound for its day. Disease was indeed the big killer of 19th-century soldiers. Respiratory and bowel ailments would slay many more rebels and bluecoats than the bullets and bolts of the enemy.

Heat would not be as much of a problem in the campaigns of 1861-65 as it was in the deserts of Mexico.

But his warning against "oily meat" and appeal for frequent bathing was sound advice in an era that had yet to grasp the germ theory of disease.

The warning against "strong coffee" must be counted the least heeded of any of the recommendations.

Civil War soldiers — both North and South — liked coffee often and in large quantities. It was the single most indispensable ration and comfort they demanded.

Having looked after the physical well-being of the departing soldiers, the Observer turned to their morale.

Music was an important part of military morale in the 19th century. Every regiment had its band of bugles, drums and horns.

And while there would later be many war songs for singing on the march or around the campfires, they were in short supply in the spring of 1861.

Nonetheless, someone had already come up with "North Carolina's War Song," new lyrics sung to the tune of a popular ballad, "Annie Laurie."

Along with the advice from "an Old Soldier," the Observer offered in the same column this first attempt at martial music:

We leave our pleasant homesteads,
We leave our smiling farms,
At the first call of duty
We rush at once to arms —
We Rush at once to arms.
To guard our roads we fly,
For the land our Mothers lived on,
Bravely to bleed or die.
Up boys, and quit your pleasure,
Up! men, and quit your toil,
The invader's foot must never
Be pressed upon our soil —
Be pressed upon our soil.
In which our fathers sleep,
Their blessed graves our care, boys,
Most sacredly must keep.
T'was in our brave old State, men
That first of all was sung
The thrilling song of Freedom
That through the land hath rung,
That through the land hath rung;
And we'll sound its notes once more
'Till our men and children shout it
From the mountains to the shore.
Sweet eyes are filled with tears, Men,
Sweet tears of love and pride,
As our wives and sweethearts bid us
Go, meet what'er betide,
Go, meet what'er betide.
And God our guide shall be,
As we drive the foe before us,
And rush to victory.

War Service Takes Toll On Militia

Originally published October 31, 1996

All through the summer and fall of 1861, columns of men in Confederate butternut uniforms tramped through Fayetteville's village streets on their way to the Cape Fear River landing.

On the river, the brand-new steamboat A.P. Hurt and others pulled up to the muddy bank to embark these new-minted soldiers for assembly camps at Raleigh or Garysburg or to "the front" in Virginia.

Since the past spring when North Carolina joined the Confederacy, more than 1,000 men from the counties of the upper Cape Fear had enlisted in old militia companies or newly formed outfits financed by local patriots.

Within a few weeks after the state's secession, Fayetteville's two volunteer militia companies, the Fayetteville Independent Light Infantry and the Lafayette Light Infantry, had marched to the river landing.

By June, they were on the Yorktown peninsula in Virginia as companies H and F, respectively, of the 1st North Carolina Infantry Regiment.

Soon, other companies came through from rural Cumberland, Robeson, Moore, Montgomery and Sampson counties.

Some marched. Others arrived on the cars of the single railroad line that angled from Fayetteville through the Sandhills to the Chatham County coalfields.

As summer came on, the rookie Rebel soldiers saw some action.

On June 10, 1861, the 1st North Carolina skirmished with Yankee columns at a Virginia country church known as Big Bethel on the Yorktown peninsula.

It was the first battlefield action of the Civil War and a North Carolinian was the first (and only) Confederate killed in the fight.

Three months later, however, the war was real.

In late July came the first big fight, the Battle of Manassas in Virginia. A Confederate win.

A month later, on Aug. 21, Union troops stormed ashore and captured Fort Hatteras on Hatteras Island on the North Carolina coast. A Confederate disaster.

A month later, Confederate troops in mountainous western Virginia were suffering from epidemic camp diseases and frustrated by terrible weather.

Grim report

In October of that autumn 135 years ago, young Capt. Jonathan Evans sat down to write a mournful report for the hometown newspaper.

Evans, only 18, had marched to the river landing with the more than 100 men of the "Cumberland Plough Boys," a company raised in what is now the Linden area of the county.

Arriving in the Virginia mountains as Company F of the 24th North Carolina, the unit was battered by illness.

Evans wrote to The Fayetteville Observer:

"The following is a list of the Cumberland Plough Boys who have died recently. Please publish for the information of their friends."

In two weeks in October, the company lost John Barnes, R.P. Freeman, James Edge, John Saw, L. Faircloth, James Collier, Daniel Bain, and James Cobb.

Evans continued:

"I am very glad to state that the health of our company is somewhat better. Although many of our men are unable to perform duty, yet we have only two or three dangerous cases.

"Still, there are many who will never open their eyes again upon their homes and the loved ones they left behind.

"The surgeon of our regiment has resigned and gone home, his failing health having rendered this course necessary.

"Dr. Millard, who was sent through the humane exertions of Mr. J.C. Blocker, arrived here yesterday and proceeded immediately to Blue Sulpher Springs to attend to our sick brothers. The regiment is to be moved tomorrow to that place, where most of our sick are."

Death arrived at Camp Fayetteville, the bivouac of the Fayetteville companies of the 1st North Carolina on the Yorktown Peninsula.

On Sept. 10, John B. Clark, 22, of the Lafayette Light Infantry (F) was reported dead of disease. On Sept. 12, 17-year-old James Weymss of the Fayetteville Independent Light Infantry (H) was reported dead of disease.

Bodies sent home

The corpses of the two young men were shipped to Fayetteville, where they were buried with military honors by the Home Guard militia company known as the Clarendon Guards.

William McKethan of Company H died a week later "at a private house in the neighborhood of 'Camp Fayetteville.' "

Then, in November, wonderful news.

The Fayetteville companies were coming home.

The 1st North Carolina, you see, was a "six-month" regiment. Its service was finished.

While many of the 200 men of the companies would go on to fight with other units, they first would see the Fayetteville they left in the spring.

The Fayetteville Observer's editor reported the homecoming of Friday, Nov. 15, 1861:

"On Friday afternoon, especially, a steam whistle from the river started thousands, a living stream of humanity, men, women, and children, white and black, in carriages, on horses, and on foot, including large numbers from the surrounding country.

"They came between 6 and 7 o'clock at night, and found the Clarendon Guards, the Ordnance Company from the Arsenal, and a large concourse of people waiting to welcome and escort them to town.

"The streets blazed with bonfires. The Town House had been decorated with great taste, under the direction of Mr. William H. Delany, Superintendent of the Gas Works. Festoons of evergreens had been arrayed from the upper part of the cupola, extending along the east front of the building and from it to corners of the streets.

"Flags floated from the buildings, and a large inscription lighted up at night with many jets of gas: 'WELCOME, HEROES OF BETHEL.'

"Under this, the Hon. Jesse G. Shepherd and William Mcl. McKay, at the request of the ladies, expressed in eloquent words what all felt, the universal joy at the return of so many of our 'best and bravest,' crowned with laurels, not the least of which had been the patient endurance of trials and hardships such as fall to the lot of few in life, returning too with the well-deserved reputation of being the model regiment, for gentlemanly deportment, in all the service."

Parade and a cake

"On afternoon the following day, the Lafayette Company was invited to parade to receive a superb cake, presented by the ladies.

"The presentation given through R.P. Buxton, Esq., who delivered an eloquent address of welcome, and read the beautiful lines addressed to the regiment by a Lady.

"Malcolm McDuffie and Thomas C. Fuller, of the company, responded.

"Then there an invitation from the Independent Company to the Lafayette Company to repair to a private house and partake of refreshment, and the march resumed, to the inspiring strains of the superb Band of the company which had been the admiration of the Peninsula.

"Due honor was paid to the refreshments, and the company returned to the Arsenal for dismissal."

In that autumn of homecoming, the unknown lady poet whose lines were read by R.P. Buxton spoke for what would turn out to be only a brief happy interlude in the darkening history of the Civil War:

The trumpets sound,
Is on the air;
They come — the brave, The young, the dear.
Their wandering over,
They come, they come;
From toil and privation
We welcome them home.

Coastal North Carolina Had A Role In Civil War

Originally published February 14, 2002

The February weather was brisk that morning 140 years ago when the gray-clad soldiers marched off the decks of the five little steamers to the docks at the coastal town of Elizabeth City.

A military band on shore came smartly to attention and then struck up a lively medley of welcoming tunes, ending with a snappy rendition of "Dixie."

It sounded like a proud moment for the year-old Confederacy.

But the reality was that these were prisoners of war, being released under parole.

They were captives from the first major Union victory of the Civil War, the capture of Roanoke Island on the North Carolina coast.

As the ill-clad Carolinians came down the gangplanks, the air was as rich with irony as it was with the music of the band of the 24th Massachusetts Regiment.

For while the 1862 history of the Civil War would soon resound with the names of battles much more famous, battles such as Shiloh and Seven Pines, Second Manassas and Antietam, the little shootout amid the sand bogs and marshes of Roanoke Island wound up being the first note of the funeral dirge of the Confederacy.

From then until the inevitable end of the war, the North Carolina coast was an often-neglected backwater of the conflict.

But that does not mean it was unimportant.

Soon after the victory at Roanoke Island, Union forces took control of the entire Carolina coast except for the port town of Wilmington.

From places such as Plymouth, Elizabeth City, Washington and Beaufort, Yankee soldiers enforced the authority of the federal government far behind the battle lines where the big armies fought in Virginia and Pennsylvania, demonstrating the hollowness of the Confederate claim to sovereignty even over its own territory.

To these towns, thousands of African-Americans fled slavery to begin experiments in self-government or to don Union army uniforms, a glaring warning to Southerners that the war they had started in defense of slavery would surely end in the demise of that "peculiar institution."

Divided loyalties

The federal presence also gave heart to nonslave North Carolinians who never accepted the Confederacy or who despaired of its future, or who simply grew weary of the neglect of the Confederate government. A study of white men from Bertie County who wore military uniforms found that while 804 joined the Confederate army, 159 joined the Union army.

Another 69 Union soldiers were once Confederates who deserted and joined the bluecoat army. And 398 African-Americans of Bertie joined the Union army.

This sort of divided loyalty mirrored North Carolina's broader experience as a Confederate state.

The federal occupation irritated relations between North Carolina's government and the government in Richmond, Va., the former demanding more help in expelling the invaders, the latter usually begging lack of resources for the task.

The result was a state where criticism of the Richmond government was often as harsh as any coming from Washington; where desertion was a constant problem for the military; and where anti-secession sentiment remained a potent political force years after the state had reluctantly joined the Confederacy.

Finally, the story of the North Carolina coast in the war would end in an appropriate crescendo almost exactly three years after the fall of Roanoke Island, when in February 1865, other Union troops marched into Wilmington, completing the capture of that "bastion of the Confederacy," closing the last port still in Rebel hands.

Personal connection

I am native to the place of this history, born only a few miles from where Union troops on Feb. 18, 1862, burned the Chowan River town of Winton.

While like all fans of the War to Suppress Southern Secession, I have read about all the big battles and in recent years reveled in Gen. William T. Sherman's adventure in the Cape Fear. This flat country of farms and forests, of rivers and swamps, sounds and islands, has always been my favorite locale for Civil War history.

When I was a boy, I first stood next to the historical marker on the bluff at Winton and looked down to where the USS Delaware and the USS Commodore Perry brought the colorfully uniformed troops of the 4th New York "Zouaves," Col. Rush Hawkins in command, up the Chowan.

They would make history by setting tar barrels ablaze and using them as fiery torches in razing the tiny courthouse town, the first such "total war" act of the conflict.

My family story includes that of a great-grandfather who joined Confederate cavalry from Bertie to defend the Chowan River line, but then found himself in Virginia and Pennsylvania, where his regiment was "ill used" by Confederate commanders in fights leading up to Gettysburg and on that famous battlefield.

Like dozens of his Hertford and Bertie neighbors, including a brother, he came home to his farm and simply quit or joined the Union army at Plymouth.

And then there is my wife's great-grandfather, first a soldier and then a sailor for the Confederacy, who was lucky to be home on furlough in 1864 when his ship, the ironclad CSS Albemarle, was blown up by a Union "spar torpedo" at its moorings at Plymouth on the Roanoke River.

When word came of his fortuitous escape from the blast, they say he went to

Browns Chapel Baptist Church, made his first connection with the Lord, and never got far from the church until he died.

A history itch

Whenever the anniversaries of these events roll around, I get that old Civil War history itch.

And this season, there is a marvelous new book that finally tells the story of all this, giving due attention to its richness and diversity.

It is "The Civil War in Coastal North Carolina" by John S. Carbone, published by the N.C. Division of Archives and History. It's a 175-page paperback. If you get the itch, too, there is no better way than this book to be introduced to this compelling story.

The war is long over, but the history is as alive as ever.

Soldiers In Gray Stand Tall In Battle

Originally published May 1, 1997

A gloomy late afternoon drizzle soaked the gray uniforms of the men of the 5th North Carolina Regiment. They were deploying into the battle line on the edge of a wheat field, in sight of the historic old town of Williamsburg, Va.

It was nearly 5 o'clock on the wet afternoon of May 5, 1862.

The 400 men of Col. Duncan Kirkland MacRae's outfit had just emerged from a half-mile tangle of trees and briers to find themselves far out on the right of their brigade line.

More than a half mile away, across the open ground, they could see the flash of a cannon's mouth. A Union battery was already firing into other Confederate troops several hundred yards to their left.

Suddenly, they were in motion. With an impetuous shout of "Follow me!" Brig. Gen. Jubal Early sent the widely separated regiments charging toward the flashing cannon.

For 43-year-old Col. MacRae and for several dozen younger men who hailed from Cumberland County, Early's command commenced a bloody mission that would win comparison to the British at the famous "Charge of the Light Brigade" and rank as a smaller version of the ill-fated "Pickett's Charge" 14 months later at the Battle of Gettysburg.

When the fight ended, nearly three out of four of them would be dead or wounded, many of the latter made prisoner.

An admiring correspondent for the New York Herald watched from behind the Union cannon and described what he saw in the wheat field:

"Still the enemy formed across the opening with admirable rapidity and precision, and as cooly as if the fire had been directed elsewhere, and then came on at the double quick step, in three distinct lines, firing as they came.

"Few brigades mentioned in history have done better than that brigade did. For a space which was generally estimated at three-quarters of a mile they had advanced under the fire of a splendidly served battery, and with a cloud of skirmishers stretched across their front, whose fire was very destructive, and after that, they had not the nerve to meet a line of bayonets that came at them like spirit of destruction incarnate."

Col. MacRae saw the fight from the other direction. His after-action report described it:

"The approach to the battery was through an open field of soft earth without any cover for my troops. I dispatched my Adjutant, Lt. (James Cameron) MacRae, and Major P.J. Sinclair to request which battery we were to charge. Major Sinclair returned to answer that I was to charge the battery that opened on us, and do it quickly. I immediately put the line in motion, and the men sprang forward at rapid pace."

"Pressing on from the first in the face of the battery, entering the plunging fire of the infantry, wading into a storm of balls which first struck the men on their feet and rose upon their nearer approach, (the regiment) steadily pressed on.

"Officers and men were falling rapidly under the withering fire of grape, canister, and musketry. Lt-Col. Badham (second-in-command) was shot in the forehead and fell dead. Major Sinclair's horse was killed and he was disabled. Captains Lea, Garrett, and Jones were all shot down, as were many of the subalterns.

"My color bearer was first struck down, and then the flag was seized by his comrade, who fell immediately. A third took it and shared the same fate, when Capt. Benj. Robinson of Co. A carried it until the shaft was shattered in his hand."

The cost: 290 lives

A regimental historian later wrote:

"Four hundred and fifteen men answered roll-call that day. Before the night, the blood of 290 fed the soil of that bleak hill."

The wheat-field charge of the 5th North Carolina that shattered against the guns and muskets of Union Gen. Winfield Scott Hancock's battery and brigade ended the Battle of Williamsburg. The battle started as a Confederate rear-guard defense against Gen. John McClellan's Army of the Potomac advancing up the Yorktown Peninsula toward Richmond.

For Col. MacRae, it was nearly his first and last action.

The Fayetteville-born lawyer's health broke a few months later, and he spent the rest of the Civil War as a Confederate emissary in Europe and as editor of a short-lived "official" newspaper, the Confederate.

After the war, he returned to lawyering, first in Memphis, Tenn., and then in Wilmington, where he died in 1888.

Three other men from Fayetteville mentioned in MacRae's battle report also survived the war.

Capt. Benjamin Robinson, the 19-year-old commander of Company A who had the regimental flag staff shattered in his hands, fought with the 5th through all the battles of the Army of Northern Virginia until he was crippled at the Battle of Spotsylvania in the spring of 1864.

He came home to be a lawyer, newspaper editor and author of a novel that included slightly fictionalized scenes of his battlefield experiences.

He died, too, in 1888.

Twenty-three-year-old Lt. MacRae, a nephew of the colonel, served throughout the war as a staff officer. He returned to Fayetteville to become a legislator, Superior Court judge and Civil War raconteur.

Maj. Sinclair, 27, son of Fayetteville's Presbyterian minister, had originally commanded Company A. Wounded again two months after Williamsburg, he resigned in December 1862, "by reason of personal difficulties in (the) regiment."

He went on to be a postwar newspaper editor in several North Carolina towns.

For the two West Pointers, Confederate Gen. Early, who ordered the 5th on its bloody charge, and Union Gen. Hancock, who commanded the guns and muskets that stopped it, Williamsburg was a prologue to Gettysburg 14 months later.

Early, who was badly wounded only minutes after shouting his order at Williamsburg, may have developed a skittishness about frontal attacks, a caution that had monumental consequences at Gettysburg.

On the first day of that battle, Gen. Robert E. Lee urged Early to attack disorganized Union forces who were retreating to Cemetery Hill. Early declined.

On the hillside, Gen. Hancock directed the regrouping of Union forces.

Two days later, on July 3, 1863, Hancock commanded another height crowned with batteries and muskets when another Confederate charge, the climactic attack of Pickett and Pettigrew's divisions, "the high tide of the Confederacy," shattered there.

Ailments Kept Many Out Of Civil War

Originally published July 18, 2002

When hundreds of presumably healthy Cumberland County men marched off to fight the Yankees in the Civil War, they left behind hundreds who were not so healthy.

By 1863, the third year of the war, the list of diseases, ailments, conditions, bodily breakdowns and general complaints that afflicted men called to duty with the county's Home Guard units read like a veritable encyclopedia of 19th-century medical misery.

Col. Walter Draughon, the 70-ish commander of the 14th and 25th Battalions of the Home Guard, would have been hard put to fend off an invasion with such troops as William Cameron, whose medical examination revealed "impaired vision & nervous affection & inguinal hernia."

But Cameron's condition was not the most desperate of the more than 120 men who had such examinations in 1863 and 1864.

The list has been published by the Cumberland County Genealogical Society from an abstract prepared by Dr. Stephen E. Bradley Jr.

Beyond help

You have to remember that Civil War medicine was distinguished more by what it could not treat than by what it could. Except for amputations, battle surgeons could offer little help to the wounded.

And back home, a home guardsman diagnosed with "organical disease of the heart, dropsical" could expect little relief, even if, as in the case of John N. Wilder, he was only 19 years old. Or similarly, in the case of D.W. Royal, who was 23 and described as "feeble, some little heart disease."

The most common medical condition of the Civil War home guardsmen had to do with their bowels and with the age-old curse of all soldiers, a case of the piles, or of hernia.

T.J. Owen, 41, suffered with "hemorrhoids (internal)." Saul Skinner, age 30, had "excessive hemorrhoids," as did J.G. Boon, age 39.

G.W. Bullard, age 44, was cursed with "double inguinal hernia," as was Willie Barefoot, age 38. R.W. Cashwell, age 35, had a "large double hernia, discharge." Isaac Hollingsworth reported "disease of the kidney and piles."

Then there were the "prolapsus anus," suffered by John McGilvary, 45, and John W. Matthews, age 44.

"Chronic diarhoea" afflicted Henry D. Brandt, who was only 23. G.W. Gainey, 43, and H.B. Ferguson, 34, reported "incontinence of urine."

Some of the conditions described by the Civil War examiners reflect medical nomenclature that is hard to translate into what we understand as modern medical conditions.

Alex McPherson, for instance, was diagnosed with "anasarcous" and was "very feeble."

Jessee Godwin had "phthisis plumo." James McAllister, age 44, had "ozena." William B. Bridgers, age 45, suffered from "coxalgia, discharge."

Pulmonary disease was a perennial health problem in the 19th century.

T.W. Fort, age 37, suffered from "hemorrhage from lungs" and was a "feeble man." W.S. Jenkins, age 35, and W.A. King, age 38, were coughing up blood. Thomas Gilmore, 34, was "consumptive." And J.H. Freeman, 42, had "a bad cough."

Heart disease was chronic in 19th-century health conditions.

Disposed to dropsy

M.J. McCoy had "asthma and disease of the heart." S.A. Phillips at 37 had "some disease of the heart with disposition to dropsy (edema)."

D.W. Royal at 23 reported "some little heart disease." N. Wall, 49, had "functional disease of the heart." W.H. Tomlinson at 40 suffered "organic disease of the heart," as did John A. Wilder, 19, who was also "dropsical." William Wade reported "some affection of the heart."

"Curvature of the spine" was the complaint of John S. Smith, age 25, and James McMillan, age 28. William Bow, 44, suffered "a nervous disease resembling epilepsy."

Some complaints seem minor, even today. M. Strickland had a "cicatrix (scar) from a burn." John S. Smith, age 19, had "an old burn on side and arm."

Abram Guin had a "cut on foot." Malcolm Fort said simply that he had "headache."

Dozens of home guardsmen were afflicted with some form of "rheumatism" and many others were "dyspeptic." Varicose veins were frequent, and Isaac Williams was to be pitied for, at age 36, "loss of teeth (entire)."

Then there were those whose medical conditions were decidedly vague, but no less distressing to read.

Neill R. Blue, 41, was simply "a feeble man," as were David Hall, 31, and Duncan Munns, age 40.

R.D. Davis was diagnosed with "general infirmity." Azal Marsh, age 47, listed "general complaint." John W. Thompson was cursed with "general debility," as was William J. Cowan, 47.

'Generally infirm'

William Wood, 49, was "generally infirm," and Joseph Piner, only 27, was also listed under "general infirmity."

B.F. Moore, 31, had a more serious condition, "the loss of right eye." And his condition led the doctor to remark: "Worthless in camp."

But perhaps the ultimate medical insult, if not a very succinct diagnosis, was that laid on 20-year-old Jere Jackson.

The examiner wrote: "Small and puny."

Volley That Felled "Stonewall" Likely Fired By Tar Heels

Originally published May 2, 1996

It was the most dolefully famous incident of "friendly fire" in U.S. military history.

It happened 133 years ago today. It was 9 p.m. May 2, 1863, when a fusillade of smoothbore musketballs crashed through a Virginia woods and found their mark.

Among five victims killed or mortally wounded by the volley was Lt. Gen. Thomas J. "Stonewall" Jackson, the South's most celebrated soldier of the early Civil War.

In a new book about aspects of the Battle of Chancellorsville, the story of the wounding and subsequent death of Jackson is characterized under a chapter title, "The Smoothbore Volley That Doomed the Confederacy."

The loss of the famous rebel commander was doubly mournful to the South because the volley that killed him wasn't fired by soldiers dressed in Union blue.

The hail of smoothbore balls that killed Stonewall Jackson and four others was unleashed by a battle line of his own men.

And all evidence, including that of the newest retelling of the incident, assigns the friendly fire to a single regiment of Confederate soldiers, the 18th North Carolina.

For the Cape Fear region, the story is especially galling: Nearly half the men of the 18th North Carolina were from Bladen, Columbus or Robeson counties.

Moreover, it is highly probable that the musketry that brought down Jackson was fired by men of Company K of the 18th, a unit enlisted in 1861 at the little courthouse village of Elizabethtown as the "Bladen Guards."

Dropped twice

The new recounting of the incident of 133 years ago is in "Chancellorsville: The Battle and Its Aftermath," a collection of essays edited by Gary Gallagher and published by UNC Press at Chapel Hill.

It doesn't alter the hundreds of previous accounts, except to make the point that Jackson might have survived the initial wounds that smashed his right arm if only they had not been badly torn open when he was accidentally dropped twice by stretcher-bearers.

A Cumberland County native who was the second in command of Company K that evening saw the shooting.

Alfred H. Tolar, later captain of the company, was a 21-year-old lieutenant at the time. Twenty-one years later, in January 1884, he wrote a detailed account for The Fayetteville Observer.

Tolar described how in the early evening his men had cheered as Jackson, accompanied by Gen. A.P. Hill and more than a dozen aides, rode their horses through the 18th's newly established battle lines.

The impatient Confederate commander was dashing forward to the skirmish line to reconnoiter what was to be a night attack against battered Union lines. The lines formed after Jackson's late-afternoon surprise attack scattered the Federals west of the crossroads of Chancellorsville.

Hill's division had missed that affair but was deployed for the after-dark action that would fulfill Jackson's repeated orders to "press on, press on!"

"A cavalry charge"

Tolar described how sometime later, as occasional musketry splattered all along the Confederate lines, his own command heard the crashing of horsemen to their front.

Rattled by a furious bombardment by Union artillery that had just ceased landing nearby, the Confederate soldiers were acutely aware of the night sounds that followed in the silence. A whippoorwill called. Then came the noise that Tolar said convinced them that "a cavalry charge" was coming at them through the darkness.

At the command of 32-year-old Col. Thomas J. Purdie, the regiment commenced firing and kept it up until, as Tolar put it, "General Hill and several staff officers were in our ranks, shouting at the top of their voices, 'Cease firing!' "

Tolar wrote, "Our regiment was fully aware of the terrible mistake that they had made, within 10 minutes after it happened."

Tolar's account differs somewhat from others who give the "commence firing" order to young Maj. John D. Barry. Barry was a former private who was actually in command of the firing line when Col. Purdie — who had also reconnoitered to the front — dashed back, yelling for his men to "fix bayonets."

The South was stunned by the loss of Stonewall. It occurred just as he had routed the Federals in a flank march that turned out to be the climax of the Battle of Chancellorsville. That battle is always considered Robert E. Lee's most brilliant victory and could rightfully be called "the high tide of the Confederacy."

For the men of the 18th North Carolina, their role in the tragic friendly fire remained forever with them. Tolar wrote that several days later, when his brigade paraded to hear the official news, "I do not think there was a dry eye in the regiment." And in a comment that might have applied to the Confederacy itself, he wrote, "Up to this time, victory had perched upon our banners in every fight, and from that time, the god of war seemed to have deserted us." Yet, that night, little time existed for reflection.

The next day, the regiment was badly bloodied on the same battlefield. Purdie was mortally wounded when Hill's division battered against the Union lines.

Dismal statistics

The 18th was destined to fight on for nearly another two years as the Army of Northern Virginia next swept north to Gettysburg, where Tolar was badly wounded, then fell back for the campaigns of 1864 north of Richmond and then hung on in the lines around the Confederate capital in the despairing spring of 1865.

In its years of campaigning, hundreds of men of the regiment — often ironically referred to as "the bloody 18th" — were killed, died of wounds or disease or did not survive captivity in Union prisoner-of-war camps.

The dismal statistics of Company K tell the story: Sixteen were killed in action, and 34 were wounded. Six of them died of their wounds. Forty-one were captured at one time or another. Twenty-one were listed as deserters.

When the Army of Northern Virginia laid down its arms at Appomattox Court House in the spring of 1865, fewer than 20 of the 18th were there to answer roll call.

Few Reasons For Thanks In 1863

Originally published November 26, 1998

On the first national observance of Thanksgiving in 1863, the United States had something to celebrate.

In the Confederate South, the story was different.

As November came in the towns and on the farms of the Cape Fear region, the autumn brought mostly gloom. Hopes that were high in the first two years of the Civil War were fading.

Newspaper columns were sad with the names of Confederate soldiers from North Carolina who had been killed or subsequently died of wounds suffered on battlefields at Chancellorsville and Gettysburg.

There was suffering on the home front as well.

Shortages of food and medicine were combined with a looming breakdown of the farming economy because so many men were absent from their farms.

Prices were skyrocketing because of scarcity and the declining buying power of Confederate money.

The county government of Cumberland was struggling to buy corn for the families of soldiers.

In early November, a committee of Cumberland County's largest planters met at the courthouse in Fayetteville and pledged $50,000 for relief.

The committee's resolution hinted at the war weariness already manifested in the third year of the conflict:

"Whereas, in this war which is waged against our lives, our property, and our liberty, and which we have neither the power to prevent nor to end if we would, and whereas through the misfortunes of war we have lost much of our productive lands, and through the partial failure of two successive crops, the scarcity of provisions in this immediate section is such that suffering will ensue among the indigent unless, a supply is procured, we resolve:

"We will assist in all manner to prosecute this war to an honorable end. We will comply with the Confederate tax law by cheerfully paying our tithe.

"We petition the county government to levy a tax to secure bread for the support of the poor of the county, especially the families of our soldiers, and that corn be given to them instead of money as heretofore."

The meeting set up committees in each of the county's townships to supervise distribution of whatever corn could be secured.

But that would not be easy.

Scarcity of goods

In prewar days, corn could be had for 90 cents a bushel.

But the weekly list of commodity prices in The Fayetteville Observer for November 1863 quoted a price of $9 to $10 a bushel and said the supply was "very scarce" even at that price.

Scarcity had sent the price of corn whiskey to $85 a barrel, from a prewar price of half that.

In the town of Fayetteville, the November relief effort started with a $42,000 subscription to provide at-cost commodities "for all classes."

Another $20,000 was pledged for relief of "the poor of the town."

The poor relief, which mainly was aimed at war widows and families of soldiers, was distributed by a committee in each of the town's election wards.

Notably, each ward was also authorized to have a committee of "four ladies of each ward" to assist in identifying the recipients.

An economic consequence of the growing despair over the future of the Confederacy was manifest in Robeson County that November by the complaint of another farmer, Speir Walker, in the matter of the forced sale of one of his slaves to satisfy a debt to Charles Ivey.

Ivey and the sheriff of Robeson were demanding payment in coin, hard cash, for the slave, rather than in Confederate paper money or a note.

Walker ran a notice in the newspaper:

"I have offered the parties double the amount without any trouble, as it is well known that gold or silver cannot be procured at a time like the present. They wish to get my property for nothing or extort from me that which it is impossible to obtain.

"Citizens are indignant. This is setting a very bad precedent and is entirely at variance with the laws and customs of the country at a time when we are struggling for our very existence against the most unprincipled and hateful nation on earth.

"We have those amongst us holding high office in the county discrediting the currency, the Government, and everything connected with it.

"Should not some stronger law be passed by the Legislature to put a stop to all such proceeding, having a tendency to cripple the Government and depreciate the currency.

"I am an old man and have two sons in the Army, am a farmer and feed the soldier's wives free of charge when they come to my house without money. Neither Sheriff King or Charles Ivey has any son in the service, nor as far as I am aware, done anything for the cause."

The legislature couldn't control the economics of increasingly worthless Confederate money.

But the Confederate government increasingly sought to augment its battered military forces through the hated Conscription Act.

Deserters and dodgers

The swamps and forests of the region were increasingly full of young men dodging the law, or soldiers who had quit the war.

The Winston-Salem newspaper reported, in a story reprinted by The Fayetteville Observer, that a five-week campaign by conscription officers, carried out by "Captain Barry and his small but efficient band of mounted men," had captured 150 deserters in the county, a third of whom chose a prison camp rather than return to the battle fronts.

The commander of the artillery company at Fort Caswell on the coast used the threat of the conscription law for recruiting.

He ran a notice pointing out that:

"Young men having to go in service soon, who have not been conscripted, will find it in their interest to come forward and volunteer where they can have comfortable quarters and be well cared for. Bounty of $100."

There was in this season a growing feeling that this was, in an already popular saying, "a rich man's war and a poor man's fight."

That was highlighted by a list of the 2,760 men in the 4th Confederate Congressional District who were exempt from the conscription law.

They included such essential workers as two apothecaries and 47 locksmiths, three telegraph operators and 24 physicians.

But there were also 26 "county and state officers," as well as several score militia officers, dozens of justices of the peace, 32 overseers, and 140 better-off men who had hired substitutes to go to war in their place.

Still in the growing despair of a Confederacy that was only 18 months from its demise, there were some thankful acts.

The Fayetteville Observer thanked farmer Harris Tyler of the Deep River district, who sent a barrel of flour to the town's relief effort.

"This is a remarkably liberal act," said the editor, "because he recently suffered loss of several thousand worth of property by incendiarism of deserters."

Civil War Brought Out Best In Folks

Originally published December 16, 1993

Santa Claus and big military events have often come together to the Cape Fear region.

In 1918, the Christmas season witnessed the essential completion of construction on the sprawling new military cantonment named Camp Bragg in the pine woods of the North Carolina Sandhills, even as the first young men from the area returned from service with the American Expeditionary Force in World War I.

In 1941, only days after the Dec. 7 attack on Pearl Harbor, civilians of the area and thousands of troops at Fort Bragg joined the nation in spending its first Christmas at war in World War II.

In 1944, families and friends of paratroopers who trained at Fort Bragg sweated out the news of the Battle of the Bulge as the 82nd and 101st airborne divisions stood across the path of a surprise German winter offensive in Belgium.

In 1989, the season was interrupted while units from Fort Bragg took part in Operation Just Cause, the U.S. excursion into Panama.

In 1990, many families of soldiers at Fort Bragg spent the season without daddy or mother as major units were deployed in the Persian Gulf.

There was a Christmas season of another sort 130 years ago, in 1863.

The little town of Fayetteville was going through its third holiday season of the Civil War.

In contrast to more hopeful times in 1861 and 1862, this season was gloomy.

Newspaper columns were thick with the names of Confederate soldiers from North Carolina who had been killed or subsequently died of wounds suffered on battlefields such as Chancellorsville, Gettysburg and Chickamauga.

And death did not always come in battle.

Death of a cousin

Malinda Ray, a Fayetteville teenager who kept a diary of the war years, reported:

"We heard today of the death of Cousin Lauchlin Ray. He was in the army in Va and was taken sick. None of his family could get to him. He was a fine young man and has fallen as many do a victim of this cruel war."

There was suffering on the home front as well.

There were shortages of food and medicine, and the farming economy was breaking down because so many men were absent from their farms.

Prices were skyrocketing, and the buying power of Confederate money was declining.

At its December meeting, the Cumberland County court appropriated no less than $100,000 for the purpose of buying corn for the families of soldiers.

The price of war

Before the war, corn could be purchased for 90 cents a bushel. At the end of 1863, the price had increased to $9 to $10 a bushel.

Scarcity sent the price of corn whiskey to $85 a barrel, from a prewar price of half that. Equally precious was salt, quoted at $17.50 per bushel, 10 times the prewar price.

So scarce was salt, in fact, that it was bought and distributed by a state government agency.

A notice in the Observer informed citizens of the Rockfish Factory Village — present-day Hope Mills — that their allocation of "State Salt" would be delivered during the week before Christmas.

"Should any remain over," said the notice, "it will be distributed to soldiers' families immediately."

The shortage of manpower was having an effect on the village's major wartime facility, the Fayetteville Arsenal and Armory, which was struggling to fulfill its quotas of arms and ammunition for Confederate forces.

Call to patriotism

On the day after Christmas 1863, Col. C.L. Childs, commanding officer of the facility, advertised for "a number of first-rate blacksmiths."

He pitched the call to patriotism.

"All such who wish to do good service to the Confederate States, now have the opportunity offered them to serve their country as acceptably as they could in Tennessee or Virginia (where the armies were) and at a same time receive a good wage for their labor."

The looming prospect that the new year might bring the war even to the upper Cape Fear River also prompted Col. Childs to gain permission to recruit "a Company of Mounted Riflemen" as a roving scouting party to watch out for Yankee raiders who might venture into the vicinity.

The company was authorized to recruit as many as 100 men, but they had to be "non-conscripts," that is, men who were not subject to the Confederate draft. Such men would thus have to be under 18 or over 45 years of age, or former soldiers discharged for wounds or other causes.

A $100 bounty

Volunteers to the company would receive a $100 bounty in Confederate money.

In the Christmas season of 1863, the highlight was a Christmas event called the "Tableaux" presented on the evening of Dec. 22, 1863, by the young ladies of the Juvenile Knitting Society in the town's largest public meeting place, known as Farmer's Hall.

The society was one of several organizations that contributed to the war effort by knitting and sewing uniforms and other clothing for Confederate soldiers.

The program of various patriotic scenes was being held "for the Benefit of the soldiers." Admission was $1, and "half price for children and servants."

An organization dedicated to helping troops, the Cumberland Hospital Association, had joined others in donating food and comforts to the Confederate hospital in Fayetteville.

The chief surgeon of "General Hospital No. 6" thanked the "Ladies of the association" and others "on behalf of the sick and wounded at the Hospital."

The hospital, located in a former school for females on Hay Street, served furloughed sick and wounded soldiers from the area.

Holidays in camps

As the people of Fayetteville struggled to make the most of Christmas, many men from the area would observe the holidays in winter camps in Virginia and on other war fronts.

Such a unit was Company K of the 18th North Carolina Regiment, which had fought in all the major battles of 1863, and which in May had the doleful distinction of mistakenly firing on and mortally wounding its own beloved commander, Gen. "Stonewall" Jackson.

From its winter quarters in Virginia, Capt. Alfred H. Tolar of the company (known as the "Bladen Guards") sent a message home, published in The Fayetteville Observer. It spoke for all the men of the area in that holiday season:

"To the citizens of Bladen County:

"We are faring finely in winter quarters, with the exception of covering. We only have drawn six blankets for 30 men this winter, and as the men all lost their blankets on the Pennsylvania campaign, I am necessarily compelled on behalf of my company to call on you to furnish us with the number required.

"You always have helped us when needed, and I do hope and believe that you will continue to bestow your kind favors on your care-worn Boys, who are willing to lay down their lives for your sakes at any moment."

Area Swamps Were Refuge During War

Originally published May 10, 2001

The swamps of southeastern North Carolina — the boggy wetlands with names such as Great Coharie Swamp, Raft Swamp, Collie Swamp, Big Marsh and the Carolina Bays — are places of mystery and beauty, today visited mostly by hunters, fishermen and outdoor enthusiasts.

During the Civil War, these same lonely lowlands were home to bands of desperate men.

When you think of the war of 1861-65 hereabouts, you usually recall the Fayetteville Arsenal, or Sherman's march through the area, the fights at Monroe's Crossroads, Averasboro and Bentonville.

But there was another side to the military history of the war.

And it is intimately connected to the swamps.

The wartime swamps in Robeson, Bladen and Sampson counties were hiding places for several classes of men who were escaping from the war.

Large numbers were eluding the Conscription Service of the Confederate government, men who either were against the war or who simply had no stomach for wearing a uniform.

Others were men who had worn the uniform, but who had had enough, deserters from Confederate service who took to the swamps with a price on their heads and the threat of hanging or a firing squad if they were captured.

Still others wore the remnants of blue uniforms. They were Union soldiers who broke out of Confederate prisoner-of-war camps, especially in the last months of the war, trying to make their way to marching bluecoat columns or to coastal towns occupied by Federal forces.

And there was that especially intrepid class, the runaway slaves. They were prized by the others because they were often the most adept at finding the deepest, and safest, parts of the swamps.

The swamps were part of their history as human chattel.

For years prior to the war, their ancestors had been fleeing bondage by plunging into the swamps, often eluding authorities for months and even years at the time.

During most of the Civil War, the nearest place to the Cape Fear region under Federal control was New Bern, occupied in 1862 and the headquarters of Union forces holding most of the North Carolina coast except Wilmington.

Thousands of runaway slaves from all over eastern North Carolina made for New Bern, where they lived in large camps.

Getting to New Bern from the upper Cape Fear meant using the lonely network of swamps as a thoroughfare to freedom.

Service hunted men

The most despised of Confederate authorities were the Conscription Service's "man hunters," who by 1864 consisted of entire companies of detached-duty soldiers and Home Guard members.

Scores of other men, many of whom had served in the Rebel army earlier but had fulfilled their term, were game for the hunters.

To elude the hunters, they took to the swamps, often gathering together in armed bands, willing to fight to stay out of Confederate uniforms.

By far the most famous fugitive from Confederate conscription was Henry Berry Lowry, celebrated as the "Robin Hood of the Swamps" because of his exploits in the lowlands of Robeson County during and after the war.

This hero of Lumbee Indian history disappeared in 1872, after more than three years of eluding first Confederates and then postwar lawmen.

He was celebrated in the popular press because, among his other exploits, he supposedly hid escaped Union prisoners of war who fled to the Lumber River wetlands after breaking out of a Confederate prisoner-of-war compound in Florence, S.C.

Confederate deserters always made up a large percentage of the Civil War swamp population.

The terrible casualties of the war by 1863 prompted thousands to desert, to risk death in order to get back to their homes.

By Christmas of that year, the problem was so prevalent that commanding officers of whole companies ran notices in The Fayetteville Observer, listing names of soldiers who were absent, and offering the standard $30 per head bounty for information about them.

Rampant desertion

The problem was so acute locally that the Observer made a suggestion.

Citing an Orange County businessman who had offered $100 each to deserters who would turn themselves in, the newspaper said:

"Is not the example worthy of imitation. What say you citizens of Fayetteville and Cumberland?

"Cannot a fund be raised among them? It might not only be well for the Army and the deserter, but save some of the plunder by which those in the woods too often live."

The approach of Gen. William Tecumseh Sherman's army in 1865 swelled the numbers of erstwhile Confederates taking up temporary residence in the swamps.

Desertion was rampant in the Rebel columns opposing Sherman.

Col. Alfred Rhett of Charleston, captured by Sherman's cavalry on the battlefield at Averasboro, boasted of his methods of fighting desertion, saying:

"I have shot 12 men myself in the last six weeks."

He said he had used hunting dogs to capture deserters in a swamp.

When the remnant of the Confederate Navy from Charleston set up a base near Fayetteville in early 1865, the commanding officer ran a notice in the Observer ordering his marching sailors to head for Richmond. The notice warned that anyone who didn't hit the road would be treated as a deserter.

Finally, of course, the Civil War swamps were refuge to another class, to escaped jailbirds and men who simply took advantage of the dislocation of wartime life to pursue their law-breaking careers.

The Observer was peppered with notices about a notorious horse thief who broke jail in Richmond County, and who then collected a gang of swamp desperadoes who raided pastures and barns throughout the area.

He made a mistake that other swamp people did not. He came to Fayetteville and was captured without a fight by the local lawmen.

The sheriff of Cumberland County ran a notice inviting other sheriffs to send in their claims for him.

May's A Month For Beauty, War, Death

Originally published June 3, 1999

The month of May brought the bright beauty of a blooming spring to the Cape Fear region.

But in 1864, May also brought dark news of wartime death and misery.

As the fourth year of the Civil War opened, hundreds of Cape Fear men were away in Confederate armies that were about to face their sternest tests in a conflict that had already brought sad news to hundreds of families in the area.

When the month opened, a huge Union army and a dwindling Army of Northern Virginia faced off across the Rapidan River north of Richmond.

The new Union commander, U.S. Grant, would shortly send his columns plunging on an offensive into the already bloody ground known as the Wilderness. His goal was the defeat of Robert E. Lee's forces, the capture of Richmond, the end of the war.

Two states away, William Tecumseh Sherman's blue columns set off at the same time, marching for Atlanta, executing the other arm of the vast Union pincer.

Once Grant's army moved, blue and gray would fight every day of the month, in the Wilderness, at Spotsylvania Courthouse, in the "Mule Shoe" salient, at Drewry's Bluff on the river gate to Richmond.

In the Cape Fear region, the month opened with some rare good war news.

Confederates had captured the town of Plymouth on the Roanoke River in eastern North Carolina, and the ironclad Albemarle was on the loose against the Union fleet.

The first news from Virginia came two days after Grant began his offensive.

Under a headline, "The Great Battle," Editor E.J. Hale wrote in the Monday, May 9, edition of The Fayetteville Observer:

"Nothing has been received here of later date than General Lee's dispatch of Friday night last (May 6), which came at noon on Saturday.

"On 2 p.m. that day it was announced that the line was down between Weldon and Petersburg, and has not since been working.

"A dispatch from the Junior Editor of the newspaper, dated on Friday, reported 'heavy fighting yesterday afternoon and this morning.' "

The junior editor was Edward Joseph Hale, a 28-year-old major on the staff of Confederate Gen. James Lane, whose brigade was in the middle of the Wilderness battle.

The ominous tone of the Saturday dispatch prompted quick reaction among workers at the Fayetteville Arsenal.

The battalion of uniformed workers known as the Arsenal Guard "handsomely tendered their services with almost unanimity to go to Virginia."

The Observer pointed out that three of the battalion's companies "were composed of artisans," who were exempt from fighting service.

The Confederate War Department turned down the offer, urging them to continue work making guns, ammunition and artillery carriages.

Even before details of the fighting in Virginia could arrive, the victory at Plymouth that raised such hopes at the beginning of the month also brought sad news.

Capt. L.R. Breece of E Company of the 8th North Carolina Regiment sent word on May 9 to C.R. McKethan of Carvers Creek in upper Cumberland County that his son, Lt. James McKethan, of the company, had been killed in the assault on the town.

McKethan had been an original volunteer in the company, which called itself the "Manchester Guardians" because the members came from the vicinity of the mill village of Manchester, today's Spring Lake.

Deluge of blood

Then came the deluge of blood.

The next three editions of the Fayetteville newspaper were gazettes of killed, wounded and missing.

Each edition carried several columns listing North Carolina casualties.

Cape Fear families read such reports as that of Lt. John F. McArthur of Company K of the 38th North Carolina Regiment:

"May 5: Pvt. N.L. Campbell, mortally wounded, died in about five hours; David Ray, mortally wounded, since died; W.S. Jackson, in the arm; J.E.J. Cain, slight in hand and knee; May 12 at Spottsylvania: Pvt. W.H. Taylor, slight in arm; Corporal William McP. Geddie, severely in the fleshy part of hip by a six-pound cannon shot."

Nearly all the Company K men came from what is now Hoke County, then the western part of Cumberland. They called themselves the "Carolina Boys."

A South Carolina regimental surgeon reported to a Bladen County family:

"Lt. John H. Tolar, 8th S.C. Volunteers of Bladen County fell bravely leading his company, pierced through the head with a minie ball. He was in 18 battles."

A Fayetteville family heard from another surgeon:

"Lt. Charles T. Haigh fell while gallantly leading his company, (B, 37th N.C.) in charging a Yankee battery. A member of the company reports he was shot through the head, the ball entering above the right eye and coming out on the left and back part of his head.

"Our men were unable to hold the ground where he fell, so I was unable to secure his remains.

"Col. Barbour has no officer whom he esteemed more highly than Charlie."

Haigh was 19, only a few months from graduation at Virginia Military Institute.

From the 52nd North Carolina Regiment, the report came that Capt. J.W. Kyle and Lt. James Huske of Fayetteville had both been wounded in the head.

Capt. Bennie Robinson of the 5th North Carolina, a veteran of every battle of the Army of Northern Virginia, was reported with a "shatttered ankle" at Spotsylvania. He was 20 years old and would go on to write a novel containing eyewitness accounts of the 1864 battles.

Young Capt. Breece had another report from the "Manchester Guardians," who had returned to Virginia from the Plymouth expedition and were skirmishing with Union forces along the James River.

May 14 shootout

Breece reported casualties in a May 14 shootout near Drewry's Bluff:

"Wounded: Sgt N.A. Gilmore, severely in foot; D.A. Cameron, in breasts (supposed mortally), and taken prisoner; Arnett Deal, in ankle, severely; Josiah Deans, in hand and back, severely; A. B. Gunter, finger shot off; H. Faircloth, in face, slight; Pvt. J.S. Bard, in both legs, seriously: Robert Gilmore, thigh, seriously; Missing: H.D.

Burns (supposed killed); Wounded, George Cox, in hand, while charging yankee works; Pvt. B.F. Ringold, in breast, supposed mortal."

The month of May in the Cape Fear region 135 years ago turned out to be more like a winter of death than a springtime.

Cape Fear Suffered In Fall Of 1864

Originally published November 24, 1994

In 1864, as the leaves fell and the third winter of the Civil War approached, home front conditions in the Cape Fear region were as bleak as the landscape.

While Civil War history is usually remembered in stories of valiant battlefield exploits, the realities of hard times back home are not so celebrated.

But in that late November 130 years ago, there was no escaping the sorrow, suffering and despair that grew darker with each passing week.

The columns of The Fayetteville Observer vividly reflected the conditions.

For despite the perennial editorial optimism of the newspaper, the news from the battle fronts was almost uniformly bad.

In Virginia, Union forces had drawn a tight cordon around Confederate armies at Petersburg. Union forces were devastating the Shenandoah Valley.

In Georgia, the bluecoats of Gen. William Tecumseh Sherman were rampaging toward the coast.

Meanwhile, the Union blockade was tightening. Only the port of Wilmington was still open to Confederate blockade-runners, and there were reports of an impending Union attempt to assault and seize it.

This action on the battle fronts meant tears back home.

After 40 months of fighting, hundreds of Cape Fear men were dead, hundreds more wounded, and hundreds were in northern prisoner-of-war stockades.

The harvest of sorrow that such losses caused is vividly described in the autumn diary of Malinda Ray, a Fayetteville teenager. In a typical entry, she wrote:

"Tommy Hybart died Sept. 16 at Smithville. He was with the 17-year-old boys, the Survivor Reserve. His mother was with him during his illness and brought his body home. James Huske was killed near Petersburg on Oct. 26. He has been in the war since commencement of it. His mother, a widow, is left desolate. She had two sons three years ago, now they are both gone. News has been received Nov. 13th of Lincoln's reelection. I know not whether to be glad or sorry. McClellan would probably have conducted the war more humanely, but perhaps Lincoln will be wearied out and make peace in the next four years, tho there is little prospect of it now."

The home front sorrow over such losses was compounded by suffering brought on by shortages, high prices and the demands of an increasingly dictatorial Confederate government.

Relief organizations trying to help fatherless families found it hard to acquire staples at any price, although government price controls ostensibly tried to make Confederate money valuable.

A relief committee in Fayetteville that had exhausted a $22,000 fund from the previous winter tried to raise another. It collected $41,000 from leading professionals and businessmen and had bought 500 cords of firewood.

But, as in nearby Orange County, the offered price of $8 for a bushel of wheat or $6 for corn was countered by a demand as high as $25 for wheat.

This problem infuriated the 68 members of the governing board of Orange County. They published a notice accusing local farmers of "making more money than you ever did in your lives," while "soldiers' wives are suffering and their children are crying for bread."

The problem of prices was not confined to civilians.

The Confederate army's commissary office in Fayetteville had to put a notice in the newspaper that it "had not impressed any bacon or corn" but was still offering cash for it.

The officer admitted, however, that he had "threatened in one or two cases where extortionate prices, above the current rates, were demanded."

The problem of providing relief for thousands of suffering families was also reflected in the problem of the Confederate government in supporting the armies.

Under new tax laws, farmers were required to make in-kind donations of their crops amounting to a "tithe" of their assessed production.

Confederate tax officers in every county were busy publishing notices of where they would be on hand to receive lists of "corn, Irish potatoes, rice, cured fodder, cured hay, buckwheat, sugar, molasses, peas, beans, ground peas and cotton."

Farmers who were "in the habit of pressing their Forage or having Cotton presses either at home or near" were required to bale their "tithe" of these commodities before delivering them to the tax agents.

Although the tax laws were widely ignored or circumvented, tax dodgers who were caught were subject to a 500 percent additional tax.

The difficulties with providing supplies and money for the army were matched with difficulties in providing manpower.

In addition to the horrendous casualties, more and more Confederate soldiers were calling it quits, leaving the battlefields, often hiding out in swamps and forests near their homes, deserting from a fight they no longer were willing to endure.

They were often joined by others resisting the Confederate Conscription Law.

The general breakdown of authority had fostered a widespread lawlessness, and gangs of thieves and robbers also roamed the countryside.

In addition, hundreds of former slaves had taken to the woods.

The problem was so acute that the Observer made a suggestion.

Citing an Orange County businessman who had offered $100 each to deserters who would turn themselves in, the newspaper said:

"Is not the example worthy of imitation. What say you citizens of Fayetteville and Cumberland?"

Meanwhile, as the Union vise tightened, military authorities struggled to fill the ranks.

Lt. W.C. Rencher, in charge of the Fayetteville enrolling office of the Conscript Service, reminded soldiers ages 18 to 45 who were home on furlough or recuperating from wounds that they were required to report on the afternoon of Oct. 10, 1864, "in every way prepared to proceed to Camp Holmes (at Wilmington)."

Only "employees of the Fayetteville Arsenal and Armory" were exempt.

Anticipating what the likely response would be, Lt. Rencher warned darkly:

"Arrangements will be made to send a guard at once after any one who may show himself recusant."

Although the Fayetteville Arsenal was thus still given priority to continue making weapons and ammunition for the war effort, that major Confederate facility was nonetheless facing growing supply and manpower difficulties.

And as a poignant footnote to a time when shortages were a way of life, when even the most humble resource was precious, and when desertion was rampant, the arsenal commandant, Lt. Col. F.L. Childs, published this notice:

"STRAYED. From the Fayetteville Arsenal and Armory, on the 9th instant, a black MARE MULE, about 8 years old, a high head in harness, slight gear marks. A liberal reward will be paid to any one finding and bringing her back. Persons are cautioned against trading for her."

Fayetteville Was Once Home To Confederate Naval Base

Originally published February 26, 1998

Sure, there's Fort Bragg. And once there was the U.S. (later Confederate) Arsenal. But a naval base in Fayetteville?

You've got to be kidding.

That was probably the reaction of Lt. Charles Porter McCorkle of the fast-disappearing Navy of the Confederate States of America in that cold January 133 years ago.

McCorkle arrived in the Cape Fear River town of Fayetteville shortly after the turn of the new year in 1865.

And, indeed, his mission was to establish a CSA naval installation on the banks of the narrow, often shallow-running Cape Fear.

On Jan. 11, 1865, he put a notice in The Fayetteville Observer with the heading:

"Office of Naval Ordnance,

"Fayetteville."

The notice solicited building supplies, "sills, plates, lifters, fencing, boarding, and 60,000 brick."

He signed the notice:

"D.P. McCorkle,

"Lt. Commanding,

"Naval Ordnance Works."

A week later, another notice solicited: "Two good Negro blacksmiths, current wages."

And so began a brief, and still obscure, naval presence in Fayetteville as the Civil War rushed to a close in North Carolina.

That presence included, in addition to McCorkle's mission, a few days when the vicinity of Fayetteville evidently was headquarters and camping ground for the officers and men of the Confederate naval base at Charleston, fleeing overland after that birthplace of secession fell to Gen. William Tecumseh Sherman's Union army in mid-February 1865.

A second article in this series will tell the story of the famous Confederate naval officer whose command camped on the banks of the Cape Fear River.

For McCorkle, trying to build a naval gun works on the Cape Fear was a distinct comedown in his long career as a leading installer of ordnance for the Confederate navy.

But he really had very little choice.

Four years earlier, he worked to arm the famous James River fleet of converted steamers. That tiny flotilla kept the Union navy at bay from 1861 until the Army of Northern Virginia abandoned Richmond and headed for surrender at Appomattox Court House in early April 1865.

Three years earlier, he worked at Norfolk, Va., in the famous Gosport Naval Yard, where the captured U.S. Navy ironclad Merrimac was fitted out for Rebel service as the CSS Virginia.

That 3,400-ton vessel, a new sort of fighting ship, was preparing for her famous shootout with the USS Monitor in March 1862. That historic fight ended in a draw.

But two months later, she was beached and burned by her own sailors as federal forces recaptured Norfolk.

That same spring of 1862, McCorkle was on duty in the yard at New Orleans when the great homemade Confederate ironclads, Louisiana and Mississippi, were under construction.

But again his duty was short-lived. Both vessels were destroyed by their own sailors when Adm. David Porter's federal fleet sailed up the Mississippi and seized New Orleans in April 1862.

So, when McCorkle got to Fayetteville, he was accustomed to fleeing from victorious Yankee enemies.

The notices that he put in the local newspaper are practically all we know so far about McCorkle's quixotic mission in Fayetteville.

Where his "ordnance works" was located is only a matter of conjecture. Or whether he ever even got it off the ground.

The Cape Fear waterfront in those days was a busy place of steamboat wharves, with one or two small boat yards where a half-dozen small river steamers were built in the antebellum years.

The work force at the Confederate Arsenal a mile away on Haymount boasted workers familiar with the manufacture of small arms, including some "Negro blacksmiths" so prized for their vital 19th-century skills.

We do know that at the time, Fayetteville was increasingly a place of refuge for the dying Confederacy.

Even as McCorkle placed his notice with editor E.J. Hale of the Observer, news came of the fall of Fort Fisher at the mouth of the Cape Fear River, the last great coastal bastion of the Rebel cause.

A month later, federal forces marched into Wilmington, and the Confederacy's last port was gone.

The surrender of Fort Fisher and Wilmington unleashed a flood of civilian refugees, wounded soldiers and deposed government officials.

This desperate tide of humanity made its way up the Cape Fear to where a Confederate flag still waved over the arsenal at Fayetteville.

By late February, that flood from the lower Cape Fear mingled with columns of Confederate soldiers retreating from Charleston.

Not all soldiers, however.

There were sailors, too.

Fall Of Fort Marked Confederacy's Collapse

Originally published February 9, 1995

The morning of Jan. 16, 1865, dawned clear over the smoking ruins of Fort Fisher, the great Confederate bastion at the mouth of the Cape Fear River that had fallen to a Yankee invasion force the night before.

Confederate Pvt. William Hooper Haigh, a 41-year-old lawyer turned soldier from Fayetteville, sat on a battered sand wall. As he waited to be shipped north as a prisoner of war, Haigh flipped through a wad of Confederate paper money that he rescued from his knapsack.

With an ironic shake of his head, he told a fellow prisoner he would keep the worthless money "in memoriam of a dead treasury and a fast-dying nation."

He was right, for the capture of Fort Fisher that night 130 years ago opened the final season in the life of the Confederacy. In three months, Robert E. Lee would surrender at Appomattox Court House, Va.

The 60 days after Fort Fisher's fall would be the last days of the Confederacy in the Cape Fear region.

On Feb. 23, 1865, Wilmington would fall to Union forces. On March 8, the bluecoat armies of Gen. William Tecumseh Sherman would cross into North Carolina near Laurel Hill. On March 11, they would seize Fayetteville, stay three days and destroy the Confederate Arsenal and Armory, newspaper offices and cotton factories. In another week, they would fight battles against Confederate forces at Averasboro and Bentonville.

When 1865 began, the Cape Fear home front had a brief moment of good news in the otherwise bleak prospects of the Confederacy.

A Christmas assault on Fort Fisher had been repulsed, and soldiers of the bastion's artillery garrison, men from Sampson and Cumberland counties, had written home with enthusiasm.

Short-lived hopes

The high hopes were short-lived. The Union forces returned Jan. 12. After a naval bombardment, troops stormed into the huge sand redoubt on the night of Jan. 15, capturing or killing most of the artillerymen who manned its guns.

Even as the Yankee fleet came into view at Wilmington, the Cape Fear River, swollen by two weeks of rain, poured out of its banks in a record flood that sent water swirling almost to the Market House in Fayetteville, overrunning mill dams and machinery. With these events, home front morale plummeted.

Editor E.J. Hale of The Fayetteville Observer, an ardent Confederate to the last, held out hope that Confederate peace commissioners who went north in late January might yet win peace and independence for the South.

The mission failed, and on Feb. 16, Hale and other political leaders called a town meeting at which speakers exhorted citizens to rally once again to the war effort.

In a final editorial on March 6, 1865, Hale could offer only the Bible as consolation: "Let us trust in God and take courage. We shall surely conquer."

A week later, Sherman's engineers destroyed the presses and offices of the newspaper.

Despite the trials of the war, commerce continued, including the sale of slaves.

The falling value of Confederate money was indicated, however, by prices paid for slaves sold in Moore County: A 27-year-old field hand sold for $11,500, a 5-year-old boy for $2,300. They were among more than a dozen slaves sold in mid-February by the estate of James Worthy of Craig's Creek.

The sagging value of Confederate money was starkly highlighted in the Observer's weekly preview of prices being paid in Fayetteville stores.

Bacon cost $6 a pound. Potatoes were $25 a bushel. Flour was $580 a barrel. Corn whiskey had gone to $125 a barrel, and corn was $90 a bushel.

Militarily, right up until Sherman marched in, Fayetteville was a center of Confederate logistical and mobilization activities.

With most able-bodied men either away in the army or hiding out as deserters, mobilization centered on home guard units and on work details of impressed slaves.

In January, Lt. W.C. Rencher, in charge of the Fayetteville enrolling office, called on wounded men, paroled men, men on furlough, free blacks and junior reserves to report for duty.

J.A. Spears, commander of the 52nd N.C. Militia Regiment in Harnett County, called on slave owners to meet to choose an overseer for slaves impressed to "work on fortifications."

In Robeson County, Maj. Charles Little called out members of the 10th Battalion of Home Guard to gather in Lumberton on Feb. 25.

Role as medical center

With the fall of Wilmington, Fayetteville's role as a military medical center grew.

Hundreds of wounded from the lower Cape Fear joined the many other wounded from other battle fronts who were authorized to seek care at Branch General Hospital No. 6. The convalescent hospital was in the handsome three-story brick building on Hay Street built as a prewar school for girls. It was just west of the present-day Arts Center.

There, surgeons Drake and Fessenden and town physicians, aided by a corps of dedicated women volunteers, tried to keep up with patients even as supplies of medicines and bandages dried up.

On March 5, the hospital was notified to expect an additional 400 patients from the Wilmington area. The ladies of the town pleaded for poultry, eggs, flour, fruit, vegetables, butter and anything else that would be "useful to the sick."

In the final days of February, as Sherman's army made its way toward the South Carolina border, Confederate military authorities in Fayetteville and vicinity were continuing their duties by the book.

On Feb. 27, Lt. Rencher, now promoted to major, commanded a mixed Home Guard unit, the "Battalion of Disabled Men," and militia. He sent word that "all detail men and light duty men not otherwise enrolled" should report to his command's headquarters in Fayetteville.

In early March, J.M. Williams, the Cumberland County "enrolling agent" in charge of collecting foodstuffs for the army, "urgently requested" farmers to bring in their "tithes" (in-kind taxables), "especially bacon, whether cured or not," to the local collection depot.

Agents in neighboring Richmond (now Scotland) and Robeson counties were asking the same.

The "impressing" agent in charge of collecting mules for army wagons called on those whose teams had been impressed to bring them in promptly, "or a guard will be sent for them."

Col. James G. Bure of the 76th North Carolina Regiment, a Home Guard outfit, ran a notice calling on "all absentees from this command" to report to its encampment in Sampson County.

Capt. Ben Robinson, a crippled veteran of the war in Virginia who was acting as town provost marshal of Fayetteville, announced office hours that stretched from before dawn to after sunset.

Defiance at arsenal

There was a note of defiance to the last public announcement from Lt. Col. Frank Childs at the Fayetteville Arsenal and Armory.

In anticipation of the arrival of the bluecoats, Childs called on "all Contractors and Employers of Contractors of this Arsenal and Armory (to) repair forthwith to this Post for duty. All who can procure horses will bring them with them for mounted service."

A week later, Union army engineers would batter down the arsenal and set it afire.

In the words of an eyewitness of that destruction, "a night made hideous by smoke" would spell the end of the Confederacy for Fayetteville.

As 1865 Begins, War Efforts Push On

Originally published January 5, 2006

Fayetteville greeted 1865, the beginning of the fifth year of the Civil War, with muted perseverance in the face of mounting gloom.

The year 1864 had seen the Confederacy steadily shrinking, with Union trenches encircling Richmond and Petersburg in Virginia, and Gen. William Tecumseh Sherman's forces marching across Georgia toward the sea.

However, the holiday season had opened in early December 1864 with a flash of good news. Confederate defenders had repulsed a Federal amphibious assault on the great Fort Fisher bastion at the mouth of the Cape Fear River.

The defense forces included local young men of the "Junior Reserves" corps, as well as men from the garrison of the Confederate Arsenal in Fayetteville.

These troops had been sent home in time for the Christmas and New Year's season. However, many others remained on duty in the trenches in Virginia and elsewhere.

The good news from Fort Fisher was bracketed by reports of further calamities: the destruction of the Confederate "Albemarle" on the Roanoke River in Plymouth, and the fall of Savannah, Ga., to Sherman's troops four days before Christmas.

In the winter of 1864-65, Fayetteville's chief contribution to the effort — beyond the hundreds of young men who had gone off to battle and the cartridge guns made at the Confederate Arsenal — was the work of women of the Cumberland Hospital Association.

Since the first year of the war, the organization had carried out a far-flung project of supplying food, clothing and household goods for military hospitals and fighting men in the field.

In December 1864, the association summed up its work for the year. It reported collections of more than $9,000 cash and disbursements of $10,318 — mostly for purchasing supplies.

The outreach of the organization was widespread. Its "recipients of aid" included:

"Hospitals in Richmond, Raleigh, Washington, and Fayetteville, Col. McKethan's

Regiment at Petersburg, Finnigan's Brigade of Florida at Petersburg, Cumberland Plow Boys, 5th and 13th Arkansas Regiments of the Army of Tennessee, Gen. Lane's Brigade of the Army of Northern Virginia, and indigent soldiers of Cumberland and other counties passing through."

The extent of a single contribution was reported in early December, in a letter from an Army doctor at the "Post Hospital" at Tarboro, which had formerly been at Washington, N.C.

He first praised them as "ministering angels of charity" who "with patriotic zeal and heartfelt sympathy have addressed themselves to their continued work of philanthropy."

The contributions included comforters, sheets and pillows for 75 beds; 40 towels; "two elegant blankets;" one bolt of cotton sheeting; 24 sets of cups; plates, knifes and forks; six coffee pots; six "Brittania table spoons;" 24 bottles and jars of jellies; a jug of scuppernong wine; six bottles of "Antimonial wine;" a keg of pickles; six bottles of wild cherry bitters; and "British ale."

In the first week in 1865, the association sent off contributions — including socks, gloves, shoes, sheeting, shirts and pieces of homespun cloth — to troops at Wilmington and Fort Fisher.

Mrs. Jessie (James) Kyle sent a piece of sheeting "directly for the benefit of the troops at Fort Fisher, a token of appreciation for their gallantry in repulsing our enemy at the Fort."

Kyle hosted the monthly meeting of the association at her home on Green Street, which today is part of St. John's Episcopal Church.

To raise money for the work, the association sponsored a benefit concert where "the young ladies and gentlemen gave a vocal and instrumental concert at Fayetteville Hall."

Crime problems

In the midst of all the goodness, crime didn't take a holiday. Horse stealing was especially rampant at a time when the demands of the military for draft and cavalry animals were heavy.

Fayetteville was able to report a victory in the war on crime with the announcement in The Fayetteville Observer of "An Important Arrest."

The note said: "The notorious William Turner Fry, who has been charged with stealing so many horses in various parts of the state, was arrested in this town on Friday last, and is now safely in jail, awaiting the applications of various jailers in other counties who have advised a desire to extend their hospitality to him."

In war as in peace, the first order of municipal business in Fayetteville was the yearly election for mayor and the seven-member board of town commissioners.

Politics was muted in 1865, this fifth year of the war.

Mayor Archibald McLean and six commissioners were re-elected. A single change was the replacement of James McKethan with Dr. K.A Black when McKethan declined to run for re-election.

The seven commissioners were elected from geographical wards. They were Black, Murdoch McKinnon, A.G. Thornton, E.L. Pemberton, R.M. Orrell, J.C. Poe, and J.R. Lee.

A key wartime government program, the "tax in kind" collections for the

Confederate Treasury, was busy in early December 1864.

G.H. Blocker and J.A. Johnson were in charge of assessing the taxes owed by individual farmers. In lieu of cash, farmers could be required to bring a portion of their annual crop to government warehouses.

The tax law covered corn, Irish potatoes, rice, cured fodder, cured hay, buckwheat, sugar, molasses, ground peas and cotton. The collectors listed 11 township locations where farmers could report "their entire crop of the above-named produce gathered or not."

The notice of the assessors concluded on a desperately hopeful note: "We hope every producer will comply promptly, and thereby save themselves and us great trouble."

Town Served Briefly As Rebel Headquarters

Originally published February 3, 2005

How did it come to pass that for three days 140 years ago, Fayetteville was second only to Richmond, Va., as a military headquarters of the Confederacy?

Like this:

Gen. Joseph E. Johnston had no illusions when he got the telegram from Richmond naming him to a new command.

His orders were, in effect, to take charge of every military formation remaining of the Confederacy, other than the Army of Northern Virginia huddled in trenches around Richmond and Petersburg, Va.

His specific mission, according to the order written by his old West Point classmate and fellow Virginian Robert E. Lee, was to stop the inexorable march of Gen. William Tecumseh Sherman's 60,000-man horde of Union soldiers.

That bluecoated juggernaut had blazed a trail of fire from Atlanta to Savannah, Ga., and across South Carolina and was headed for North Carolina with the ultimate objective of winning the Civil War by squeezing the last life out of a dying Confederate States of America.

Johnston saw his orders in a realistic light.

He told a friend that he only got the command "to be the one who will surrender."

By late February, Sherman's columns were closing up to the North Carolina border.

On the North Carolina coast, Fort Fisher at the mouth of the Cape Fear River had fallen in January.

And on the very day that Lee made his decision to call on Johnston, Gen. Braxton Bragg pulled his ragtag defensive forces out of the port town of Wilmington, ordered the destruction of huge supplies of cotton, tar and military stores, and retreated toward Goldsboro.

But Johnston was a consummate soldier. He would follow orders. At 57, he was the South's most trusted military luminary next to Lee and a veteran of 35 years in uniform.

In the "old army," Johnston had been the highest-ranking officer, a brigadier general, to resign his U.S. commission and put on the uniform of the Confederacy.

If he had not been wounded in a battle outside Richmond in the spring of 1862 and replaced by Lee, he might have been the South's top commander throughout the war.

But his fate was to command in the western theater, to pick up the pieces when other generals failed to stop the advances, first of U.S. Grant and then of Sherman.

When he could not hold Atlanta against Sherman in the summer of 1864, Confederate President Jefferson Davis, never a fan of the prickly Johnston, relieved him of command.

Since then, Johnston had been staying in the Carolinas to keep tabs on Sherman's advance.

So it was on Feb. 23, 1865, when Johnston received the telegram from Richmond ordering him to take on Sherman once again, he immediately boarded the train to Charlotte.

There, he met a few thousand Confederates, the remnants of once-great armies he had commanded as the Confederacy's top general in western theaters of war.

To get to Charlotte, these soldiers had ridden trains from Mississippi, Tennessee, Georgia and Alabama on trips that took as long as two weeks.

Ghost units

In their depleted ranks, a corps was no larger than a brigade. Regiments often consisted of a few officers and some color-bearers.

Nonetheless, on Feb. 24 Johnston took time for a "grand review" of the troops under his command.

Then he went about trying to organize the forces to throw across Sherman's path.

In addition to the troops at Charlotte and others who straggled in by rail from the west, he could count on a corps commanded by veteran Lt. Gen. W.J. Hardee that had been falling back before Sherman all across South Carolina, giving up Charleston, the Cradle of the Confederacy, without a fight.

Hardee's few thousand men, most of whom had been artillerymen in coastal forts until Sherman's march made them irrelevant, were first ordered toward Charlotte.

Then there was the collection of remnants of Rebel corps and divisions comprising Bragg's command from Wilmington.

But this force had more than Sherman to worry about.

A large federal formation of more than 20,000 was congregating outside Wilmington and New Bern under orders to march on Goldsboro to rendezvous with Sherman as soon as his army had passed through Fayetteville.

As he surveyed the situation from a headquarters in Charlotte and then in Raleigh, Johnston typically wanted to get closer to the scene of likely collision between his forces and those of Sherman.

Hotel headquarters

So, on Monday, March 6, 1865, in the late afternoon, the commander of all Confederate forces south of the Virginia line arrived in Fayetteville with his staff.

A local observer wrote later that the general's entourage included an ambulance containing his "camp equippage" as well as his own "camp wagon with a splendid pair of bays & 2 led horses."

Johnston and his staff set up headquarters in the Fayetteville Hotel.

That three-story hostelry stood at the eastern corner of Hay and Donaldson streets, on the south side of the 100 block of Hay Street. Across Hay Street were the offices and press rooms of The Fayetteville Observer.

For the next 72 hours, Johnston used Fayetteville as his base for directing deployment of his scattered forces, trying to find a place to make a stand against

Sherman's masses.

Units from Charlotte were ordered east along the railroad toward Smithfield, where they would be in easy distance of Bragg's forces in the vicinity of Goldsboro.

Hardee, whose vagabond command was gathered at Cheraw, S.C., was already on the march toward Fayetteville, where Hardee thought Johnston might make his stand.

But on Thursday, March 9, 1865, when Hardee rode into Fayetteville ahead of his marching corps, his orders from Johnston were to keep right on going across the Cape Fear River.

By then, the vanguard of Sherman's army was barely a day behind Hardee's rear guard.

Hardee's new orders were the last issued from the Fayetteville Hotel, as Johnston closed his headquarters there that evening and headed for Raleigh.

Fayetteville's brief history as a Confederate command center was followed by four days of occupation by Sherman's troops, beginning March 11, 1865.

Ten days after leaving Fayetteville, "Old Joe" Johnston took his stand at the Battle of Bentonville in Johnston County, the last hurrah of southern arms outside Virginia.

Johnston went on to surrender his tiny forces to Sherman at the Bennett House near Durham.

Twenty-six years later, the South's premier general after Lee died of pneumonia contracted while standing in bare-headed tribute beside the coffin of none other than William Tecumseh Sherman.

Kilpatrick's "Skedaddle" Part Of Civil War Lore

Originally published March 11, 1993

Old-timers say that at dawn on a crisp March morning, you can still hear the wild yap of the Rebel yell echoing through the boggy swamps near Monroe's Crossroads, a lonely pineywoods place now on the Fort Bragg reservation a few miles east of Southern Pines.

One hundred and twenty-eight years ago at the spot, the loom of morning was just showing in the eastern sky when that yell burst over the sleeping forms of a brigade of Union cavalrymen whose camp encircled the angular frame farmhouse of the Monroe family. It was March 10, 1865.

Two minutes after the first high-pitched yell, hundreds of Confederate cavalrymen burst into the camp, sweeping the Union men out of their blankets, cutting the horse lines, swinging sabers at the heads of the startled enemy. Many surrendered on the spot. But many others dashed for the haven of nearby swamps where the Rebel horses could not navigate.

As startled as his men by the wild charge was the Union commander, 29-year-old Maj. Gen. Judson Kilpatrick.

"Little Kil" was not sleeping among his men, however. His night had been spent in a bed in the Monroe house. Beside him, according to well-established tradition, was the lovely form of a young lady from Columbia, S.C., who had been his traveling companion — night and day — since the Union army of Gen. William Tecumseh Sherman roared through the South Carolina capital three weeks earlier.

Accounts differ, but all agree that the agile young general escaped into the swamp with only seconds to spare.

Some say he galloped away on a horse with his nightshirt flying. He said he had on his trousers, but not his boots, and that he made it to safety on foot.

Whatever the true details, the scene was indelibly engraved in the memories of many of the participants in the affair. And the circumstances attached an enduring name to the fight.

In official reports, it was the Skirmish at Monroe's Crossroads.

But it has come down in history as "Kilpatrick's Shirttail Skedaddle."

The word "skedaddle" was a favorite of soldiers on both sides of the Civil War. It was an informal word, but with a vivid meaning.

Generals might write in their reports about "retiring" or "falling back" or even "retreating." But bluebellies and rebels alike knew that a "skedaddle" meant to get out in a hurry, and the devil take the hindmost.

As the Union men of Kilpatrick's force retired at high speed toward the nearby bogs, the ragged men of Lt. Gen. Wade Hampton's tiny Confederate brigades swirled through their camp, capturing men, horses and a parked battery of artillery.

Even though it lasted only a few minutes, this was the first success at arms the rebels had known in weeks as they retreated before the blue columns of the Union armies tramping through the Carolinas.

There were official reports of the attack. But the most memorable account of those few minutes was written years later, in a poem, by professor Steve Smith of Sandhills Community College in Southern Pines.

It is in the form of a reminiscence of a great-grandpappy who was there, recalling it for the storyteller, Bushnell Hamp.

As it was told in poetry:

> *"Bush, there ain't never been a bunch of fellas surprised like them bluebellies was when we went whipping through the blackjack and turkey oak right on top of them — Boots & blankets, mules & muskets gone all which ways, us graybacks giving them the old yip-yip & the Yankees scrambling in their scanties, hell it was like falling through the roof of a henhouse: everywhere confusion — horses screaming, Captain Bostic yelling orders there weren't no hearing for the racket, some fool puffing a bugle. I seen a belly-shot trooper double over in the saddle & and that big-talking Texas boy went down sprawling & and it occured to me maybe I oughta ease on over beside the Monroe house where there less of a rumpus."*

The poem goes on to describe how great-grandpappy then observed Gen. Kilpatrick's lady friend standing in her night dress on the porch of the Monroe house, and concludes:

> *"Bush, if ever a man was struck stupid it was your great-grandpappy as he stared at that handsome hardbodied woman."*

The wild scene around the Monroe house was not over with the Confederate attack, however.

Little Kil and his men rallied in the swamp. The Confederates could not penetrate the wet ground. And although they had skedaddled for the low ground, the Union troopers quickly recovered their cool. They were hardbitten veterans from Pennsylvania, Kentucky and Illinois who weren't used to losing a fight. There were as well loyalist southerners of the 1st Alabama Cavalry who didn't cotton to being made prisoner by fellow southerners.

While many had ducked out of camp without boots or trousers, nearly all had grabbed their weapon, the deadly Spencer repeating carbine, which had already broken many a Confederate attack.

With Little Kil at their head, they charged out of the swamp, joining with others who had formed a sketchy firing line. The tide of battle reversed. It was the Confederates' turn to be pressed out of the camp they had so recently seized. The artillery was recaptured and turned on the disorganized enemy. The Rebel cavalry pulled out of the fight, dashed off toward Fayetteville, 20 miles away down the rain-soaked "Morganton Road."

By 8 o'clock the camp was clear and a brigade of Union infantry had posting on the road to make sure Hampton's horsemen did not return.

The skirmish at Monroe's (also called Solemn Grove) had enabled the Confederate cavalry to slip through a tangle of Union columns moving from Lumberton toward Fayetteville and to make it to the latter place that evening, a few hours ahead of the bluecoat army.

The skirmish caught the attention of the nation through a report in Harper's Weekly, with an illustration depicting a mounted Gen. Kilpatrick leading a charge of his troops onto the grounds of the Monroe house.

It was left for Gen. Sherman, Kilpatrick's commander, to sum up the affair. In his memoirs, he wrote:

"(Wade Hampton's cavalry) captured the house in which Gen. Kilpatrick and the brigade-commander, Gen. Spencer, were and for a time held possession of the camp and the artllery of the brigade. However, Gen. Kilpatrick and most of his men escaped into the swamp with their arms, reorganized and returned, catching Hampton's men in turn, scattered and drove them away, recovering most of his camp and artillery; but Hampton got off with Kilpatrick's private horses and a couple hundred of prisoners, of which he boasted much in passing through Fayetteville."

The official North Carolina historical marker for the event, located on the U.S. 401 bypass at Raeford, states: "Monroe's Crossroads: General Kilpatrick's Union cavalry repulsed Gen. Hampton's Confederate cavalry there, March 10, 1865, ten miles north. Now in Fort Bragg area."

Repulsed, perhaps. But for a few glorious minutes at dawn, it was a southern triumph in a war that was soon to end with the South overwhelmed.

Even Gen. Kilpatrick, a veteran of every cavalry battle of the Civil War, gave them their due.

In his official report, Little Kil called the attack on his camp "the most formidable cavalry charge I ever witnessed."

*A full account of the Battle at Monroe's Crossroads is in "The Civil War In North Carolina" by John G. Barrett (UNC Press, 1963). The poem is from "The Complete Bushnell Hamp Poems" by Stephen Smith (Briarpatch Press, 1991).

A Curious Tale Amid Chaos At War's End

Originally published February 27, 2003

When March came to the Cape Fear region in 1865, the situation could only be described as chaotic.

The Union army of Gen. William Tecumseh Sherman, 60,000 strong, was pouring across the North Carolina border, headed for Fayetteville.

The port of Wilmington was in federal hands. Refugees were taking to the roads made muddy by days of torrential rains. The river itself was rising to "freshet" stage, overrunning some low-lying water wheel mills.

Small columns of Confederate soldiers straggled north just ahead of the Union horde, retreating from Charleston, from Wilmington, from Cheraw, S.C., burning bridges and military supplies along the way.

As they made their way through the lowlands of present-day Sampson and Robeson counties, many of these men of the dying Confederacy dropped away, deserting the already-thin ranks.

A Confederate colonel would soon boast of shooting a dozen of his own men who were "recovered" from the swamps.

And in the swamps, these just-arrived dropouts from the Civil War found many others.

There were earlier deserters from the Confederate ranks. There were outlaws. Plus runaway slaves and men hiding from the Confederate Conscription Service.

And there were even Union soldiers, escapees from Confederate prisoner-of-war pens in Charleston, S.C., or in Florence, S.C.

In the face of this growing gathering of the lawless and the desperate, civilian law and order was strained to the breaking point.

Country homesteads and plantations were largely at the mercy of marauding bands, even before the foragers and "bummers," the Union army, showed up to add to the misery.

Taxing as usual

Amid the confusion and despite the looming occupation by Sherman's troops, local Confederate authorities still tried to collect commodities under the "tax-in-kind" law.

Patrols of the Conscription Service were still chasing down likely recruits, sometimes with fatal results. According to a story in The Fayetteville Observer, "the artful deserter named Cain" was mortally wounded in a shootout with Capt. Williams of the Conscription Service.

In Fayetteville, soldiers of the garrison of the Confederate Arsenal advertised for a "shovel army" to construct defensive earthworks around the town, even as they packed arsenal machinery to be shipped out of the grasp of the federals.

In this increasingly gloomy time, just five days before Sherman's army marched into the town, the Fayetteville Telegraph found a bit of news meant to lift Confederate spirits:

The headline read:

"Thrilling Affair in Robeson County:

A Lady Uses a Repeater with Fearful Effect."

The story went on:

"On Tuesday about one o'clock, a band of deserters and escaped Yankee prisoners, about 15 in number, and led by a Yankee naval officer, attacked the house of Mrs. Dr.

McNair in Robeson County and after spending nearly an hour in the attempt, forced an entrance by breaking down the front door.

"Mr. William Stanton of this county was staying with Mrs. McNair on the night in question, and being called by the lady to her assistance soon after the attack commenced promptly reported for duty.

"An exciting contest followed.

"Mrs. McNair and Mr. Stanton both being provided with Colt repeaters (pistols) and a rifle, which they used with fatal effect, one of the attacking parties being killed and one wounded in the breast.

"We regret to learn that Mrs. McNair was badly burned and her face otherwise injured by fragments of glass and powder. Mr. Stanton was also slightly wounded.

"Mrs. McNair and Mr. Stanton having exhausted their ammunition, the latter, at the urgent solicitation of Mrs. McNair, who feared the desperados would wreak vengeance on him for the death of their comrade, retired, and she remained alone to brave their fury."

Plot twist

"On approaching her, the Yankee officer in command was so struck with her bold and determined mien that instead of harassing her, he complimented her warmly on heroic conduct, stating that she was the first person, man or woman, who had offered any resistance to his band.

"They carried off nearly all the valuables they could find, including silverware and jewelry, six thousand dollars in Confederate notes, and two breastpins containing pictures of Mrs. McNair's husband and son, the latter of whom fell in defense of his country on one of the battlefields of Virginia."

In a final bit of patriotic editorializing in a war that was soon to be lost, the Telegraph said: "The courage and determination of this noble woman savors more of romance than of reality. But the instances given in this article are literally true.

"It carries one back to the days of Revolution, when the fearless women of the Colonies wonder, and noteworthy to be imitated by many of the male sex whose knees are smiting together, like Bellshazzars, lest they be attacked by deserters and other roving bands."

The story of Mrs. McNair and the "Yankee naval officer" is a rarity even among the many stories of such encounters in the final days of the Civil War.

And a lot of mystery.

Mrs. McNair is not otherwise identified.

And where did he come from, "the Yankee naval officer" who was commanding what must have been a rather disciplined mixed crew of deserters and escaped POWs?

There were few opportunities for Confederates to capture Union naval officers.

Perhaps he was from some rare event during the blockade of Charleston.

When that "cradle of the Confederacy" fell to Sherman's army without firing a shot, perhaps retreating Rebels marched him from Charleston to the POW pen at Florence, where several dozen Union army officers are known to have been held briefly before they escaped to the swamps of eastern North Carolina.

Whatever the answers, the story of Mrs. McNair and the mysterious Yankee officer who so admired her pluck comes down among the more diverting tales of the last chaotic days of the Civil War along the Lumber and the Cape Fear.

HOME GUARD DEFENDED 'TIL THE END

Originally published August 21, 2003

What do you do when an enemy army of 60,000 is inexorably moving toward your little town while local law and order is breaking down around you?

It is always a fascinating picture, how the Fayetteville area strove for business as usual even as Union Gen. William Tecumseh Sherman's army of federals bore down in the last days of the Civil War.

It was a time when the crumbling Confederate government still made its demands on an area where there obviously would not be a Confederate government in a few weeks.

Where the woods and swamps were full of deserters, conscripts running from the law, escaped slaves, and even escaped Yankee prisoners of war.

Where civilians suffered shortages while desperately carrying on amid a deluge of rumors, and while wrestling with how to absorb a tide of refugees from as far away as Charleston and Wilmington.

There are many sources of information about the times from late 1864 until Sherman marched into Fayetteville on March 11, 1865.

Newspapers, official accounts, federal diaries and local reminiscences all tell stories.

A recently published collection of the records of the stay-at-home military forces of North Carolina, the militia and "Home Guards," provide more colorful insight into the times.

The records are in the form of the scores of orders issued from Raleigh by the adjutant general of the state. He was the commander of the organizations of part-time soldiers, the underage and overage, the infirm and crippled, who remained at home while the regular Confederate armies campaigned elsewhere.

Even as Sherman's army completed its invasion of South Carolina, North Carolina authorities were called on to supply slave labor for defensive works under construction far to the north on the Roanoke River near Weldon, about 10 miles south of the Virginia border.

On the day before Christmas in 1864, the commanding officer of the Cumberland Home Guard, the venerable Col. Walter Draughon, received orders to "forward 40 able-bodied slaves" to the Weldon project.

A month later, the Home Guard of Bladen and Columbus counties was called on to raise 20-man detachments, commanded by an officer, to carry out the most ubiquitous work of the stay-at-home soldiers.

They were to "proceed to the West Prong of the Big Swamp to search for deserters who are committing serious depredations upon inhabitants of the region." The detachments were to "capture, destroy or drive these depredators."

Majs. James Monroe at Elizabethtown and M. Smith at Fair Bluff received the orders. Smith was directed to name Capt. James Mears to command the Columbus detachment.

The swarm of deserters from Confederate ranks who congregated in the vast swamps of the Cape Fear region grew even larger in February 1865 as Sherman's troops seized Columbia and Charleston and aimed for Fayetteville. The stay-at-home soldiers were the main line of defense against the tide of desertion.

When David Kivett offered to raise a volunteer company of locals at the textile village of Manchester near present-day Hope Mills, the adjutant general said he would be welcome if he would accept assignment from Draughon "to guard bridges for purpose of apprehending deserters."

On March 1, the 59th Battalion of local militia in Moore County got its orders from Maj. Charles Dowd to "go to the field for 30 days to hunt deserters in Moore County."

Sherman nears N.C.

A few days later, as Sherman's army approached the North Carolina line south of Laurel Hill, the Home Guard soldiers "of the second class" in Moore, Montgomery, and Richmond counties were called up "to aid Lt. L.P. Leach of the 23rd Regiment in arresting deserters and recrusant conscripts."

On March 8, three days before Sherman's army marched into Fayetteville, the adjutant general replied to a letter from the commanding officer of the most unusual military organizations ever to be in the town.

It was the "Office of Naval Ordnance," commanded by Lt. David McCorkle of the Confederate Navy.

McCorkle had been in the area since just after the turn of the year, a refugee from installations in the Deep South that were in Yankee hands.

And despite the looming presence of Sherman's huge land forces, he meant to establish a naval gun factory on the Cape Fear.

In his letter to McCorkle, the adjutant general addressed him at the "N.C. Works" and gave permission for a "Mr. McArthur" to escape Home Guard detail so he could remain with McCorkle's command.

By the middle of the week prior to Sherman's arrival on Saturday, March 11, columns of retreating Confederate troops moved through Fayetteville and camped in fields and streets.

On March 8, the adjutant general sent a terse command to Draughon, commanding officer of the Cumberland Home Guard: "Call out the Home Guard. See that the waggons are loaded only with state property. It is understood that the commandant has impressed them for Confederate use."

The "commandant" was Lt. Col. Fred Childs, commanding officer of the Fayetteville Arsenal, the prewar federal facility that had served to make guns and bullets for the Confederate army.

On this day when Sherman's army began crossing the North Carolina line near Laurel Hill in Scotland County, Childs was preparing to dismantle his machinery and ship it out away from the Yankee line of march.

Last message

The next day, March 9, the adjutant general sent what would turn out to be the last message to the Home Guard of Fayetteville.

It was addressed to Capt. Robert Mitchell, who headed the Fayetteville company of the Cumberland Home Guard.

In the midst of the general confusion, Mitchell was attempting to scrape together a command, but he obviously was having problems. He tried to enlist the "town guards and police," but they apparently declined. He wired the adjutant general on March 7.

The adjutant general's March 9 telegram informed Mitchell that such men were not exempt from Home Guard duty. Only "millers, public and private, blacksmiths, tanners, wheelwrights and certain others" could claim such exemption.

Forty-eight hours later, the question was moot. Sherman had arrived.

Gen. Sherman's Army Cut A Large, Destructive Swath Through Region

Originally published March 9, 1995

The morning of March 8, 1865, dawned in a downpour that beat down most of the day on the long columns of blue-coated soldiers snaking through the rural Carolina countryside.

At the end of the day, in a letter dated from "Laurel Hill, Richmond Co., N.C.," Maj. Thomas Osborn, chief artilleryman of the right wing of Gen. William Tecumseh Sherman's army, reported:

"We crossed the South Carolina and North Carolina line at 8 o'clock this a.m. and we felt relieved in getting out of the most contemptible state in the Union. Laurel Hill is a village consisting of a fine church, an old church not so fine, no other houses in town."

The thousands of soggy soldiers who were huddled around the church buildings at Laurel Hill that night were opening a two-week chapter of the Civil War that would see them march through the Cape Fear region, occupy Fayetteville, fight battles at Monroe's Crossroads, Averasboro and Bentonville, and then move toward final victory during the last week in April 1865.

The saga of 130 years ago followed a five-week march that began in Savannah, Ga., and carried the 60,000-man Union army across South Carolina.

They were headed via Fayetteville to Goldsboro, where they expected to rendezvous with federal columns moving inland from the port city of Wilmington, which had fallen to federal invaders on Feb. 23, 1865.

By the time the columns converged on Laurel Hill, some bluecoats had been across the Carolina border.

On March 7, a detachment of mounted scouts from the right wing had demonstrated Sherman's scorched earth tactics by destroying railroad shops at the little village of Laurinburg, only five miles from Laurel Hill.

Leaving Laurel Hill on the morning of March 9, the main columns pressed on through the rain, heading for Fayetteville. By nightfall, many had crossed the swollen Lumber River.

Sherman traveled with troops who that night camped on the grounds of Bethel Presbyterian Church in what is now Hoke County. Under a "terrible storm of rain," the commanding general slept dry on a church pew.

That morning, a regiment of horse-riding soldiers skilled in scorched earth practices — the 9th Illinois Mounted Infantry — veered east from Campbell's Bridge on the Lumber (present-day N.C. 71) to the Robeson County seat of Lumberton, where two bridges and a mile of railroad track were destroyed.

On that same day, other bluecoats marched through the church crossroads of Philadelphus in Robeson County, where a fire-eating Southern lady recalled: "They visited us in torrents ... acting like escaped fiends from the lower regions."

As Sherman bedded down at Bethel church on the evening of March 9, 10 miles to the north in the sandy pine woods of what is now the Fort Bragg reservation, the curtain was about to rise on a military melodrama.

Moving east

For two days out on the far left of Sherman's infantry columns, Union and Confederate cavalry forces had been bumping against each other, skirmishing as they ranged eastward toward Fayetteville.

By nightfall on March 9, Gen. Wade Hampton's rebel troopers were ready to pull a trap on the Union forces of 29-year-old Union Brig. Gen. Judson Kilpatrick, who had stopped for the night at the deserted crossroads house of Charles Monroe.

In the dawn of March 10, Rebel horsemen pounded into the yard surrounding the farmhouse, scattering horses and men of the Union squadrons.

Kilpatrick, a veteran of a hundred actions in the saddle, rushed out into the night in his nightclothes, leaped on a nearby horse and rallied his men. In a ragged charge, they recaptured their camp as the Confederates scampered off with their wounded and several score of prisoners.

Officially listed as the Skirmish at Monroe's Crossroads, the melodrama soon earned an enduring nickname: "Kilpatrick's Shirttail Skedaddle."

As the Union columns converged on Fayetteville, a key Confederate place with its Fayetteville Arsenal and Armory, the town waited for the inevitable.

At a converted school on Hay Street, wounded were treated from the fighting in Wilmington.

Soldiers and slaves enlarged a formidable earthen redoubt that had been thrown up 19 months earlier at the southeastern foot of the Clarendon Bridge over the Cape Fear River.

Operations ceased

On March 9, the Fayetteville Arsenal ceased operations. Much of the machinery — lathes, belts, steam engines — was removed and loaded on dilapidated railway cars that during the night were hitched to a locomotive and hauled away to be lost in the confusion of the end of the Confederacy.

By midday of March 10, 1865, federal soldiers were only hours from Fayetteville. That afternoon, riders of the ubiquitous 9th Illinois Mounted Infantry pounded into Rockfish Factory Village, present-day Hope Mills, and proceeded to burn the 25-year-old Rockfish cotton factory.

As Saturday, March 11, dawned, there was shooting within sight of the Market House in Fayetteville as a patrol of Union cavalry cantered into the town, surprising Confederate officers — including Gen. Hampton — who were stirring at breakfast in a Hay Street hotel.

The Confederates in turn surprised the Union horsemen. In a sharp running fight that included dashes through narrow garden lanes, they killed 19 of the bluecoat horsemen and captured a dozen, including young Capt. William Duncan, leader of the scouting party.

In a few hours, however, Hampton was gone, burning the Clarendon Bridge over the Cape Fear.

The main body of Sherman's army trooped in, the 14th Corps of the Left Wing winding in from the west on present-day Morganton Road.

The scene in the little town of a few thousand was told well in the recollections of former slave Sarah Louise Augustus. Sixty years later she told an interviewer: "The Yankees came through Fayetteville wearing large blue coats with capes on them. Lots of them were mounted, and there were thousands of foot soldiers. The Southern soldiers retreated, and then in a few hours the Yankees covered the town.

"They busted into the smokehouse on Marster's, took the meat, meal, and other provisions. Grandmother pleaded with them, but it did no good. They took all they wanted. They told us we were all free. The Negroes began visiting each other in the cabins and became so excited they began to shout and pray."

By nightfall, Mayor Archibald McLean had formally surrendered the town, and Sherman's thousands were camped on the grounds of the arsenal, in fields, yards, streets, even in gardens and orchards.

Sherman set up his headquarters on the grounds of the arsenal and was busy issuing orders and making reports.

Fire, destruction

On Sunday, March 12, the town, which was not unaccustomed to spectacular fires and destruction, saw it again.

That morning, in a letter written from his headquarters on the arsenal grounds, Sherman informed Grant of his intentions for the arsenal: "We cannot afford to leave detachments, and I shall therefore destroy this valuable arsenal, so the enemy shall not have its use; and the United States should never again confide such valuable property to a people who have betrayed its trust. Since I cannot leave a guard to hold it, I therefore shall burn it, blow it up with gunpowder, and then with rams knock down its walls."

And so, wrote eyewitness Alice Campbell years later: "The nights were made hideous with smoke. The crowning point to this nightmare of destruction was the burning and battering down of our beautiful and grandly magnificent Arsenal, which was our pride, and the showplace of our town."

More destruction

As the troops moved out on Monday and Tuesday, March 13 and March 14, there was other destruction, in keeping with Sherman's orders to lay waste to buildings that could provide provisions or equipment for the collapsing Confederate military forces.

Brig. Gen. Absolam Baird, military governor of the town during the army's stay, later reported: "Before leaving the town, I destroyed 2 foundries of some importance, 4 cotton factories, and the printing establishments of 3 rebel newspapers."

By Tuesday, March 14, the Union columns were on the move, crossing the Cape Fear on a temporary pontoon bridge. By Wednesday evening they were gone.

The Day The Arsenal — And Sherman — Blew Up

Originally published March 10, 1994

Gen. William Tecumseh Sherman was in as good humor as he had ever been in his life that March day in 1865 when he and his entourage rode into Fayetteville.

Before his four-day visit was complete, something he found in the old town on the Cape Fear would provoke a display of rage that would forever be remembered by those who witnessed it.

It is one of the most unusual of the thousands of tales growing out of those few hours when the Civil War came to the Cape Fear.

The commanding general of the Union army that had just occupied the town had been known throughout his 45 years for a temper as fiery as his red hair.

In the early years of the Civil War, his rages were so uncontrollable that he feared for his own sanity. For a time, he even gave up his military duties in order to get a grip on himself.

But for the six months before he rode into Fayetteville on March 11, 1865, Sherman had been in what everyone agreed was the best mood of his life.

Professionally, he had reason to be. His army of 60,000 hard-bitten bluecoats was coming to the climax of a triumphant campaign that had begun in the fall of 1864.

Marching through the heart of the Confederacy, he "made Georgia howl" and then laid waste to a wide swath of South Carolina, vowing, "this is where the war started, and this by God is where it will end!"

Sherman, chomping on an endless succession of cigars, joined the exuberance and satisfaction of his troops in a military accomplishment that had the Union shouting its praises.

Members of his staff with his traveling headquarters noted the equanimity of their chief. His mellowness was demonstrated in a practical way when the army crossed into North Carolina. He issued a general order calling on the troops to treat Tar Heels more kindly than they had the Palmetto state rebels. He believed there was abiding Union sentiment in the state and that many might flock to the Old Flag once it flew there.

Gen. Sherman's good mood was still evident in his first hours in Fayetteville when, as he wrote later in his memoirs:

"I took up quarters at the old United States Arsenal, which was in fine order, and had been much enlarged by the Confederate authorities, who never dreamed that an invading army would reach it from the west."

But then, in the very next phrase, the fiery general, writing 20 years after the fact, recalled what would provoke him:

"And I also found in Fayetteville the widow and daughter of my first captain (later general) Thomas Childs, of the Third Artillery, learned that her son Fred had been the ordnance-officer in charge of the arsenal, and had of course fled with Hardee's army."

On that same March day, Sherman had written to U.S. Grant, his commander, that:

"Here we find about 20 guns and the magnificent United States arsenal. I shall therefore destroy this valuable arsenal so the enemy shall not have its use, and the United States should never confide such valuable property to a people who have betrayed a trust."

By then, Sherman had called on Mrs. Thomas Childs, widow of Brig. Gen. Childs. Mrs. Childs apparently lived in the arsenal quarters of her son, Confederate Lt. Col. Frederick Lynn Childs, who had indeed skipped out with Hardee's forces on March 10.

Sherman had a long association with the family. As a new second lieutenant just out of West Point in the summer of 1840, he met the wife of his company commander, Capt. Childs, West Point class of 1813.

He perhaps had something to do with encouraging Fred Childs to become a West Point cadet. Childs graduated in the class of 1851 and was a career officer Army until he resigned in March 1861 to join the Confederacy.

But what Sherman found in Fayetteville astonished and enraged him.

Twenty-nine years later, in 1894, when Fred Childs died in Charleston, S.C., an account in The Fayetteville Observer said his passing "makes us remember the story of Gen. Sherman's rudeness to the venerable mother."

According to the story, when Sherman called, and "upon being informed that Fred was not at hand, his smile turned to a scowl, and is reported to have exclaimed:

"I wish I had him here now. I would hang him!"

The scene in which the general, to put it mildly, lost his cool over the perfidy of Fred Childs was apparently first recorded in an 1880s history of the Fayetteville Arsenal by S.A. Ashe, who had been a young officer on Childs' staff.

If that weren't enough, however, it turned out that yet another early acquaintance of Sherman's was also on the arsenal staff and was still in town.

We have several eyewitness accounts of what happened when they got together.

Maj. Thomas P. Osborn, a staff officer, wrote home about it from Fayetteville:

"There is a fine old U.S. Arsenal here and one for which the old officers seem to have a warm attachment. It is in good repair. It is built on the general plan of the one in Washington, but smaller, and with more taste. General Sherman says he 'will not leave two bricks of it together.' The military store-keeper at the arsenal formerly belonged to General Sherman's regiment in the old army, and appealed to the General for assistance or protection. The general gave him the most severe reproof I have ever heard one man give another, and sent him out of his sight, after telling him that 'a man who had turned traitor to the flag he had served under was too vile a thing to be in his presence.' "

Another staff officer wrote that after the meeting, "The staff never saw him under such emotion; the corners of his mouth twitched. He could scarcely eat ... his hand trembled as he raised the bread to his mouth ... there were tears in his eyes."

The object of Sherman's unforgettable wrath has been identified by historians as Edward Monagan, who was apparently a civilian ordnance official first for the prewar U.S. Army and then for the Confederate arsenal.

With his fiery temper rekindled, William Tecumseh Sherman was no longer in any mood to be nicer to North Carolina rebels than he had been to those in South Carolina.

In accordance with long-standing plans, the arsenal was indeed destroyed, as were cotton factories, foundries and flour mills.

Also razed, of course, were the offices and shops of the Observer, whose editor, E.J. Hale, had been a prewar acquaintance and political ally of the Sherman family (the general's brother, William Sherman, was a congressman and U.S. senator, a leader of the Whig Party in Ohio when Hale was the leading voice of that party in North Carolina).

Like so many others, however, Hale had become an ardent Confederate. His editorial columns breathed rhetorical fire at the bluecoats up until the final hours before they marched into Fayetteville.

In view of the vividly etched personal change that came over Gen. Sherman as a consequence of his Fayetteville experiences, it is fun to speculate what might have been.

If, for instance, Fred Childs had stayed behind and surrendered the arsenal. Or if Ed Monagan had fled! Or if, after Sherman had crossed into North Carolina, E.J. Hale had returned editorially to his strong prewar Unionist sentiments, thus justifying Sherman's optimistic perspective expressed in his general order.

Would the arsenal have been spared, or the Observer?

Who knows? As it is, history records that Cump Sherman showed his famous temper, in spades, when he came to Fayetteville.

Union Scout A Success

Originally published January 5, 1995

On that late winter day 130 years ago, a rain-swollen sky hung over the tiny Scotland County crossroads of Laurel Hill near the South Carolina line.

It was March 6, 1865, and a squad of bedraggled Confederate militiamen stood picket at the crossing where one road came in from Camden, S.C., headed northeast for Fayetteville and intersected with another heading west toward Charlotte.

Theirs was an outpost for Confederate forces gathering mostly to their west whose task was to get in front of the relentlessly moving bluecoat horde that was the Union army of Gen. William Tecumseh Sherman.

The squad was brought to alert by a party of horse soldiers pounding in from Camden.

As the riders reined in, the pickets saw they were wearing the gray uniform of Confederate Gen. Wade Hampton's cavalry.

The young captain in charge of the horsemen greeted the pickets, asked a few questions and then told them that Sherman's huge force was indeed only hours behind. It was moving, he assured them, in a northwesterly direction, apparently heading for Charlotte.

A detail of the militiamen quickly moved out immediately to relay the information. The horsemen wheeled away and headed south.

Forty-eight hours later, Sherman himself was in Laurel Hill, traveling with his bluecoat columns as they poured into North Carolina, heading for Fayetteville after a march that had brought them first from Atlanta to Savannah in Georgia and now through South Carolina.

While Sherman camped overnight in a church near Laurel Hill, the Confederate forces were hastily boarding dilapidated cars of the North Carolina Railroad, heading east in a frantic rush to pull across the path of Sherman's advance.

But for a day or more, they had hesitated as their commanders pondered the warnings of danger in the direction of Charlotte.

What about those "Confederate" cavalrymen who had put out that false information?

Soldiers of today's special operations forces will recognize the Civil War handiwork of one of their own kind.

For, of course, they weren't Confederates at all.

Daring escapades

They were Capt. William Duncan (1840-1925) and his scouts, men of Company F of the 11th Illinois Cavalry, a swashbuckling crew that served as the out-front eyes and ears of Gen. O.O. Howard's 14th Corps of Sherman's army.

At 24, Duncan was a battle-hardened veteran of Civil War fighting, much of it in the improvisational style of the scout. Born in Edinburgh, Scotland, he had emigrated to the United States only a few years before the start of the Civil War. By 1865, he had nearly four years of war under his belt, including such battles as Shiloh and Chickamauga.

On that March day, he risked his life in tricking the Confederate pickets. If caught in the gray uniform, he would have been hanged as a spy.

But that was just a day's work for Duncan, the latest in a series of spectacular exploits and escapades.

On the celebrated march through Georgia, Duncan and a command of exactly four men had bluffed a Confederate commander into surrendering the town of Milledgeville, then the capital of the state.

A few weeks later, he had floated in a rowboat down a sluggish Georgia river and past a Confederate fort to reach a ship of the Union blockading fleet off Savannah, bringing the news that Sherman's army was nearing the port.

Just before that, he had derailed the last train trying to leave the port by the simple expedient of shooting a mule and dumping its carcass across the tracks even as the engine came around the bend.

In South Carolina in late February, he had burned the bridge at Camden, capturing and destroying Confederate supplies and horses. Among the latter was a blooded thoroughbred owned by none other than Jefferson Davis, president of the Confederacy.

Night fighting

The week before his ruse with the militiamen at Laurel Hill, Duncan's company of fewer than 70 men had clashed with the better part of a depleted regiment of Hampton's cavalry at the little town of Mount Elon, S.C. In a nighttime fight at close quarters, Duncan's company had driven off the 6th South Carolina Cavalry, inflicting more casualties than he had soldiers, mortally wounding Col. Hugh Aiken, commander of the Confederates. Duncan's casualties included a severely wounded officer and two others.

For all these "hairbreadth escapes" and bold actions, Duncan had earned a reputation admired throughout Sherman's command.

Howard said of him: "He was invaluable to me in the way of scouting and reconnaissance."

Sherman called for him personally to pick scouts to send across country to link with Union forces at Wilmington.

Duncan was described by another young fighting veteran on Howard's staff, Maj. Thomas Osborn, chief of artillery:

"Captain Duncan is 26 years of age (actually only 24 at the time), medium size, fair hair and complexion, moderate ability, great energy, and wonderful personal recklessness."

Of course, if you know the history of Sherman's march into Fayetteville on March 11, 1865, you know the rest of the story of this intrepid young precursor of the Special Forces soldier.

Leading his troops into the early-morning streets of the town, Duncan became embroiled in a pistol-popping battle on horseback with none other than Hampton and his personal staff and was captured.

Prisoner of war

A few moments before the shootout, Duncan had been sitting on his horse in front of a hotel on Hay Street, unaware that Hampton and his staff and several other Rebel generals had just dashed away from the hotel breakfast table and out the back door, hastily mounted their horses, and were organizing a counterattack.

As Duncan put it in well-written war memoirs 30 years later, if he had but known, he could have made prisoners of them all.

"But now for the first time in nearly four years service, I was a prisoner of war myself," he wrote.

For once, the young scout had been the victim of a ruse. Hampton had sent an old black man from the hotel out to tell Duncan that the Confederate forces were up on the hill to the west of town.

After surrendering, Duncan was stripped of the new uniform he had acquired while on leave in New York following his Savannah exploit and confined as a war prisoner in a rude cabin east of the Cape Fear River.

On March 15, 1865, the day Sherman's army crossed the river, Duncan and another officer escaped out the chimney of the cabin.

Sherman described the scene. The Union commander wrote:

"As I sat on my horse, I was approached by a man on foot, without shoes or a coat, and his head bandaged by a handkerchief. He announced himself as the Capt. Duncan who had been captured by Wade Hampton in Fayetteville, but had escaped. He explained that when he was a prisoner, Wade Hampton's men made him 'get out of his coat, hat, and shoes,' which they appropriated for themselves. He said Wade Hampton had seen them do it, and he had appealed to him personally for protection as an officer, but Hampton answered him with a curse."

The last laugh

Duncan nonetheless got the last laugh, and provided one last example of scouting skills, by being able to warn Sherman that the Confederate forces gathering before him were aware of how the wings of the Union forces were now a day's march from each other, and were hoping to strike one of them. This they did two days later at Bentonville in the last major battle of the war in North Carolina.

Duncan picked up that information from listening in on mess table conversation while languishing in a circle of prisoners camped with the headquarters of Confederate Gen. J.W. Hardee.

For Duncan, the Civil War came to an end three months later when his mounted troop led Howard's corps down Pennsylvania Avenue in the big Grand Parade of the Union army. He had been given the honorary "brevet" rank of major "for gallant and meritorious service." He never again was called on to practice those martial skills that won him such renown.

Back home in Illinois, the celebrated young scout felt the tug of the American frontier. He soon emigrated to the Dakota Territory, where he served several terms in the territorial legislature.

A few years later, however, he moved to Minneapolis, Minn., where for more than 30 years he was foreman of buildings for the city school system.

He and his wife died a few hours apart at their retirement home in Cannon Falls, Minn. The local newspaper covered its entire front page with his biography and with eulogies.

The Day They Traded Swords For Pens

Originally published February 10, 2005

As we approach the 140th anniversary of Gen. William Tecumseh Sherman's Civil War march through Cumberland County in March 1865, there are many breathtaking tales to tell.

Sherman is the man who made it plain that "war is hell," and you can find plenty of evidence of that in the record of those days.

There were nighttime shootouts, fighting in the streets of Fayetteville, an arsenal razed to the ground as the night "was made hideous with smoke" of the destruction. And so on.

But then there is also the prosaic story of mail call in the midst of the martial din.

Even in these days of real-time communication, "mail call" continues as a favorite time for people in uniform.

In the Civil War, mail moved more slowly. But it was just as important.

And so it is a great story how one day in March 1865, the opposing Union and Confederate soldiers in and near Fayetteville took time out from war to write home.

When Sherman's army got to Fayetteville in early March, many of the 60,000 men of the Union columns had not seen a letter from home for months, or written one either.

So you can imagine the excitement when on the second day in town, a Sunday morning, word went out from Sherman's headquarters on the arsenal grounds that the tugboat Davidson, which had just arrived from Wilmington at the wharf on the Cape Fear, was also a mail boat!

If you could get your letter written by afternoon, you could tell the folks where you were and what you had been doing.

Hasty writing

Civil War soldiers were the letter-writingest fighting men in history, up to that time.

So you can imagine the scene as thousands of writers took pen or pencil in hand. Sherman joined in by penning detailed reports of the exploits of his army since it set out from Savannah, Ga., on Feb. 1.

But while the Davidson was taking on outgoing mail, there was huge disappointment when it was learned that it had no incoming mail on board.

It turned out that thousands of letters, dispatches and newspapers waiting to be delivered to the men of Sherman's column were sent to the coastal village of Beaufort instead of Wilmington.

Beaufort had been under federal control for years and Wilmington for only three weeks. The Union commander in Wilmington hoped soon to control the railroad that made its way from just east of Beaufort to New Bern and on to Goldsboro and Raleigh.

Where did the Union soldiers finally pick up their mail? Most likely at Goldsboro, which Sherman's forces occupied after the nearby Battle of Bentonville on March 19-21, 1865.

Southern missives

And what about the Confederate letter-writers?

Would you believe it, at the very time the Union soldiers in Fayetteville were writing their first mail in weeks, soldiers of the little Confederate army of Gen. William Joseph

Hardee were doing the same thing.

At the time, the two armies were less than 20 miles apart.

Hardee's forces had retreated across the Cape Fear River in the 24 hours before Sherman arrived on March 11, 1865.

Hardee stopped after a day's march at the Cape Fear village known as Smithville — near the border of Cumberland and Harnett counties — the ancestral home of a gaggle of Smith families.

And there, the weary Southerners, who had been marching in retreat for nearly a month since Sherman's army chivvied them out of Charleston, S.C., were told that mail would be collected, to be forwarded by the Confederate postal service (which, by the way, was the only bureaucracy in the Confederacy that during its brief existence actually turned a profit).

Did the mail get delivered to Rebel homes before the war ended barely a month later?

Many of the soldiers in Hardee's command were South Carolinians or from Georgia, from along the route of Sherman's notorious "March to the Sea" in Georgia and his monthlong push through South Carolina, neither places where the Confederate postal bureaucracy was likely to be in very good shape.

Whatever the final story of the great flood of mail from that day 140 years ago, this letter-writing spree in the midst of the military maneuvering resulted in literally thousands of eyewitness views of those few days.

Where are they now?

Thousands of the letters no doubt have long disappeared as attics were cleaned out over the years.

But there are hundreds of letters in public archives throughout the country.

The extent of this treasure trove of historical material is illustrated in the bibliography of many books, such as Mark Bradley's monumental retelling of the Battle of Bentonville and the campaign that led to it.

He lists no fewer than 90 collections of personal letters, as well as scores of printed diaries and recollections.

And 140 years later, that is the enduring historical significance and the legacy of the day of mail call in 1865.

Union Troops Mixed On Fayetteville

Originally published September 6, 2001

The never-ending issue of Fayetteville's image, whether it is a great place or Fayettenam, has an ancient history.

Is the place "old and rusty"? Or is it the Paris of the South?

The question was raised around 136 years ago when Gen. William Tecumseh Sherman's army rolled into Fayetteville on a soggy weekend in early March 1865, only weeks before the close of the Civil War.

Many of the military visitors, soldiers in the bluecoat column, took time to record their impressions of the town, which, small as it was, was nonetheless the largest place they had been since leaving Columbia, S.C.

Two officers of the bluecoat Union army saw the place from decidedly differing perspectives.

Maj. Thomas Osborn, commander of artillery for the corps of the Right Wing, was not flattering.

He wrote:

"The town contains perhaps 3,000 people, is old and timeworn. It has probably been a fashionable town, but now looks old and rusty. A portion of the town is built on a series of little hills, perhaps a third of it on a plain. The country surrounding the town is as poor as the Lord could well make it."

Paris of the South

On the other hand, Maj. George W. Nichols, a member of Sherman's personal staff, saw it differently.

In his book, "The Story of the Great March," published within a few months of his visit, he quoted his diary of the time:

"The city of Fayetteville is beautiful.

"The arsenal buildings are situated upon a commanding eminence at the west end of the city, and from every point they present an exceedingly picturesque appearance.

"Taken together with the old buildings buried among the trees, which are just putting on their livery green, they give to the place the romantic air of some of the old towns in the vicinity of Paris.

"An ancient market-house, of very tasteful architecture, stands in the center of the main street, which is a wide avenue, lined on either sides with substantial stores and dwelling-houses.

"Toward the river there are mills and manufactories, and on its banks, strongly-constructed steamboat piers, all showing evidence of the trade and commerce belonging to the river navigation, although there is not depth of water sufficient for any but light-draught steamers, except at certain seasons of the year.

"The people generally are of the better class. They do not all profess to be original Unionists, but they do not disguise their hostility to Jeff Davis and his despotism."

Nichols was keenly eager to get a glimpse of "the Peculiar Institution," that is, chattel slavery of black people, the underlying cause of the war. It was a system in its death throes.

He wrote:

"The slave population is not large, and is composed chiefly of women and children.

"As in other parts of the South which we have visited, the masters have run away; taking with them all the able-bodied slaves; but the negroes who were able to escaped, and have returned to join our column.

"It is generally understood among these colored men that the Rebel Government intended to put them in the army to fight against the 'Yankees.'

"The infatuation of the slaveholders upon this point seems to me one of the most singular of all self-deceptions; for among the hundreds of black with whom I have conversed during the progress of this campaign, I do not remember one who did not possess a better understanding of the merits of the questions at issue than the master who claimed to own him."

Slaves won't join rebels

"While the masters still have faith that the slaves will fight for them, and offer the additional inducement of freedom if they come safe out of battle, the slaves distrust them, and understand that their bondage was one of the principal questions involved in the rebellion.

"An intelligent old quadroon woman, whose mother, 86 years of age, sat near, and who was surrounded by her daughters and grandchildren, four generations in one group, said today:

" 'There, sire, are my two sons-in-law. Yesterday morning, their master tried to take them away, offering them their freedom if they would go in the army voluntarily; but they knew better than that. They never would fire a gun against the Federals.'

" 'No,' interposed one of the young men: 'I would not fight for the man who is my master and master of my father at the same time. If they had forced me into the army, I would have shot the officer they put over me the first time I got a chance.' "

Arsenal demolished

But if Nichols admired the early springtime look of the town, he painted a dark and balefully accurate picture of what was about to happen.

Looking out from the tents among Sherman's headquarters on the grounds of the former U.S. Arsenal, now a Confederate arsenal, he wrote:

"We take possession of this property by a double right. It was originally the property of the United States, paid for by the general government, and was stolen from us; and again it is ours by right of conquest."

And then Nichols paraphrased the grim orders that Sherman had already issued:

"We shall destroy it utterly. There is not a piece of this costly machinery that will not be broken in fragments; not a stick of timber that will not be burned to ashes; not one stone or brick of these beautiful buildings will be left standing upon another.

"By Monday night, that which should have been the pride and honor of the state and the country will be a shapeless mass of ruins."

Sure enough, efficient army engineers, former iron miners from the state of Michigan, took only half a day to knock down the buildings and set fire to the rubble.

The next day, Osborn and Nichols and all of Sherman's army departed from the "old and rusty" town (or was it a Paris of the South?).

They left indelible memories for those who remained.

And they left, too, unique eyewitness descriptions of where they had been.

Civil War Trenches Are Part Of The Landscape

Originally published February 8, 2007

This time of year, historians around here always look back to the first weeks of 1865, when Gen. William Tecumseh Sherman's federal bluebellies were marching hard toward Fayetteville.

The federal army swept through South Carolina and tramped inexorably on toward its goal of squeezing the life out of the fast-fading Confederacy.

The end of the Civil War was only weeks away.

This is a report on one aspect of Fayetteville's response as Sherman's hard-bitten veterans loomed over the horizon.

A customary response to an invading army was to dig — that is, to shovel out trenches and to throw up earthworks.

By the fifth year of the Civil War, digging was as much a part of fighting as shooting. Given a few hours, most armies could and would shovel elaborate trenches and artillery gun positions.

Today you can see the remains of three military digging projects in and near Fayetteville.

Only one has a historical marker and is in a public place. So start with it.

Go out Ramsey Street where the Veterans Affairs Medical Center is on the east side of the highway and Lafayette Memorial Gardens is across the road to the west.

On the grounds of the hospital, a state historical marker tells of the Confederate breastworks dug in 1865 in direct response to the threat of Sherman's invasion. The remnants of the lines are carefully preserved by personnel and patients of the hospital.

The trenches were dug in a few days, probably by slave labor, under the direction of officers from the garrison of the Fayetteville Arsenal.

The Confederates weren't sure which way the federal army would come.

In early 1865, the thinking was that the Union juggernaut would go west of Fayetteville, aiming at Charlotte.

And Sherman didn't disabuse that notion. Right up until his army marched into Fayetteville on March 11, 1865, he sent scouting parties out west to confuse Confederate outposts.

Of course, Sherman's army rolled into Fayetteville from the southwest and south, and the Raleigh Road lines in a sense were sited the wrong direction.

The line of earthworks across Ramsey Street, then known as Raleigh Road, would stretch from the high river bluff just east of the veterans' hospital to the swampy lowlands of Cross Creek a half-mile to the west. The line was just north of the grounds of present-day Lafayette Memorial Gardens. Some remnants of the line are still visible on the west side of Ramsey Street.

Two other military digging sites in and around Fayetteville date to earlier years of the Civil War. And like the Raleigh Road trenches, they apparently never were used.

The earliest is at the mouth of Rockfish Creek, just off N.C. 87 south of Fayetteville, now on the grounds of the Public Works Commission's waterworks.

The well-defined "fort" commands a high bluff overlooking a stretch of the Cape Fear River.

The little earthwork was created in the early months of 1862 by soldiers of the arsenal garrison. It was built as an artillery emplacement, and was meant as a defense against any federal gunboats that might get past the Wilmington waterfront and make their way up the river.

I have never found evidence that any guns were ever placed in the little earthwork, which at times was grandly called "Fort Booth," after an early commandant of the arsenal who died in September 1862.

The third well-preserved Civil War earthwork is tucked back in the woods on the east side of the Cape Fear just north of where Sol Rose is completing his new Lord's Mill Restaurant at the east end of the Person Street bridge.

The earthwork was sited to command approaches to the bridge that spanned the river there in 1865: the 45-year-old covered bridge known as the Clarendon Bridge.

The story of the earthwork goes back to the summer of 1863 when new alarms of a Union invasion of the Carolina coast followed the great battle of Gettysburg far off in Pennsylvania.

Two weeks after that battle, the commandant of the Fayetteville Arsenal called for major new military initiatives in Fayetteville.

On July 16, he announced that he was recruiting 100 additional men for the garrison.

Meanwhile, Fayetteville Mayor Archibald McLean published a notice under the head: "Laborers for the Defenses" appealing for 50 to 75 laborers "so Major Childs can complete the defenses of the Clarendon Bridge."

The appeal urged slave owners to donate the labor of their slaves to dig large earthworks to guard the bridge and the arsenal. The mayor said the number of workers sought could complete the fortifications in three days.

But the appeal evidently was not totally successful.

In early August, the arsenal commander, newly promoted Lt. Col. Fred Childs, called on the general commanding Confederate forces in North Carolina to supply engineers with authority to hire 75 to 100 workers at $20 a month to "fortify the town."

The alarm subsided, however, and three weeks later, Childs reported that workers were "still needed to work on fortifications."

Sixty workers had been furnished, he said, but 100 were needed. If workers were not provided voluntarily, he warned that he would use military authority to impress them into shovel service.

Once again, fortifications never saw action.

When Sherman's army arrived, Confederates burned the Clarendon Bridge and retreated north into the Cumberland County countryside.

Sherman's pontoniers — military engineers who build pontoon bridges — took only a couple of hours to throw a bridge across the Cape Fear next to the smoldering ruins of the Clarendon. The earthworks overlooking the bridge were largely forgotten. Obscurity has helped preserve them. Sol Rose has owned the site for many year and has lovingly guarded it.

Flag Born In War Was Burned Before Surrender

Originally published January 1, 2004

The contrast could not be more stark in the two scenes from the Civil War.

First scene:

The late summer of 1861, a parlor full of enthusiastic women in crinoline and lace, their fingers busy with scissors, needle and thread.

The result of their handiwork, a large red, white and blue flag, a star in the left field. On the bottom right white panel in capital letters is the word BETHEL.

The ladies of Fayetteville were making a banner for the first regiment of North Carolina soldiers to fight the Yankees.

Second scene:

In April 1865, in a patch of woods a few miles from the village of Appomattox, Va., a group of ragged officers of the doomed Confederacy gather sticks and leaves for a small fire.

When the flames rise, they quickly shred the bunting of a well-worn banner and feed it into the flames.

A few hours later, with other survivors of the Army of Northern Virginia, they surrender to the Union forces of Gen. Ulysses S. Grant and turn toward the homes that some left as far back as that spring four years earlier.

In the voluminous bibliography of North Carolina military history of the American Civil War, few units have had more words written about it, more stories told, than the 11th North Carolina Regiment, which began as the First North Carolina Regiment of Volunteers, and which is ever known as "the Bethel Regiment."

However, perhaps not lately have you heard the story of the regimental flag and the role played by "the ladies of Fayetteville."

Why did the ladies of Fayetteville have the distinction of making the flag that, you guessed it, would be hastily burned in the little fire near Appomattox 40 months later?

Two of the First North Carolina's 10 companies were Fayetteville outfits, former volunteer militia units.

The Lafayette Light Infantry, commanded by Capt. Joseph B. Starr, was the first to leave Fayetteville for the regiment's training ground. It was followed shortly by the Fayetteville Independent Light Infantry, commanded by Wright Huske.

They were in the confused skirmish on June 10, 1861, at Big Bethel Church in Virginia when the North Carolinians drew first blood in what would be the terrible and disastrous conflict that wrecked the South while preserving the Union.

Henry Lawson Wyatt, 19, of Tarboro was the only casualty, since mourned as the first to die in the war. A statue of Wyatt is on the Capitol grounds in Raleigh.

Special banner

To commemorate the fight, the legislature adopted a resolution calling for a special flag for the First Regiment.

The ladies of Fayetteville drew the assignment to make the flag.

It was presented to the regiment in camp at a bivouac grandly named "Camp Fayetteville" near Yorktown, Va., on Sept. 9, 1861.

A history of the regiment describes the scene:

"Col. Lee called the regiment to dress parade for the presentation by John W. Baker Jr. of Fayetteville.

"Mr. Baker, although impaired by some 'Virginia Tangle-leg,' spoke 'in behalf of the ladies of Fayetteville.' The standard-bearer grasped the colors, while Charles Lee expressed the gratitude of the regiment for 'the kind remembrance in which the ladies of Fayetteville have held them.' "

That September day may have been the most festive in the regiment's history, although even by then more than 40 young men had died of camp diseases. In its subsequent history, it earned the sobriquet, "the bloody Bethel regiment."

A few weeks after the flag ceremony, the regiment raised for a six-month enlistment would disband, shortly to be resurrected as the 11th North Carolina under the scheme of new military numbering.

Although the number changed, many men of the First came back to the 11th. Dozens of others went into other units of the Confederate army.

A regimental history concluded that the ranks of the original regiment were a veritable academy for those who would go on to higher rank. By the end of the war, the list included "one lieutenant general, one major general, two brigadier generals, 14 colonels, ten lieutenant colonels, eight majors, and 57 captains."

By the end of the war, 492 soldiers of the original First and the 11th were dead, 164 killed in fighting, most of them at the Battle of Gettysburg in 1863, the others dead of disease in camp or prisoner-of-war cages.

When the time came to surrender at Appomattox, seven officers and 92 enlisted men were left to sign paroles.

Before then, however, a final scene, told in the regimental history by Capt. Edward R. Outlaw, a veteran of every day of the regiment and then a company officer.

"When General Lee rode to the front and through the lines to meet General Grant, everyone knew that the surrender had come.

"The officers present with the regiment at once retired to a secluded thicket, and taking up a pile of twigs and leaves, committed the flag to the flames. Before burning it, Captains Outlaw and J.M. Young tore out pieces of each color.

"It had been given by the legislature of North Carolina to the Bethel regiment, and then committed to the keeping of the 11th. It had never been dishonored and they could not bear to see it a trophy of an anonym."

Bunting from flag

Outlaw, an original soldier of the First North Carolina, kept his piece of bunting from the Bethel flag until his death in 1921, and it was presumably buried with him.

And what of the ladies of Fayetteville. Who were they?

As far as my research goes, they will remain anonymous.

No doubt some of the area's oldest families may have stories handed down by ancestors. If so, now is the time to tell them.

Cape Fear Steamboats Once A Southern Lifeline

Originally published February 19, 1998

For four years, Fayetteville's doughty little fleet of Cape Fear River steamboats was a loyal lifeline of the Confederacy.

In March of 1865, Union invaders swept into the region, and disaster threatened the small stern-wheeler vessels.

The threat was not from the federals, however.

Instead, they were fleeing their own Confederate army.

How two notable stern-wheeler steamboats survived the threat is a little-known story from the last desperate days of the Civil War in North Carolina.

The two vessels were the A.P Hurt of the Cape Fear Steamship Co. and the North Carolina of Thomas Lutterloh's extensive shipping and turpentine-distilling operation.

Not so fortunate was the Kate McLaurin of the Cape Fear company.

In early 1861, the Hurt made Civil War history ferrying the town's first contingent of soldiers to training camps in Wilmington.

From then on, the little river steamers of the Cape Fear company joined the North Carolina and other vessels of Lutterloh's line plying the river to Wilmington, hauling large quantities of cotton and turpentine raised in the upper river valley.

These sought-after commodities were the cargo on bold blockade-running Confederate ships that dashed from the Wilmington waterfront, hoping to slip through the U.S. Navy's cordon and make their way to markets in Europe.

But in early 1865, the end was near for the Confederacy and perhaps for the steamboats, too.

Wilmington fell to federal invaders on February 22, 1865.

Fires blazed along the wharves as huge stores of cotton, turpentine, rosin and other commodities were destroyed.

The burning was not the work of Union troops.

It was carried out by retreating Confederates, under orders from their commander, Gen. Braxton Bragg.

"Scorched earth" was a favorite tactic of Bragg.

Eighteen months earlier at Chattanooga in Tennessee, Bragg ordered the torching of tons of bacon, flour and other quartermaster stores.

His half-starved soldiers retreated by the light of the fires.

The Fayetteville steamboat fleet, loaded with cotton and refugees, fled up the Cape Fear from Wilmington, hoping to evade the same fate.

Years later, Robert Orrell of a Cape Fear line vessel said he steamed away to "keep her out of the way of Confederate authorities." Thomas Lutterloh said it was to "evade the Confederate forces."

Directed by Archy White, a free black man who was her pilot, the North Carolina slipped into the well-concealed mouth of Rockfish Creek, seven miles below Fayetteville.

The shallow-draft vessel hugged the creek bank under the high breastwork of an abandoned Confederate artillery emplacement thrown up years before when alarms of federal invasion periodically swept the area.

Today, a private residence on the south side of the creek looks over to the remains of the breastwork on the north side.

With 30-year-old pilot Dan Buxton, also a free black man, directing, the Hurt apparently hid in the mouth of Carver's Creek on the west side of the Cape Fear.

The Kate McLaurin went another few miles, probably to Tranthams Creek or McKays Creek.

Their havens wouldn't be safe for long.

As March of 1865 came, columns of bluecoats came boiling up from the south.

Gen. William Tecumseh Sherman's army poured into North Carolina, headed for Fayetteville and for a linkup with federals in Wilmington and Kinston.

Confederate forces, perfecting their well-honed talent for retreating, known as "skedaddling" by Civil War soldiers, stayed a day ahead of them.

Hot tactics

Perfecting Bragg's scorched earth tactic, Confederate soldiers caught the Kate McLaurin in her creekside hideout.

And even in their haste to evade the oncoming Yankees, they always had time to burn cotton.

On March 10, 1865, the McLaurin and her cargo went up in flames.

South of Fayetteville, however, Sherman's columns were too quick for the Confederate cotton-bummers.

On the same day, bluecoats swept across Rockfish Creek a few miles above the North Carolina's hiding place.

They didn't pause but did send a detachment of "mounted infantry" veering off further up the creek on its own mission of destruction.

It knocked down and burned the cotton textile factory of the Rockfish Manufacturing Co. at Rockfish Factory Village, present-day Hope Mills.

The next morning, Sherman's hard-bitten veterans streamed into Fayetteville.

Some time during that day, W.S. Cook, agent for the Kate McLaurin, got a message to Thomas Lutterloh that a Union quartermaster officer said steamers running away from the Confederate army would be returned to their owners.

Robert Orrell later swore: "I sent word if he came up(river) our army would burn his boat. The Yankees might do the same thing the next morning, but they might not after explanation, and if they did he might get pay for her out of the U.S. The Kate was burned by our own people, so I had no hope for getting anything for her."

Within a day, Lutterloh met with Capt. Samuel Lamb, a quartermaster on Sherman's staff, and revealed that the North Carolina was smuggled up Rockfish Creek. The Hurt was meanwhile on her way to her Fayetteville landing. Lamb formally took charge of the two vessels on March 14.

For the next five and a half months, the A.P. Hurt and the North Carolina became the workhorses of a busy river trade run by the U.S. government.

As in wartime, the boats were loaded with cotton and turpentine, some from captured Confederate stocks, but much of it from civilians who had waited years to find a market.

For several weeks also, the boats did a brisk passenger business and hundreds of refugees from the lower Cape Fear headed back home.

The U.S. Army collected the freight tolls and passenger fares.

In mid-August of 1865, the steamers were returned to their owners.

For lost revenues

Twenty-five years later, in 1890, the owners entered a claim before the Committee on War Claims of the U.S. House of Representatives, seeking $40,000 each for the lost revenues of that summer.

The committee took affidavits from pilots Archy White and Dan Buxton, from Lutterloh, Cook, and Orrell, and from various Army officers.

In 1896, the committee recommended payments of $18,750 to each of them.

By then, Thomas Lutterloh was an important patriarch of the Republican Party in North Carolina. He served in the General Assembly for a half-dozen terms just after the end of the war and into the early 1880s.

The two little stern-wheelers saved from the rebel firebrands continued on the dwindling Cape Fear River trade for many years, the A.P. Hurt being the last of its kind. It burned in 1923 at its dock in Wilmington.

Records Offer A Sketch Of Black Militias

Originally published May 9, 1996

They were short-lived but significant players in Cumberland County's military heritage, the black men who served in 19th-century state militia companies.

Composed of former slaves or sons of slaves, these official units of the N.C. State Guard flourished briefly in the three decades after the Civil War.

They could be found taking part in Fourth of July celebrations and parading at the annual Black State Fair in Raleigh. They provided an outlet for young black men to exercise leadership and often acted as semiofficial political organizations in election seasons.

By the 1890s, the black militias, along with other black organizations, were hard put to survive in the tightening era of Jim Crow laws and rigid racial segregation.

After a brief and often tragic heyday during the Spanish-American War in 1898, they disappeared.

The earliest mention of a black militia company in Fayetteville occurred in 1876, in a report of participants in Fourth of July festivities for that year.

The unit was referred to as the "Fayetteville Rifle Guards (colored)."

At the time, the two white military companies in Fayetteville were not seen at Fourth of July parades. They saved their marches for Confederate Memorial Day in May. Both these units retained pre-Civil War names, the Fayetteville Independent Light Infantry and the Lafayette Light Infantry.

At least a decade before the 1876 account, black soldiers, or at least black men in blue Union army uniforms, were conspicuous at Fourth of July celebrations around the Market House in Fayetteville.

These were local men who were veterans of service in black regiments during the Civil War.

Detail in archives

Records are incomplete, but a list in the State Archives in Raleigh names officers of black State Guard units and provides some detail of the history of the Fayetteville units.

The list contains names, rank, unit, and service years of 19 black men from Fayetteville who held State Guard commissions between 1877 and the mid-1880s.

Among them is Abram Halliday (1835-1915), who became the highest-ranking black man in the State Guard until the Spanish-American War.

First as a major and then as a lieutenant colonel, he was the commanding officer of the statewide black battalion of the Guard, which included units from Raleigh, New Bern and several other towns.

The records identify Fayetteville's first official unit in 1877 as Company C of the 5th Battalion of the State Guard.

The sparse record does not make clear when this company took a new name, as the "Howard Light Infantry."

By 1883, however, the name was well established.

The name was that of a famous general of the Union army in the Civil War, O.O. Howard, who after the war was director of the Freedman's Bureau.

Gen. Howard saw Fayetteville only once. That was in March 1865, when he commanded a wing of Gen. William Tecumseh Sherman's army as it marched through the town.

The military company was not the only black institution honoring Howard. A subscription school started in 1865 under the auspices of the Bureau was renamed the Howard School in 1867. It was the forerunner of Fayetteville State University.

In 1883, the General Assembly reorganized the State Guard, authorizing 25 companies in four regiments of white militiamen and a separate "1st Battalion (colored)" for black militia.

By September of that year, the State Adjutant General had inspected and approved small appropriations for the two white companies in Fayetteville and for Company C, the Howard Light Infantry, of the black battalion.

The only other black unit approved so far was Company A, the "Oak City Blues" of Raleigh.

In that year, both the Howard Light Infantry and the Oak City Blues demonstrated their parading skills at the Black Fair days of the annual North Carolina Exposition in Raleigh.

1877 Officers list

In an 1877 list of officers of black State Guard units, five men from Fayetteville held commissions.

Halliday was listed as major of the 5th Battalion. In 1882-83, he was a lieutenant colonel.

Lewis Smith was listed as captain of the company in 1877-78.

Samuel Buxton was first lieutenant. In 1878, he would be promoted to captain and become a battalion officer through 1880. His family included several pilots on Cape Fear River steamboats.

Matthew N. Leary and Dallas Chesnutt were second lieutenants. Chesnutt served only that year. Leary served through 1878.

Chesnutt was a mail clerk on steamboats plying the Cape Fear River. Leary was a member of a free black family. During Reconstruction years just after the Civil War, Chesnutt's father was a county commissioner and his brother a member of the General Assembly.

Other officers came on board in 1878.

S.H. Cuningham was captain of Company C later that year. Robert Barefoot was first lieutenant and would hold that post until 1884.

James R. Deale and L.W. Levy were second lieutenants. Deale served through 1883. Levy only in 1878.

Also in 1878, William Mastiller of Fayetteville was commissioned a captain and commissary officer on the staff of Halliday of the 5th Battalion.

He served through 1880.

Six new names appear in the list in 1879.

George T. Potts was a second lieutenant through 1882 and then became captain of Company C, serving through 1886.

Lewis A. Barge was a first lieutenant who served through 1884. Daniel White was a first lieutenant for 1879.

James Pearce and Henry Gee were second lieutenants for 1879.

That year also, the Rev. F.G. Fry joined the 5th Battalion staff as battalion chaplain with the rank of captain.

The list in the state archives concludes with Robert Berry, a lieutenant in 1881; Jonathan Bayless, a lieutenant in 1883-84; and Benjamin McKethian, a second lieutenant in the 1st Battalion (colored) in 1885.

CHAPTER FIVE

World War I

"Great War" Was Merely The Beginning

Originally published August 19, 1999

The end of the 20th century, 100 years drenched in war's bloodshed, will be here in a few months.

But the violence that gives the century its mordant historical legacy began 85 years ago this month.

On Aug. 4, 1914, German troops poured across the border of Belgium, beginning at 8 a.m. at the tiny village of Gemmenich. Belgian frontier forces engaged them in a brief firefight.

The four-year conflict that started at Gemmenich, known as the Great War, later as World War I, was the opening act of the age of worldwide conflict, spawning World War II and practically every other great bloodletting since.

Within a month after the border crossing, German armies were moving across northern France, descending on Paris.

By the end of August, hundreds of thousands on both sides were dead, wounded or captured.

By the time the Great War was over in November 1918, at least 13 million men, women and children were dead, including at least seven million soldiers.

Yet, in that summer when it began, the world seemed especially bright and cheerful.

As the armies of Europe mobilized that August, the great capitals — Paris, Berlin, London, St. Petersburg — were basking in especially fine weather. Parks were filled with summer vacationers.

Even when soldiers marched away to the mobilization trains, many wearing nosegays of summer flowers tucked into tunics by well-wishers in the cheering crowds, the generals as well as the public expected a quick war, almost certainly to be over by autumn.

On the move

And what about across the Atlantic, in the Gemmenich-sized town of Fayetteville, N.C.?

Well, it so happened that troops were on the march there, too.

On the bright Monday morning of Aug. 5, 1914, the officers and enlisted men of Company F of the North Carolina National Guard, Fayetteville's own Fayetteville Independent Light Infantry, gathered at the train depot.

The 57 civilian soldiers were on their way to summer training camp, this year at Camp Wheeler in Georgia, where they would join other Guardsmen of the 9th Division from Georgia and Florida.

The Fayetteville Observer reported that "Camp Wheeler is situated on a high hill overlooking the city of Augusta, and is said to be a delightful resort."

The reporter for the hometown newspaper concluded his report:

"We trust that the boys of Company F will have a fine time, and we believe they will."

If the trip to Augusta sounded as much a holiday lark as a serious military exercise, the men of Company F were no different from the rest of the community that August.

Fayetteville was paying more attention to baseball, the cotton crop, and watermelons than to the headline on Page 2 of the Aug. 5 edition of the Observer:

"War Clouds Gathering in Europe."

A front page story in that same edition was headlined, "Snappy Baseball," and reported the 7-4 victory of the "Eutaw Team of 71st" over the nine from the Holt-Williamson cotton factory in Massey Hill.

The Eutaw team was a new one but was proving formidable.

"The Eutaw boys are hustlers," said the account of the game. "The aggregation is composed of fine young men and should be encouraged."

In more serious matters, folks were happy to hear that the cotton crop and the watermelon crop were in fine shape, despite dry weather.

And several thousand folks from counties surrounding the Robeson County town of St. Pauls were still recalling the good time there a week earlier at the annual "veterans reunion and farmers picnic," held under the "big oak grove" at the town's Presbyterian church.

The crowd had arrived by excursion train, by wagon, and by newfangled automobiles.

The latter, more than a hundred of them, were used for the first time to transport the veterans in the parade.

As August 1914 ended, hopes for a quick end to war in Europe were dimming, although President Woodrow Wilson was said to be engaged in a behind-the-scenes peace initiative.

But even if it did not end soon, the editor of the Observer found a silver lining in the war clouds.

In an editorial headlined, "A Great Opportunity Before Us," he suggested that the war in Europe affords a great opportunity for the United States to "make its jack" by expanding its textile chemicals industry to fill the void left as Germany's industry concentrated on chemicals for making war.

War on the horizon

And as they returned from summer camp and returned to their peaceful pursuits, several men of Company F could hardly anticipate that three years later they would be fighting in the trenches of northern France.

Lt. Robert Lamb would be Capt. Lamb in 1918 and take the company to the battlefields of France.

Quartermaster Daniel Byrd of 1914 would be second in command as a lieutenant in 1918.

Both would win the Distinguished Service Cross for heroism in battle with the 119th Infantry Regiment of the 30th Division.

Cpl. Ted Frye of 1914 would become Capt. Frye in 1918, also winning the

Distinguished Service Cross in action with Company H of the 30th.

Pvt. Gerald Erambert of 1914 would be Cpl. Erambert of Battery B of the Trench Mortar Unit of the 119th Infantry.

For Lamb, Byrd, Frye and Erambert, as for their generation everywhere, the Great War that began in that August 85 years ago would be the defining moment of their life, as it would also be the dark door into the 20th century.

National Guard On Mexican Border: Deja Vu

Originally published June 24, 2006

National Guard on the Mexican border?

Big news in 2006 but deja vu in Fayetteville — a story that is 90 years old.

Nearly 100 men and boys from Fayetteville went with the National Guard to the Mexican border in 1916.

Among them were several teenagers eagerly recruited to fill the ranks of F Company of the 2nd Regiment of the North Carolina National Guard, a unit which before it was "federalized" was the 123-year-old volunteer militia company known as the Fayetteville Independent Light Infantry.

Among the teenagers, George Ward of Fayetteville was only 16 years old. Two years later, he would would go on to win the Distinguished Service Cross for fierce gallantry in the trenches of World War I.

Also among the ranks was 16-year-old Robert Porcelli, the popular and strappingly handsome Italian immigrant with a melodious singing voice and great talent with a bugle.

In July 1918 in France, "Mike" Porcelli died when a German shell collapsed the trench where he was training with British soldiers. He was the first soldier from Fayetteville to die in the Great War of 1914-18, now known as World War I.

In the early summer of 1916, neither Ward nor Porcelli nor any of the 100 or so members of the Fayetteville Independent Light Infantry were thinking military thoughts.

But the week of July 4, 1916, the company learned that it was to go into federal service as F Company and go to the Mexican border.

A federal medical inspection caused a manpower crisis for the former militia company. The medics disqualified 27 men, including four sergeants, leaving only 61 fit for the mission. A hasty recruiting drive brought in Ward and Porcelli and several others.

On July 6, 1916, the company — along with companies from Lumber Bridge and Raeford — took trains from Fayetteville to a little city of white tents near the coastal village of Morehead City.

The grandly named Camp Glenn on the sandy plain was definitely not like back home. The men were now full-fledged soldiers with a good chance of being shipped out to dangerous locales, specifically to the border between the U.S. and Mexico.

Little remembered now, the U.S. "Punitive Expedition" into Mexico, chasing after irregulars under the notorious Pancho Villa, became a tiny warm-up by the Army for the huge World War I mobilization that would soon follow.

It also was a warm-up for the Cumberland County men, many of whom would fight in the trenches of France in the Big War that was just over the horizon.

A bit of luck

In 1916, however, they were lucky.

The fighting in Mexico was largely over before they embarked that autumn for what turned out to be a six-month stay in a prairie camp near the border town of El Paso. Two North Carolina Guard regiments boarded 74 Pullman cars at Camp Glenn on a five-day train trip to the border camp where thousands of other guardsmen, most from eastern states, were gathered.

An early letter from Sgt. B.M. McFadyen describes Camp Stewart, the tent city just outside El Paso:

"The camp presents a busy spectacle. Everywhere one looks, the prairie is covered with tents and motor trucks, baggage in trains and the cavalry is ever on the move.

"So far we have done nothing but hunt prairie dogs, horned toads and rattlesnakes. The dust is in motion all the time and that is the thing we have to contend with."

In October, F Company took part in war games organized by the regiments. The company won a commendation for its soldierly appearance, and more seriously, for its skill at digging trenches. Its work was "held up as a model for other companies."

The company mascot — a Cumberland County goat named Woodrow in tribute to President Woodrow Wilson — survived a desert crisis. He tried to eat a cactus. The thorns were removed, carefully, by his handler.

By Thanksgiving, the question was being raised openly in the Observer: "When are they coming home?"

On Thanksgiving Day, F Company sat down to a turkey dinner, capped by cigars, cigarettes and fruit, all prepared under the direction of mess Sgt. Ross Jones.

In early December, the company learned it was to "winter over" at Camp Stewart. Tents were to get wooden floors. They already had stoves to ward off the chill of the desert nighttime.

At Christmas, the company had another typical North American feast of roast turkey and all the trimmings, ending with fruit and mince pie. Each soldier got a present of four linen handkerchiefs — gifts from the Stein Brothers Store in Fayetteville.

At New Year, according to McFadyen, bands serenaded the lines of tents for 20 minutes after 1917 came in. Resolutions were made, "the universal one seeming to be 'never to enlist in anything again, the National Guard not excepted!'"

The Cumberland soldiers waited nearly three months into the new year before they finally got back home on March 27, 1917. In less than a month, the United Sates entered World War I.

Within weeks, many men of F Company were back in federal service as soldiers of the 30th Division — which soon left for Camp Sevier, S.C. — and were trained for World War I fighting in France.

In that war, others who had been "on the border" joined Porcelli as special heroes.

Cyrus Adcock was only 20 when he was killed in action in May 1918, the first Cumberland County man to die in the war. William McLaurin, John McPhail and Charles Cornwell Hall also were killed in action or died of wounds.

Sgts. McFadyen and Jones won officers' commissions as wartime captains. The former hash slinger Jones came home to briefly serve as police chief of Fayetteville.

Call To War Mobilizes The Region

Originally published June 5, 1997

Eighty years ago today, thousands of young men, both white and black, lined up at the county courthouse and at voting precincts across the Cape Fear region and the country to sign up for "the war to end all wars."

Congress had put the United States into World War on April 6, four days after President Wilson issued a "war message" calling for a declaration of war against Germany in the conflict that had raged in Europe since 1914.

The nation's first universal "draft law" was soon passed, and on June 5, 1917, young men ages 21 to 30 were answering the call.

Before nightfall, more than 2,500 had registered in Cumberland County, including 900 blacks, the first time in American history they were treated equally in at least one phase of military mobilization.

As the country went to war that bright day in April 1917, more than 70 Cumberland County young men were already wearing Army khaki, and they had just returned from a "war zone."

They were members of Company F of the 2nd North Carolina Regiment of the North Carolina National Guard who had been in service for six months, stationed in El Paso, Texas, as part of the U.S. Expeditionary Force on the Mexican Border.

Many had been members of Fayetteville's ancient militia company, the Fayetteville Independent Light Infantry, when the unit was called into Guard service.

They had returned to Fayetteville from Texas on March 27, 1917.

Soon after war was declared, the National Guard was mobilized and gathered at several former summer training camps across the state.

Company F and Company L from Lumber Bridge in Robeson County, a militia company known as the "Scotch Guards," went to Goldsboro and were assigned guard duty along railroad lines all over North Carolina.

Two weeks after the declaration of war, Capt. R.J. Lamb of Fayetteville wrote from "Camp Royster" in Goldsboro that Company F had been divided into seven train-guard detachments, all under Lt. Dan Byrd, and were mostly stationed in western North Carolina.

Lamb called for volunteers to fill the ranks of the understrength Guard companies.

That same week in Lumber Bridge, the town mourned the death of Capt. Joseph L. Shaw, who had commanded Company L in El Paso. He had contracted pneumonia there, recovered, but fallen ill again at Camp Royster and died in a Goldsboro hospital.

In June, Company F suffered its first casualty when 25-year-old Pvt. Walter W. Cook accidentally shot himself in the left foot while on guard duty in the little mountain town of Murphy. He was reported "getting along nicely" in a Mission Hospital in Asheville.

Other military services quickly began recruiting. On April 9, a Navy recruiter arrived and set up a "recruiting sub-station" in the Fayetteville Post Office on Hay Street.

Others volunteered for the Marines, while several college-age men were accepted at officer training schools.

Among those who quickly signed up for military service was George Myrover Jr., who by mid-May was training with a cavalry unit at Fort Ethan Allen in Vermont.

Closer to home at the officers training camp at Fort Oglethorpe, Ga., were six

Fayetteville men, Dr. E.J. Carson, J.S. Huske, Terry A. Lyon, E.D. Orrell and Donald F. Ray.

In the first days after Wilson's war call, everybody wanted to be in on the action.

Cheers for FILI

In a throwback to the past when a call to arms brought out men of all ages, more than 150 white men of Fayetteville gathered at the tiny armory of the Fayetteville Independent Light Infantry only two days after Wilson sent his war message.

They spent an evening in military drill. "They marched through the streets and were greeted with cheers from crowds on the sidewalks," The Fayetteville Observer assured its readers.

In the textile factory village of Massey Hill, another such "patriotic rally" on the next evening brought out 100 men, and 40 of them signed up for a "home protection company."

Sgt. J.B. McNeill, a National Guardsman, conducted a drill, and it was noted that several of the men, veterans of Company F's duty in El Paso, already knew their lefts and rights.

What was to be different this time was that black men also were in on the act.

A day before the "minute men" rallies, Professor E.E. Smith, superintendent of the Colored Normal School, predecessor of today's Fayetteville State University, called all 450 students, men and women, to a convocation and "made a patriotic talk."

Of the 110 male students, "16 signified his willingness to enlist in defense of the flag." They were the first men in Cumberland County to take that vow.

And also for the first time in wartime, there was official work for women.

Eunice Sinclair of Fayetteville, who already was a career woman as a child welfare worker, was the second woman from North Carolina to join the Navy League's Service School, spending the month of May 1917 in Washington, training to be a Red Cross worker with Navy units.

A more practical approach to war service was taken in May when several dozen farmers participated in a "food rally."

They signed a pledge to speed up cultivation of "idle lands," to plant crops "other than corn for feeding of hogs and stock," and even to work "part of Saturdays."

Rallying citizens

In a series of rallies, 258 citizens pledged to buy $115,000 in the newfangled "Liberty Bonds."

By mid-June, however, rallies were losing their appeal. When Mayor McNeill called for a gathering at the courthouse to promote Liberty Bond sales, the editor of the Observer regretted "that it was not well attended."

At the same time, however, a meeting of the women of the Fayetteville Red Cross Chapter at the library in the Market House was well attended.

The meeting heard a talk by Eunice Sinclair, just back from her training in Washington. The chapter organized for its major role of preparing surgical bandages by dividing into teams under the general chairman, Mrs. H.W. Lilly.

And perhaps the most significant early wartime contribution of Fayetteville was launched by Mrs. John Underwood.

She proposed that the city raise money for an ambulance, to be named the

"Lafayette Ambulance," for use by the armed services or the Red Cross.

By early June, fundraising was well under way, with a limit of $5 per donor. The United Daughters of the Confederacy got into the act by being among the first to give to the ambulance fund, as well as $25 to the surgical bandage fund of the Red Cross.

Meanwhile, a young Fayetteville man away at college, Noel Paton, signed up for an "ambulance corps" at Sewanee, Tenn. He wrote home that they would be inducted, "and then on to France where we will train."

2,600 Register

By mid-June, the records of 2,600 men who had registered for the draft had been copied by a team of volunteers that included 20 men and 13 women.

Clerk of Court W.M. Walker thanked the group for "a hard week's work," and listed their names, saying, "I want the public to know who these patriotic men and women are."

The Cumberland County draft board that would hear appeals when call-ups began was named by Gov. William Y. Bickett. It was composed of Dr. J.W. McNeill, N.A. Sinclair, and "Colonel" George McNeill.

As the first 90 days of the war passed and the country waited for the first draft numbers to be chosen, Cumberland County heard of a local man who was already overseas.

Curtis Vinson of Beaverdam Township was reported to have arrived in France on June 29.

He was the vanguard of hundreds of others who would go "over there" in the war that started that bright April morning 80 years ago.

County Residents Entered Great War Early

Originally published July 7, 2005

Young men from the land between the Cape Fear and Lumber rivers who served in France during the Great War of 1914-18, now known as World War I, had short military careers.

They typically arrived overseas in the spring or summer of 1918.

If they were in battle, the fighting occurred in late summer and autumn, up until the armistice that ended the war Nov. 11, 1918.

But at least two young men from the region got to France in 1917.

One was killed in battle just as the United States entered the war in April 1917. The other arrived in the summer of the year and was severely wounded early in 1918.

When Gerald Marsh, a Cumberland County native who entered the U.S. Army from Parkton, died in early April 1917, he was wearing the uniform of the 5th Battalion of the Canadian army.

He was in the great wave of Canadians who stormed a heavily defended German position known as Vimy Ridge.

More than 14,000 were killed, wounded or missing in the assault, which was the single bloodiest action in the World War I history of the Canadian army in France.

In early June, Dan Marsh of Parkton received word from the war department of Canada that his son was missing in action on the Western Front.

Not until March 15, 1918, did word come that Marsh had been killed in action, probably on April 28, 1917.

Marsh thus became one of the heroes of a defining moment in Canadian military history.

The assault on Vimy Ridge is sadly honored in many Canadian towns and cities by memorials listing the names of local men who died in the action against what was considered the most stubborn of all German defensive positions on the Western Front in World War I.

Honored by Canadians

"Vimy" remains a word honored by succeeding generations of Canadians.

Before the Canadian assault in 1917, more than 150,000 French soldiers had been killed or wounded in unsuccessful attacks along the Vimy portion of the German defenses.

By 1919, when Robeson County veterans were compiling lists of servicemen from the county, Gerald Marsh was credited with being the first county casualty of the war, killed in action only a few days after the United States entered World War I.

How did a Parkton man find his way to the battlefield as a soldier of Canada?

Gerald Marsh was a professional soldier. He joined the U.S. Army in 1914.

In the winter of 1916, he was a gunner in an artillery unit stationed on the coast of Maine.

He and several of his buddies were eager to get in the fight of the big war in Europe.

They put on civilian clothes, crossed the undefended Canadian border and enlisted in the Canadian army.

Only after donning their new uniforms did they reveal that they were fully trained soldiers. They were sent overseas within two months.

Son tells story

Dan Marsh told his son's story in an interview with The Fayetteville Observer:

"He enlisted in a Canadian regiment about two months ago and left for the battlefronts soon after. It is supposed he fell fighting in the fierce combat of Vimy Ridge, which was finally captured by Canadian forces."

Among Cumberland County soldiers in World War I, Pvt. Curtis Vinson is among the earliest to die in combat.

Vinson's story is told in the Observer in July of 1918, under a headline "First Soldier Back."

"The people of Fayetteville had the pleasure recently of shaking the hand of one of our Cumberland boys who has been in service in France.

"This Cumberland soldier went over on June 10, 1917, and was wounded January 1, 1918, in the trenches.

"He has been in Walter Reed Hospital, where he had one of his lower limbs removed."

Vinson was from the Beaverdam section east of the Cape Fear River. His father, E.M.Vinson, died a few weeks before Vinson was greeted in Fayetteville.

He was on furlough from Walter Reed, visiting his brother, also E.M. Vinson, of Fayetteville.

Pvt. Vinson told the Observer that he was being "well cared for by the United States" at the famous military hospital in Washington.

Clouded history

Vinson's military history is hard to find. None of the stories indicate what unit he was in or when he enlisted.

But it seems likely that he was a soldier in the famous 1st Infantry Division, or "the Big Red One."

The 1st Infantry went to France in early June 1917 and for several months was the only large unit in the American Expeditionary Force.

The dates listed by his brother for Vinson fit the sailing of the 1st Infantry.

By contrast, the 30th Division, in which a large percentage of Cape Fear men served, didn't arrive in France until May 1918.

Curtis Vinson, like Gerald Marsh, evidently was eager to get in the fight.

Vinson's brother heard from him in October 1917, and he told the Observer:

"I had a letter from him this week saying he would advise American boys to get the chance to get over to France. It is a fine country and he is getting plenty to eat and enjoying it fine."

Area Figured Heavily In Two WWI Companies

Originally published February 12, 1998

Camp Sevier was a hastily built World War I training camp in the foothills near Greenville, S.C.

For men from the Carolinas and Tennessee who arrived there in late summer of 1917, the area was a familiar scene of cotton fields, small towns and scruffy forests on rolling hills. It was hot in summer and chilly in winter.

By February 1918, the 20,000 new doughboys had been in training for nearly five months. They were getting anxious to see another place, preferably the battlefields of France, where World War I had raged for three years. This was especially so among the dozens of men from the Cape Fear River and Lumber River areas, notably Company F and Company L of the 119th Infantry Regiment of the 60th Infantry Brigade. Both units had high concentrations of soldiers from the same counties of North Carolina, mainly because they had been local volunteer militia companies converted to National Guard outfits.

Company F had been the Fayetteville Independent Light Infantry, which dated to 1793.

Company L from Lumber Bridge in Robeson County dated from 1849 as the Lumber Bridge Light Infantry. Since the Civil War it was proudly known as "the Scotch Tigers."

Few military units preparing for France in early 1918 had been in uniform as long as these companies of the former 2nd Regiment of the North Carolina National Guard.

They were called into federal service in the autumn of 1916 to serve on the Mexican border, where a U.S. "punitive expedition" had chased rebel Pancho Villa.

After posting at El Paso, Texas, the companies came back home in the spring of 1917 just as President Wilson called for war with Germany and the U.S. formally joined

World War I.

For a time, the companies were on federal service, sent in small detachments to guard railroads from Goldsboro to Asheville.

Then in late summer of 1918, the War Department "drafted" the National Guard units of the Carolinas and Tennessee to form the 30th Infantry Division and ordered them to assemble at Camp Sevier.

Following its service in France, the 30th would take the nickname "Old Hickory," honoring President Andrew Jackson, born either in North or South Carolina, and elected president from Tennessee.

But at Camp Sevier in early 1918, they were still debating an appropriate nickname.

The more refined among them were proposing "the Blue Ridge Division."

But the favorite of many of the men was less refined. They wanted a name fitting an industry that was popular in all three states.

They wanted "the Moonshine Division."

The much-traveled men of Company F found a surprise present from back home waiting for them when they pulled into the raw cantonment in cotton fields of western South Carolina.

Among Capt. Robert Lamb's first correspondence was a letter to Mrs. Lucy Wooten of the Fayetteville chapter of the Red Cross, thanking the ladies of the town for the "comfort bags" containing all sorts of goodies as well as hard-to-get toilet articles.

Lamb wrote: "While they are seldom blue or homesick, I can see in each face more sunshine now."

And Lamb gave his first assessment of the new home of Company F, writing:

"After 16 months of moving about, Greenville is the most hospitable place we have been."

Before they could even think about fighting the Germans, the thousands of young Southerners at Camp Sevier had to fight illness that rampaged through the tents and wooden mess halls.

The camp diseases were described by young H.N. McLaughlin, a member of the Machine Gun Company of the 119th Infantry, who listed them as "measles, mumps, pneumonia, and spinal meningitis."

Among the civilian volunteers was Dr. Robert McGeachy, a brother of Cumberland County Sheriff Hector McGeachy. Robert was an older physician who came to treat soldiers and then, in a burst of patriotic fervor, sought to enlist.

When a military doctor told him he had a "defect" that needed fixing (probably a hernia), McGeachy became a patient in the camp hospital and wrote that he was "trying to go overseas."

The doctor's son, Robert Jr., was already at Sevier, a private in the 30th Division. Robert Jr., said the report, was also fervently patriotic, having enlisted even though he was "under age for the draft."

The cantonment that was battered by disease in the autumn was battered by cold weather that winter.

In mid-January, Capt. Lamb wrote that training was practically suspended for two weeks because of "the ice and snow-covered ground," as the Piedmont of the Carolinas suffered the most frigid winter in years.

Lamb said the men had held up well. "The thing that suffered most," he wrote home, "was the old saying, 'the Sunny South.' "

As the 30th trained in the South Carolina foothills and looked forward to boarding ships for the battlefields, other men from the Cape fear region already were seeing France.

James McDaniel of Fayetteville went over in November of 1917 in an engineer unit.

Lt. D.P. Shaw from Cumberland County's Seventy-First Township, a brand-new officer from officer training school, sailed in December.

An early veteran of the Atlantic crossing was J. Kidwell Grannis from Flea Hill (now Eastover) in Cumberland County.

He was a merchant seaman on the government freighter Montosa and by January had made two round trips with shiploads of engineer supplies.

Grannis liked what he saw, even in war-weary France. He told his mother: "This country is peopled by fine men and lovely, attractive women!"

And already a combat veteran, to hear him tell it, was Seaman Cory P. Cain, whose address was Canal Street in Fayetteville.

Cain was in the crew of a U.S. Navy ship that used 6-inch guns to sink a surfaced German submarine "just before Christmas" in 1917.

Just after that Christmas, young Alphonso Newton, known to his Fayetteville buddies as "Newton," landed in France with an advance party of U.S. artillerymen.

He wrote that they were "well-housed, well-fed, and well cared for."

At Camp Sevier in that February 80 years ago, the men of the 30th Division were wishing they could see it for themselves.

The 1st Americans Arrive In World War I Europe

Originally published March 19, 1998

To a small-town man such as Pvt. Joseph D. Breece of Fayetteville, the rolling fields and forest copses of northern France were a new world when he first saw them in March 1918.

And yet, there were familiar touches.

And as soon as he arrived on March 26, he wrote home to his father J.D. Breece:

"This is really pretty country. It is spring weather here. We see mules working with yokes, and horses plowing with the traces on the horses. Good luck to all."

Breece was among the early arrivals of the American Expeditionary Force on the battlefields of World War I. He was a volunteer who didn't wait for the draft to call him up after the United States entered the war 11 months earlier.

He was in a service unit, the 13th Company of the 2nd Motor Mechanic Regiment, sent to France to teach repair and maintenance techniques to British soldiers who had been fighting in scarred hills and ruined villages for three and a half years.

For Joseph Breece and for hundreds of thousands like him, the early spring of 1918 was a time of waiting before they joined the bloody conflict raging along 1,000 miles of trenches from the North Sea to Switzerland.

Exhausted by a war that had lasted since August 1914, the British and French armies were counting on a flood of American manpower promised by President Wilson.

Meanwhile, across the deadly no man's land, German armies were pulling themselves together for one last great spring offensive.

But in March, the promise of American manpower was still mostly a promise. Only a few understrength fighting units and a growing array of service and construction troops had so far arrived in France.

Most of the newly minted soldiers of the U.S. Army were still training in hastily built camps in the United States.

Carolinians in the 30th

Hundreds of soldiers from the Cape Fear region were at Camp Sevier in the South Carolina foothills, where the 25,000 men of the 30th Division were preparing for war.

The division included National Guard outfits from the Carolinas and Tennessee.

It had been training since the previous summer. Plagued by illness and bad weather, the division was behind in its training schedule.

Even if it had been up to snuff, there was little prospect of an early departure. A severe shortage of shipping was slowing deployment of the million-man army that the United States planned to send to France.

Meanwhile, Sgt. Bernice M. McFayden of Fayetteville received his commission as a second lieutenant in the 119th Infantry Regiment of the 30th Division.

McFayden, like dozens of others, was a member of the former militia company known as the Fayetteville Independent Light Infantry, which was mobilized as a National Guard unit in 1916 to go to the Mexican border

It was now Company F of the 119th Infantry.

Pvt. Edgar McBride wrote his nephew, Vincent McBride, that he was settling in with Battery B of the 53rd Artillery Regiment.

The unit was typical of early arrivals of the American Expeditionary Force. It had no guns, and was attached for training to a British artillery regiment.

The largest group of Cape Fear men in France in that spring 80 years ago was engineers, members of the 17th Engineer Regiment.

That unit had gone to France in February to build port facilities and improve railroads, preparing for the flood of American men and material scheduled to come in the summer and fall of 1918.

Pvt. Jesse Williams wrote his sister Roxana that he felt right at home, because Fayetteville friends Ed Owen and J.P. Ewing were also in his company.

The regiment already suffered a casualty. An "Asheville boy" in their company was killed in an accident.

Also in training with a British unit, learning how to use a machine gun, was Pvt. Paul Amos Hatch of Fayetteville, a member of a "motor machine gun company."

Hatch was said to be only 16 years old, "the youngest man in his regiment." But like the others who were in France at this early date, he had volunteered for service rather than wait for the draft.

Sgt. Lauchlin McNeill was also engineer. He wrote his parents, Dr. and Mrs. J.W. McNeill, that he was "safe in France."

These early arrivals on the battlefields of France actually were preceded by several Cape Fear men who managed to join either the Canadian or British armies earlier in the war.

And as spring of 1918 arrived, sad news came of the fate of one of them.

It happened a year earlier, in April 1917, when the Canadians stormed a bloody height near Arras known as Vimy Ridge.

For the small bit of battlefield real estate, the Canadians suffered 14,000 casualties in two days.

Among them was Gerald Henry Marsh, son of Dan Marsh of Parkton. He was killed in action April 28, 1917.

The terrible losses at Vimy Ridge were just a single moment's cost of a conflict that by March 1918 had killed millions of men on both sides.

It was an ominous harbinger of what the future might hold for men such as Pvt. Joseph Breece as he enjoyed the coming of spring to the French countryside.

Black Soldiers Hold Impressive Record

Originally published February 1, 1996

Black men in Civil War days overcame early opposition to become soldiers in the U.S. Army. They sang a song that said, "We'll fight for the Union, if we only get our chance."

Fifty years later, black men were wearing khaki and World War I soldiers were still rigidly segregated into units commanded by white officers.

They were often scorned by their own commanders and seldom called on to fight with the white American Expeditionary Force in France.

Despite the barriers and when given a chance, men such as Sgt. Walter Richardson of Cumberland County became heroes of the "war to end all wars."

Richardson won the country's second highest military award, the Distinguished Service Cross, as an orderly in the Medical Detachment of the 371st Infantry Regiment.

Like most such awards for medical orderlies, it was given for his bravery in rescuing wounded comrades from the bullet-swept battlefields of France.

The 371st was a unit of the famous 93rd Division, one of two World War I divisions composed of black enlisted men under white officers.

While many of the black units of the 93rd Division were comprised of men from New England and New York, including hundreds who had been in all-black National Guard units, the men of the 371st were largely from the South, particularly the Carolinas. Most of the white junior officers of the regiment also were from the South.

Aligned with French

Sent to France in April 1918, the regiments of the division were attached to French army divisions. They even wore French helmets rather than the "tin hat" trench headgear of the American Expeditionary Force. The division's shoulder-patch insignia bore the French helmet as its design.

The 371st Infantry Regiment was in combat almost from its first week in France.

Before the war ended in November, it had suffered casualties of more than 30 percent and had become one of the most decorated outfits in the Great War.

Richardson was among 60 officers and 124 men of the 371st who were decorated with the Distinguished Service Cross or the French Croix de Guerre, or War Cross. The entire regiment was awarded the Croix de Guerre with palm, the World War I equivalent of the Presidential Unit Citation of World War II.

A historian of black soldiers in World War I wrote of the 371st Regiment:

"Perhaps because of its humble background, the 371st was practically ignored in the writings of black commentators of the times; there were no poets to sing the praises of the Carolina 'darkies.'

"The regiment is not often mentioned by white observers either, because they tended to focus on poor performance by blacks, and the 371st gave little opportunity for that."

First black volunteer

E.C. Wright, an 18-year-old student at the State Normal School, now Fayetteville State University, is credited with being the first black volunteer from Cumberland County in World War I. He also served in the 93rd Division. Wright enlisted in October 1917.

Before the war was over, Wright became a sergeant and spent more than six months in front-line sectors with that famous unit.

Another Cumberland County man who saw combat was William Henry McDuffie, from the Flea Hill (now Eastover) section, who served in Company B of the 368th Infantry Regiment, a unit in the 92nd Division.

The division was composed largely of draftees with a leavening of old Army regulars and National Guardsmen.

It was known as the "Buffalo" division, a legacy of the old Regular Army black cavalry units nicknamed the "buffalo soldiers" by the Indians who fought them.

A newspaper note about McDuffie said "he wears a 'buffaloe' on his sleeve," a reference to the divisional shoulder patch that depicted an American bison.

More typical of the black men who wore World War I khaki was the first soldier from Cumberland County, black or white, to die in the war.

Thomas H. Kirkpatrick Jr., a member of a quartermaster labor unit, succumbed to pneumonia in France on Feb. 10, 1918.

Kirkpatrick's unit had been one of the first in France.

At the port of St. Nazaire, thousands of black men in labor and quartermaster battalions built the docks, barracks and warehouses that would serve the millions of men of the American Expeditionary Force who began to arrive that spring.

Kirkpatrick was typical of the scores of black men from Cumberland County who eagerly sought to enter military service but were denied combat roles and were used essentially as laborers.

Legacy of pride

Nonetheless, their service in World War I gave the black veterans a sense of belonging and a legacy of pride that served them into the years to follow.

One who did survive was Abednego Purdie of the Swann's Creek community. He died at 61 in 1956, proudly remembered for his service as a private in Company C of the 349th Quartermaster Labor Battalion in 1918.

Buried in the same graveyard with Purdie is Ella M. Burns, a private in the 431st Labor Battalion. He died in 1934.

Buried at Savannah Baptist Church is Douglas Williams, who died at 63 in 1954, a member of the 829th Company of the Transportation Corps.

Eastover story

In another graveyard in Eastover, a handsome stone tells the story of a young black man from Cumberland County who did make it into a combat unit and was killed in action in France.

Pvt. Nathan Byrd, 23, of the Flea Hill community went to France with the headquarters company of the 367th Infantry of the 92nd Division.

He died in action Nov. 7, 1918, only four days before the Armistice that ended the war.

His regiment was attacking German positions to extricate units of the U.S. 7th Division and French forces pinned down by heavy fire near the Moselle River.

The action redeemed a unit that was regularly maligned by its viciously racist American Army commander, Gen. Robert Bullard.

The grateful French awarded the unit the Croix de Guerre.

The inscription on Nathan Byrd's tombstone in the cemetery of Mount Zion AME Church speaks for all the young men, black and white, who, as it says, "Died In France."

It reads:

"Just in the morning of his day, in youth and love he died."

Region Troops Embark For Trenches Of France

Originally published April 23, 1998

April 1918 came to northern France with blossoms.

And with another season of blood and destruction.

Behind the jagged scar of trenches that stretched from the North Sea to Switzerland, poppies bloomed in Flanders, roses in Picardy and apple trees in the lovely valley of the Marne River.

But along the lines, there was only more fighting and more killing.

In the fourth spring of World War I, known then as the "Great War," German armies launched a final massive offensive.

They hoped to crush weary British and French armies before hundreds of thousands of American soldiers, the "doughboys," could arrive to turn the tide of a struggle that had gone on since the summer of 1914.

Now a year after the United States entered the war, the young men of the country were heading for France in a growing flood. Only a few had yet to see the trenches, but their hour was fast approaching.

Back home, thousands of newly minted soldiers were preparing or leaving for France.

April in the Cape Fear region opened 80 years ago with a mass rally of Cumberland County's black community at the Lafayette Auditorium on Person Street in Fayetteville.

Hundreds of men, women and children turned out to bid goodbye to more than three dozen men leaving for military training camps.

The Fayetteville Observer reported:

"The drafted men from all over the county were there ready to take the train for camp, and there their friends and neighbors had given them a farewell ovation."

The event was arranged by Dr. E.E. Smith, principal of the Fayetteville Normal School. The men and the crowd heard an inspirational address by "Dr. Cotton of Henderson."

A few days later, hundreds turned out for a big Saturday parade in Fayetteville, after which Gov. William Y. Bickett made an impassioned call for the success of the fundraising campaign for the "Third Liberty Loan."

Even before spring came to the battlefields, two families from the region had received telegrams announcing the death of a loved one in France.

Thomas H. Kirkpatrick Jr., son of a tenant farmer in Cumberland County, died of pneumonia Feb. 10, 1918, while serving with a quartermaster labor battalion building supply warehouses for the American Expeditionary Force.

Kilpatrick was the first black soldier from Cumberland to die in World War I.

On Feb. 27, Pvt. George E. Galloway of Fairmont in Robeson County died near Toul in a field hospital of the 1st Infantry Division.

He and 70 other doughboys were victims of a poison gas attack on the trenches where they were learning the ropes of war in what was considered a "quiet sector" of the Western Front.

In April, the news columns of the Observer were increasingly sprinkled with reports about men from the area arriving in France.

At Camp Sevier, an Army training camp near Greenville, S.C., thousands of men of the 30th Infantry Division were told that they would soon be leaving the red-clay foothills for the landscapes of poppies, roses and apple blossoms.

After eight months of training, the untested Doughboys of Company F and Company L of the 119th Infantry Regiment, mostly from Cumberland and Robeson counties, were eager to experience what real war was like.

An early eyewitness description of the fighting, although seen from a distance, came in on April 8 from a Fayetteville man already in France.

Second Lt. James C. Cooper was an infantry platoon leader in the 5th Regiment of U.S. Marines. The two-regiment Marine brigade was part of the 2nd Infantry Division of the U.S. Army.

Separate units that formed the 2nd Division had arrived in the fall of 1917, following the arrival of the 1st Infantry Division. They were the early vanguard of the flood of soldiers now coming to France.

In October 1917, Cooper's outfit first entered the trenches in the "quiet sector" near Toul, where Pvt. Galloway would die in February.

Letter home

In early April, the lieutenant wrote his "Mamay," Mrs. C.J. Cooper in Fayetteville:

"Lately we've been doing the kind of work that doesn't get in the newspapers or win you the Croix de Guerre.

"For several nights we have been working all night in the rain and about six inches of mud, constructing a communications trench behind the lines.

"It has to be done at night because the place is exposed to artillery and machine gun fire during the day.

"One night, we were wonderfully repaid.

"We witnessed a night attack at a distance.

"First came the artillery barrage with a roar and flashes lighting up the rainy night and cloudy sky.

"The innumerable rockets began to go up all along the line, lighting flares illuminating the air for miles, then a variety of signal colors and stars that shower over the scene of attack.

"Machine guns added their rat-tat-tat to the chorus and the display lasted over an hour.

"We have not had permission or time to write, but you probably guessed from the field card that I sent that we have been in action, but I hope you were not excited or worried.

"Our sector of the front is comparatively quiet and although anti-aircraft guns, airplanes and artillery are continuously in evidence, it is not strenuous comparatively.

"We had a gas alarm one night and it served to show what real men I have in my platoon, and when they get under fire they don't have to be told anything twice, but are 'right there' with that 'here I am' demeanor.

"Have been getting good food lately and expect to gain weight.

"Do hope and pray that this big drive will be the last real battle of the war."

This vivid word picture of what war was like along the trenches in Flanders, Picardy and the Marne would soon became the real thing for Cooper and for hundreds of others from the fields and towns between the Cape Fear and Lumber rivers.

Reality Of World War I Hits Home Fast For Soldiers

Originally published May 21, 1998

The month of May in 1918 was a season first of high adventure and then of sober reality for men from farms and towns of Cumberland County who filled the ranks of the brand-new World War I army.

In company with thousands of Carolina and Tennessee comrades of the 30th ("Old Hickory") Infantry Division, they left the red clay hills of Camp Sevier near Greenville, S.C., riding trains to Hoboken, N.Y., and Boston.

There they boarded troopships for a trip across the Atlantic Ocean.

They arrived at Liverpool, the great seaport of northwest England. Within a few more days they were landing at the ancient port of Calais in northern France.

They were in a great stream of doughboys headed for northeastern France, a bloody battleground where soldiers of Britain and France and Germany were locked in desperate struggle in the fourth spring of the Great War.

Despite what must have been excitement at their ocean crossing and arrival in new lands, some letters home were calm.

"We had a pleasant and uneventful trip," wrote Capt. Robert J. Lamb, commander of Company F of the 119th Infantry Regiment, the 30th Division unit comprised largely of men from Cumberland and surrounding counties.

But Pvt. Kenneth McNeill of Company F probably expressed more candidly the real excitement of the small-town and farm men arriving in a new world.

"We are in fine feather," he wrote. "We had a little fun with a submarine. I got two sweet kisses when we came through the streets." He finished with a boast:

"I won't be back until I get the Kaiser's helmet!"

The tingle of the new was still there in the last days of May when they moved into training areas near British trenches in the blood-soaked "Flanders Fields" of Belgium. There, battle-savvy veterans of the famous Black Watch Scottish Highlander regiment of the 39th Division of the British army took them in hand for training before they would take their place in the trenches.

McNeill was thrilled:

"I have met a Scotchman," he wrote to his mother. "They are a bunch of fine fellows and I am proud of my ancestors."

But then, within hours after their arrival, the terrible human sacrifice of World War I became real to the Cumberland County doughboys.

It was May 28.

In a trench looking out across the shattered no man's land between British and German lines, 28-year-old Pvt. Cyrus Adcox was getting first instructions in machine-gun operations.

In civilian life, Adcox was a worker at Victory Mills in the Massey Hill textile village of Cumberland County. Now, he was in the machine gun battalion of the 30th Division.

This was a so-called "quiet sector" of the Western Front. But it was also deadly.

A German artillery round exploded nearby.

Adcox went down, badly wounded by a shell fragment.

A day later, the War Department sent the sad news that he had died of his wounds in a field hospital.

Adcox was the first Cumberland County soldier to die in combat in a war that had already claimed hundreds of thousands and would claim many more before the armistice on Nov. 11.

Another Cumberland soldier, Pvt. Daniel Edwards, wrote to Adcox's mother:

"He was wounded in battle; yet his last words to me before he was carried back on the litter were: 'write to my folks and tell them.' Then: 'put my canteen on the litter, boys, I may want a drink on the way back.' "

Edwards continued:

"Now I have been with Cyrus since leaving Texas (they had served together on the Mexican border in 1916) and I have never known a better or more conscientious boy. His was of a nature which made him friends wherever he went, and his death was regretted by our entire company, who through me send their regard and sympathy. You should be proud of such a son, and no doubt you are, whatever your sorrow at his untimely end."

In Massey Hill, the whole textile village community gathered a few days later for a memorial service.

It was held in the white-framed Methodist Church (now at the intersection of Whitfield Street and Southern Avenue). The hymns were sung by the choir of the Tolar-Hart cotton factory.

The death of Adcox came as the American Expeditionary Force in France swelled to over 1 million men at a time when the British and French were fighting "with their backs to the wall" against enormous German offensives launched in hopes of winning the war before the Americans could make a difference.

Almost on the day that Adcox died in Flanders, the doughboys were going into action in large numbers for the first time.

Responding to desperate pleas from the Allied High Command, Gen. John J. Pershing ordered five of his AEF divisions away from quiet sectors and toward the valley of the Marne River where rampaging German divisions were threatening to sweep all the way to Paris, barely 50 miles way.

On June 1, Lt. J.C. Cooper of Fayetteville and his platoon of the 6th Regiment, U.S. Marines, attached to the Army's 2nd Infantry Division, unloaded from a truck convoy near the banks of the Marne.

They looked out across rolling countryside dotted with tiny French villages and patches of forest.

Directly to the front a few kilometers away was a village named Bouresches.

To the left of the village was a leafy glen known as Belleau Wood.

From the right came sounds of fighting in the riverside village of Chateau-Thierry. There, units of the U.S. 3rd Infantry Division had just destroyed a bridge over the Marne, blowing it up literally in the faces of probing columns of German troops.

The lieutenant and his men knew this was the moment they had prepared for.

But even then they could not have expected that in a few days the names of Bouresches, Belleau Wood and Chateau-Thierry would be immortalized in U.S. military history, and their own names as well.

Soldiers' Letters Home During The Great War

Originally published July 19, 2001

As the old songs put it, the roses were blooming in Picardy and the poppies were blowing in the fields of Flanders.

It was France in May 1918.

And men from the small towns and cotton fields of North Carolina were glimpsing the peaceful roses and poppies on their way to the trenches of the Great War, joining the bloody conflict that had raged for four years across northern France and in Flanders, which stretches into the area of Belgium near the North Sea.

In that springtime 83 years ago, men from the Cape Fear region wrote home to wives and parents describing the scenes and the life so different from what they had ever seen or known.

Among them was Harry Johnson of Fayetteville, top sergeant of Company F of the 119 Infantry Regiment of the 30th Division, who had arrived in England in early May 1918 and then gone across the English Channel to join British forces in Flanders.

Johnson wrote to his wife in June.

After describing the 16-day voyage across the Atlantic, he wrote of France:

"France beautiful with farms as neat and beautifully kept as gardens in America. The houses are all made of brick, and in American eyes, curiously constructed.

"It is an interesting country, but a thousand years behind the times.

"The sun rises at 3:30 a.m. and sets at 9:30, but the boys are busy with their duties and giving a good account of themselves, and they are all well and in high spirits."

David Owen, a farm boy from Seventy-First Township, arrived in France with the 323rd Ambulance Company of the 81st Division, and wrote to his parents, Mr. and Mrs. J.M. Owen:

"We have been having some fine time since we crossed over. Have been in three camps since arriving in France. The scenery over here is beautiful beyond description. We are no where near the front, 300 or 400 miles away.

"Our drinking water is not very good. Our camp is near a beach and we go in the surf sometimes."

Pvt. Kenneth McNeill of Company F expressed his excitement.

"We are in fine feather," he wrote.

"We had a little fun with a submarine. I got two sweet kisses when we came through the streets."

He finished with a boast: "I won't be back until I get the Kaiser's helmet!"

A month later, McNeill wrote his mother:

"I am machine gun sergeant of my company and I have under me the best bunch of men I ever had. I know they will stick with me until the last man falls."

McNeill's brash eloquence was tinged with yearning for home in a letter written after the armistice on Nov. 11, 1918, ended the war, now known as World War I.

To his father, Dr. J.W. McNeill, he wrote:

"When the bullets were whizzing around and falling fast around me, I thought how I would love just to walk around the old market house in Fayetteville again.

"They can talk all they want about a glimpse of the Statue of Liberty, but all I wanted to see was our old market house once more."

Sgt. Josey D. Shaw of Fayetteville, a veteran of the Spanish-American War and top sergeant of the supply company of the 119th Infantry, echoed others, but also reminded the folks of things to come.

He wrote:

"France is a beautiful country. The weather is fine and my health is splendid. I am under the sound of the big guns now, and expect to be up closer in a few days. The customs of the country are different than America, but the boys are all cheerful and well provided for and in fine spirits. They are of the belief that the strife will not be in duration much longer.

"Some of my command are already in the trenches and others will follow in a few days."

For some who went into the trenches, that May would be their last.

Cpl. Robert Porcelli came over with Company H of the 119th Infantry.

He was killed by a German shell burst on July 12, 1918.

Porcelli was the "first man from the municipal limits of Fayetteville" killed in the Great War. He was 19 years old and a volunteer.

Cpl. Archie Rasperry of Company F wrote to his sister, Mrs. W.L. Starling of 229 C Street in Fayetteville:

"I am well, enjoying life and having a good time. I have seen some fine country since reaching France.

"Don't expect many letters, since soldiers are often sent to places where they cannot reach post offices."

On Sept. 29, 1918, Rasperry was severely wounded when the 30th Division attacked German fortifications known as the Hindenburg Line.

Hopeful letter

In January 1919, Rasperry wrote hopefully from a hospital in Salisbury Court in England that "my wounds are healing nicely and my general health is good."

But in March he died of "empyema." He was buried in the American cemetery at Winchester.

Cpl. Wayman Thaggard of Company F, a farm boy from Cedar Creek in Cumberland County, dutifully wrote his mother, Mrs. W.A. Thaggard, almost as soon as he arrived among the Flanders poppies:

"My Dear Mother, I will write you just a line to let you hear from me. I am well and getting on all right.

"Mamma, there is a lot of pretty country over here. The climate is warmer than it is back home.

"I hope you are not worrying about me. I am getting along all right. I know it is harder for you than it is for me. You ask the Lord to give you strength to bear it, and all will be well. You all write just as often as you can. Don't you worry when you don't hear from me. I will write more when I can.

"I close with love to you all. Your son, Wayman Thaggard."

On Sept. 19, 1918, Thaggard was killed instantly when the men of Company F charged across the poppy fields toward the Hindenburg Line.

The Other Historic 6th Of June

Originally published June 4, 1998

In 20th century military history, 6th of June is always D-Day, the 1944 invasion of Normandy, France, by the Allies in World War II.

It was the day of fierce fighting on Omaha Beach and of the heroics of paratroopers of the 82nd Airborne Division as they captured the village of St. Mere Eglise.

But there is another 6th of June. And it is also an immortal date in the annals of U.S. military heritage.

Eighty years ago, in the early dawn of June 6, 1918, soldiers and Marines of the U.S. 2nd Infantry Division charged out of their scratched-out "foxholes" (a new name) near the Marne River in France.

They surged toward German positions in a patch of French forest known as Belleau Wood.

It was the fifth year of World War I, and these newly minted American warriors were being thrown into the path of an onrushing German offensive that had shattered French lines and French morale. The enemy was only 40 miles from Paris.

Although an Army division, the 2nd Infantry Division included more than 5,000 Marines of the 5th and 6th Marine Regiments, grouped in the 4th Marine Brigade.

Yet, the commander of the Marine brigade was an Army brigadier general, James Harbord, recently chief of staff of the entire American Expeditionary Force in France.

Firing in support of the Marine advance were guns of the Army's 2nd Artillery Brigade, including the 17th Field Artillery Regiment.

Commanding the 17th was Col. A.J. Bowley. After the war, he would become longtime commander of a new artillery post in the Carolina Sandhills, Fort Bragg.

The 17th would be stationed at Bragg from 1921 until 1942, when it left for a World War II combat career beginning in North Africa.

In the story of Belleau Wood and Bouresches, the U.S. Marines would win the starring role.

Hammered by fire from an array of machine gun nests, the onrushing Americans went down "like flies," in the words of one officer.

But they began to make headway along the enemy's first line. Late in the day, as other units plunged into the rocky interior of Belleau Wood, a battalion of the 6th Marines commanded by Maj. Thomas Holcombe rushed toward Bouresches.

Capt. Donald F. Duncan led his 96th Company in a running charge.

But machine gun fire virtually wiped out his line.

Among the dead were Duncan himself and the Navy dentist, 25-year-old Lt. j.g.

Weedon E. Osborne, acting combat medical officer for the battalion. He was killed by a machine gun round while carrying a gravely wounded comrade to the rear.

Duncan would be posthumously awarded the Distinguished Service Cross. Osborne was awarded the Medal of Honor.

But others kept going.

As night fell, Lt. James F. Robertson and 21 of his men burst into the village streets, chased off the last German machine gunners and set up a defensive line.

Dozens of other men of the 6th Marines managed to filter into the village, among them Lt. C.J.C. Cooper, a Fayetteville native.

In the hours after midnight, they beat off a German counterattack, and Bouresches was thereafter in American hands.

The battle for Belleau Wood went on.

Not until June 15 did the report go back: "Belleau Wood now exclusively U.S. Marine Corps."

When it was over, the cost had been high.

Between June 4 and July 1, nearly 10,000 young men of the 2nd Division were casualties, including 1,811 killed in action.

Among the dead was Fayetteville native David Graham, the oldest son of Alexander Graham (1844-1934), the pioneering public education superintendent in Fayetteville and Charlotte.

David's younger brother, Frank Porter Graham, also born in Fayetteville, was later in the Marine brigade, after receiving a commission as a second lieutenant.

Frank Porter Graham would go on to be president of the University of North Carolina and a U.S. senator.

For Lt. Cooper of Fayetteville, the savagery of Bouresches would turn out to be his first and last combat in World War I.

On June 10, the Paris edition of the New York Herald reported:

"Other lieutenants who played a courageous and spectacular part in the battle of Bouresches were J.C. Cooper of Fayetteville, N.C., V.H. Hampton of Charlerri, Michigan, J.M. Bars of Mississippi.

"They are all to be sent to America at once to picture to the people of America some of the sterner things connected with the war to impress upon the tens of millions in the East, West, North, and South what a gigantic task and duty lies upon the United States."

On July 12, The Fayetteville Observer carried the news:

"Lt. James Crawford Cooper, who has served gallantly in France being in some of the fighting by the Marine Corps, arrived in the city last evening. He has received honorable mention for conduct on the field.

"Lt. Cooper has been sent over here to give instructions to troops preparing for active duty."

The scale of the fighting forces in those June days 80 years ago at Bouresches and in Belleau Wood was small compared with the struggles that marked the Western Front in previous years.

But the French military writer Jeane de Pierrefeu saw an impact far larger than the numbers.

He wrote of the arrival of the 2nd Division:

"Just at this instant, a perfect cloud of Americans swept over the country. They were crowded as close as possible on immense auto trucks, perched in all sorts of grotesque positions, bareheaded, with their shirts open at the throat.

"They roared out the songs of their country at the top of their voices, while the people along the way greeted them with indescribable enthusiasm.

"The sight of these splendid youths from across the ocean, these smooth-shaven 20-year-old youngsters with their spik-span new equipment, their vigor, and their health wrought a miraculous change in our feelings.

"We felt we were witnessing the magic transfusion of new blood flowing in warm waves through the exhausted body of France; and when in those days of heaviest trial, when the enemy stood again on the banks of the Marne and fancied that we had lost all hope, the hearts of the French were filled with a confidence which it is impossible for me to describe."

This eloquent contemporary tribute is repeated by historians today.

After the war, an American military cemetery and battle monument were established near the shattered remains of the old hunting forest of Belleau Wood.

Among those under the crosses is Lt. j.g. Osborne, the young Navy dentist who died on the field near Bouresches.

Today, modern maps still refer to the Bois de Belleau by its old name.

But 80 years ago, soon after the fighting ended there, the commander of the French 6th Army issued an official order giving it a new name.

In military history it comes down the years by that name: Bois de la Brigade de Marine.

Hero Of Great War Survived Gas Attack

Originally published January 31, 2002

When James Crawford Cooper died in September 1939 after a lingering bout with tuberculosis, his obituary celebrated him as "a well-known Fayetteville salesman."

The 52-year-old widower, who lived quietly with his mother and two of his sisters at a home on Green Street, was mourned as "a man of friendly disposition with a wide circle of acquaintances who were saddened by his death."

He is buried in Cross Creek Cemetery No. 2 under a headstone that lists only his birth and death dates.

From that simple record, you wouldn't know that Cooper was a hero of the first desperate struggles of the American Expeditionary Force on the battlefields of France in the Great War of 1914-18, now known as World War I.

Cooper's name is linked with battle honors that are hallowed in the military history of that war, with Chateau Thierry, Bouresches and Bois de Belleau.

While he was in the maelstrom of the first fierce fighting in that fateful June of 1918, Cooper's service was, like that of many Americans in World War I, as brief as it was incandescent.

Cooper first saw action June 6, 1918, as an Army officer, a first lieutenant in the 6th Marine Regiment of the Army's 2nd Division.

Less than a month later, Cooper was on his way home, suffering the effects of a German gas attack that virtually wiped out his company.

The Fayettevlle Observer reported on July 12, 1918:

"Lt. James Crawford Cooper, who has served gallantly in France being in some of the fighting by the Marine Corps, arrived in the city last evening.

"He has received honorable mention for conduct on the field.

"Lt. Cooper has been sent over here to give instructions to troops preparing for active duty."

Honor, bravery

The first notice that Fayetteville got of Cooper's exploits was a story reprinted on that day from the Paris edition of the New York Herald, dated June 10, 1918:

"Other lieutenants who played a courageous and spectacular part in the battle of Bouresches were J.C. Cooper of Fayetteville, N.C., V.H. Hampton of Charlerri, Michigan, J.M. Bars of Mississippi.

"They are all to be sent to America at once to picture to the people of America some of the sterner things connected with the war to impress upon the tens of millions in the East, West, North, and South what a gigantic task and duty lies upon the United States."

In October, another story said:

"The first of Cumberland boys sent back for patriotic duty was Lt. James Cooper, who was three times cited for bravery at Chateau-Thierry, Bouresches and Bois de Belleau."

A key description of Cooper's riveting combat career came in an article in the Saturday Evening Post published four months after the June struggles in the valley of the Marne River:

"Lt. Cooper was transferred to three different companies during a few weeks time and was with the Marines with their fine work at Belleau. He was put in a company that was made up of five men left over after the rest had been wiped out. In one company he belonged to 187 men were knocked out by mustard gas, 30 of them dying from it."

While there is not enough evidence to flesh out Crawford Cooper's personal battle experience beyond these contemporary reports, a Marine historian concludes that Cooper had to have served for a while in the 96th Marine Company of the 6th Marine Regiment.

That company suffered "horrific" casualties; on June 6, 1918, when it attacked across a wheat field to capture the French village of Bouresches.

On June 13, after days of fierce fighting in the adjoining Belleau Wood, the 96th was again virtually wiped out in a gas attack, which killed 34 and sent more than 120 to aid stations.

A leading historian of the Marines in World War I, George B. Clark, says that if Cooper was in a company that was largely wiped out by gas attack, it would have to be the 96th.

The designation of the other two companies referred to in the magazine article is unclear. Or how, as the Saturday Evening Post article asserts, he was also involved at Chateau Thierry.

That town on the Marne River became immortal in U.S. Army history because of its defense on the last days in May of 1918 by the 3rd Infantry Division, specifically the 38th Infantry Regiment. The 3rd Division, now based at Fort Stewart, Ga., took as its nickname, the "Rock of the Marne."

German offensive

The battles of the Marines of the 2nd Division at Bouresches and Belleau Wood, a few miles west of Chateau Thierry, got much of the publicity when these two U.S.

divisions joined the fray against the sputtering German offensive in that summer of 1918.

How did an Army officer get in the Marines, and how did the Marines get in an Army division?

When World War I began for the U.S. in 1917, the Marines were organized into two regiments, the 5th and 6th Marines, and sent to France. Because they were ready for combat, they were formed into a brigade and made part of the Army's 2nd Division there.

Clark says that because of a shortage of trained officers, the Marines "accepted" as many as 50 young Army officers, mostly lieutenants fresh out of officers training school.

Cooper was in this group. He apparently attended officer training school near the headquarters of the American Expeditionary Force.

A letter to his father in Fayetteville, written in October 1917, reported that he was "stationed at an infantry officers school near a small French town."

Cooper said that "we are the first Americans the citizens of the town had seen, and there is much excitement."

He described his life:

"We have large barracks and are comfortably settled and eat at the cafes, served from French menu cards. American boys are learning the French language and gestures, but claim they will spread the English language while they are learning French. Some of the boys said they had a brush with a German submarine, but I was satisfied we did not see one."

The 32-year-old veteran of the first furious American fight in the Great War apparently spent several months in uniform after returning to the United States.

But by 1919, he was back in Fayetteville, working in the family banking business.

For the next 20 years, he led his quiet life in the midst of his family and lifelong friends.

His wife, Georgia Dixon Cooper, died in 1937. They had married in the autumn 20 years earlier when he prepared to go overseas.

Cooper was born in 1886. He got his first taste of military training as a student at the Donaldson Military Academy in Fayetteville, where many of his classmates would also serve in the Great War.

He apparently attended the University of North Carolina in 1904-05, but by 1906 was back home working as cashier of the National Bank of Fayetteville.

Large family

Cooper's family was large and notable.

His father, Sol Cooper, was a leading banker and businessman. Two of his sisters married other local businessmen, John Rose and Rhodes Moffitt. His two unmarried sisters lived with Mrs. Sol Rose. His sister, Aline "Teeny" Cooper Moffitt, became a noted "society" reporter for The Fayetteville Observer and a beloved community activist who survived him by more than 20 years.

She and her sisters apparently followed the wishes of their brother in not even mentioning in his obituary that 21 years earlier he so notably shared the searing experience of the first major American baptism of fire in the Great War.

Attempts to locate a photograph of James Crawford Cooper have been unsuccessful. If anybody has a picture, please let me know.

Above: Lts. Mallory and Cox are seen here with a plane nose down at Pope Field in 1919.

Above: An airplane does a nose dive at Aviation Field on Camp Bragg in 1919.

Above: An unknown soldier labeled this photo in September 1919, pointing out his room in the Medical Officers Barracks at Camp Bragg.

Right: A group of black men shoot dice at Camp Bragg in April 1919.

Left: Percy Rockefeller, nephew of John D. Rockefeller, is seen here at the Overhills Fox Hunt Club. Percy had been a key player in buying the land near Fort Bragg for the Rockefeller family. Fort Bragg purchased more than 10,000 acres of the property in 1997.

Above: Col. McEnery is seen here on "Princess Pat" at Camp Bragg around 1920.

Left: Baseball was a hit on Fort Bragg just as it was across the rest of the country. Pictured here is the Bragg baseball team in 1919.

ABOVE: A group of men pose in front of a plane at Pope Field in 1919.

ABOVE LEFT: Lt. Potter and Lt. Bigalow sit near their plane at Pope Field in 1919.

ABOVE RIGHT: Lt. Black, flight surgeon, Lt. Hopkins, observer, and Lt. Mallory, pilot, show off their plane in 1919 at Pope Field.

THIS PAGE: Just months after being named General of the Armies, Gen. J.J. Pershing is seen here inspecting the grounds at Pope Field in 1919.

Above: This is one of the first units assigned to Camp Bragg. It is likely the 83rd Field Artillery Regiment, which came to Bragg in 1923. On the horizon behind the snaking column of horse-drawn artillery is an experimental observation balloon. Records show that the first Bragg parachute jump was made from a balloon on July 4, 1923. This picture was provided by Capt. Pete Esbach of the 18th Airborne Corps Public Information Office.

Above & right: Horse-drawn caissons prepare for action with the 6th Field Artillery Battalion at Fort Bragg sometime around 1920.

Above: Planes sit outside a hangar at Pope Field on Fort Bragg in the 1930s.

Left: This aerial shot shows the landing strip at Pope Field near Fayetteville. In an Army magazine, the caption reads: "According to a notice in the 'Airmen's Guide,' a publication of airfields and airways during the 20's, pilots were advised to 'buzz' the area from low altitude before landing to chase away grazing deer."

Right: The main barracks building for Air Corps troops at Pope Field in the 1930s.

Right: Secretary of War G.G. Dern and Brig. Gen. Manus McCloskey stand with their staff at Fort Bragg on Jan. 26, 1934.

Above: A parade of soldiers marches through downtown Fayetteville near the Market House around 1935.

Right: An unknown couple stands next to a plane at Fort Bragg in March 1936.

Right: The Champion of 1936 was Battery A, 17th Field Artillery at Fort Bragg.

Above: Pvt. 2nd Class George Remaley and Pvt. 2nd Class Orr pose with a gun at Fort Bragg in 1940.

Left: An unnamed soldier sent this photo home with a label of "How am I doing brother?" on the back. It's from the 4th Field Artillery at Fort Bragg in 1941.

Left: Lt. Gladys Irene Gilds, left, Lt. Ruth H. Briggs, center, and Lt. Della Raney stand in front of the "colored ward" of the No. 2 hospital at Fort Bragg.

Below: The Soldier's Town Home on Old Street in Fayetteville in 1941.

Above: Soldiers at the opening of the train depot in Fayetteville in March 1942.

Right: Men of the 16th Battalion Crack All-Negro Training Unit at the Field Artillery Replacement Training Center at Fort Bragg in 1943.

Below: Soldiers from the 41st Engineers march through the streets of Fayetteville in 1945.

Left: Pictured are members of the 13th Field Artillery Brigade, which was formerly stationed at Fort Bragg and commanded by Brig. Gen. John Crane. The photo was submitted by Lawrence Presnell of Fayetteville, a former member of the 1st F.A. Obsn. Battalion.

Above: This huge hangar housed balloons that were among the earliest aircraft based at Pope Field.

Right: This aerial photo provided by U.S. Army Photography from 1968 shows the "old and new hospital" at Fort Bragg.

Left: Pope Field at Pope Air Force Base is seen in this 1968 aerial photo provided by U.S. Army Photography.

Above left: An 82nd Airborne Division paratrooper with camoflauge.

Above right: A paratrooper struggles to pull in his parachute at a Fort Bragg drop zone.

Above: A group of soldiers rides through Fort Bragg in this undated photo.

Above: A group from the Braxton Bragg Chapter of the Association of the United States Army (an Army support group) stands on a balcony in Washington, D.C.

Above left: Maj. Gen. Sidney Marks and Lt. Gen. Richard Seitz walk across Fort Bragg in January of 1975.

Above right: This historical marker was put up in 1986 to honor Fort Bragg.

ABOVE: The C-130 was a popular plane used by the Air Force. Many of the planes sit on base now, unused.

ABOVE & LEFT: Planes at an air show at Pope Air Force Base outside Fayetteville.

Above: Three C-130s taxi into place at Pope Air Force Base in August 2004 after a flyover to mark the 50th anniversary of the first C-130 flight.

Right: Maj. Gen. Charles Swannack Jr. leads 82nd Airborne Division paratroopers from a plane March 27, 2004, at Pope Air Force Base's Green Ramp after returning from a deployment to Iraq.

Bottom: Paratroopers from the 1st Battalion, 504th Parachute Infantry Regiment, 82nd Airborne Division wait in a C-17 Globemaster III that will take them for a night jump at Sicily Drop Zone on Fort Bragg.

38th Infantry Stood Against German Onslaught

Originally published January 8, 2004

U.S. Army history boasts many victories on the offensive.

There also are stout defenses that make the honor roll, such as Andy Jackson at New Orleans in 1814, John Gibbon's division defending the "little clump of trees" on the ridge on the third day at Gettysburg and the 101st Airborne surrounded in Bastogne in World War II.

And then there was the 38th Infantry Regiment on France's Marne River in July of 1918, blunting the spearhead of the last great German offensive of World War I.

The defensive feat earned the regiment its enduring nickname: "the Rock of the Marne."

The price was high.

Capt. Jesse Woolridge reported of his front-line company:

"We had started with 251 men and 5 lieutenants. I had left 51 men and 2 second lieutenants."

Among the dead that day was a local man.

Pvt. Charles Cornwall Hall of Red Springs and Fayetteville was only 20 when the German storm burst over the 38th in the early hours of July 15, 1918.

He apparently was killed in the artillery barrage that opened the German attack.

His remains were not found, and his name is engraved on the monument to the missing and presumed dead in the beautiful Aisne-Marne American Military Cemetery at Belleau, France.

Charlie Hall came from a military line.

Military heritage

His Cumberland County family included his father, a major in the Confederate army. An ancestor fought in the American War of Independence.

He was only 17 when he joined the Lumber Bridge militia company, the famed "Scotch Tigers," a Robeson County volunteer outfit with a long history. He went with the company for service on the Mexican Border in 1916.

An article about his death in The Fayetteville Observer carried on his story:

"Volunteering with the National Guard in 1916, he went overseas when 20 years old, refusing an honorable discharge last winter.

"He was an unusual personality, handsome, a fine physique and possessed of a rare spirit that radiated happiness and good will. He did his part in a way that is a credit to his ancestors. So this blood shed in the cause of humanity was of the stain of true patriots."

The article ended with a poem:

How better could a man die
Than facing fearful odds
In defense of his country,
His home and his God.

The article was illustrated with one of the rare portrait-photographs of fighting men that the Observer published during World War I.

It depicts a slim, pleasant-faced young man in a summer Army blouse with suspenders, a campaign hat pushed back on his head, and his Springfield rifle at right shoulder arms.

It was probably taken during the summer of 1917, perhaps between Hall's Mexican Border service and his service with the 38th Regiment of the 3rd Infantry Division.

It is interesting that Hall didn't go to France with the 30th Infantry Division, to which most North Carolina National Guard companies were assigned and which trained at Camp Sevier in South Carolina.

In time for battle

He apparently joined the 3rd Division in the winter of 1917 when it was training at Camp Green in Charlotte.

As fate would have it, the 3rd Division sailed for France in March 1918 and was thus available in early June when a German offensive threatened to crumble Allied lines northeast of Paris. The 30th Division arrived in May.

The 2nd and 3rd divisions rushed straight from training grounds behind the lines in France to the front lines along the Marne River.

In June fighting at Chateau-Thierry and Belleau Wood, the 2nd and 3rd divisions wrote celebrated chapters in U.S. Army history by checking this first German assault against the river line.

The battle in which Charlie Hall died came five weeks later, along the same Marne line.

The fight was a baptism of fire for the 38th, which had largely missed the early June battles.

Known as the Second Battle of the Marne, the German attack of July 15 quickly fizzled and would become the last offensive of Germany's four-year war in France.

On July 18, the Allies opened their own offensive, the first of a drumbeat that would last until the Great War of 1914-18 ended with the armistice on Nov. 11, 1918.

The role of the 38th Infantry was told by no less a correspondent than Gen. John J. Pershing, commander of the American Army in France, in his official announcement:

"On this occasion, a single regiment of the 3rd Division wrote one of the most brilliant pages in our military annals.

"It prevented the crossing of at certain points on its front, while on either flank the Germans who had gained a footing pressed forward.

"Our men, firing in three sections, met the German attacks with counterattacks at critical points and succeeded in throwing back two German divisions into complete confusion, capturing 600 prisoners."

Weeks of fighting

The 3rd Infantry Division was in active sectors on the Western Front for 86 days, second among all U.S. divisions in the American Expeditionary Force.

Charlie Hall was among 3,277 men of the 3rd who died in battle in World War I.

The Rock of the Marne was a celebrated regiment in a celebrated division.

A quarter century later in World War II, the 3rd Division fought across Africa, Sicily, Italy and into central Europe when my friend, Fayetteville's own Jimmy Gilbert, was in the ranks.

It was in the thick of the long wars in Korea and Vietnam.

The 3rd Infantry Division (Mechanized) is now based at Fort Stewart, Ga.

And, of course, it was the spearhead of the U.S. Army's latest offensive, the invasion of Iraq.

In the annals of the division, and especially of the 38th Infantry, the defense of the Marne remains the most honored moment. The motto is emblazoned on the regimental crest, and in the name of the 3rd: the "Rock of the Marne."

Young Staff Officer Describes Furious Attack By The "Boche"

Originally published September 17, 1998

The most difficult writing of military history is the eyewitness story by a participant.

As those who have seen the movie "Saving Private Ryan" realized, combat is such a chaotic experience that even remembering what happened may be confused.

And so when survivors sit down to write about what happened, they often fall back on descriptive language that is heavy on general terms but short on detail.

But there are exceptions.

This military history consists mostly of a letter written from the 1918 World War I trenches of France by Lt. William O. Huske of Fayetteville. The 24-year-old Huske was a staff officer in a battalion of the 149th Field Artillery Regiment, so he was not in the deadly maelstrom of the infantry attacks across no man's land that were such a feature of that war.

But he does manage a straightforward yet vividly descriptive word picture of what a great World War I attack was like, with huge artillery duels followed by rushes of infantry and the desperate fighting if attackers got to the trenches of defenders.

Huske's letter appeared in The Fayetteville Observer in September 1918.

He described an offensive that was the final last-gasp effort of the Germans (referred to by the disparaging name of "Boche") to win the four-year war, an attack toward Paris from the front near the city of Rheims.

Huske volunteered soon after the U.S. entered the war in the spring of 1917. Eighty years ago in July, he had been in France for seven months in the 42nd "Rainbow" Division. The division served under French command in the action that he described in his letter to his mother, dated July 23, 1918.

He wrote:

"On July 14 the colonel phoned the major and being on the major's staff and in the same dugout with him, I heard orders that 'the German attack comes tonight. Have every one up and on the alert.'

"From that time on until the 23rd, we did not have our clothes off.

"The first 86 hours after that I got six hours of sleep, and have been up every night since then.

"At 12 o'clock the Boche started, but our heavy guns had been booming away for half an hour, firing on roadways, trenches, and points of assembling.

"When we opened up in return, the sky for miles was a deep red glow in every direction, and the great flashes of the guns on both sides and the bursting of projectiles looked like great flashlights.

"The roar of the shells sounded like a thousand men ratting a piece of 20-20 tin.

"Then came the great explosion that shook underneath you as munition dump after munition dump was struck, and the flames leaped heavenwards, seeming to disappear in the clouds overhead.

"The preparation lasted until 4:15, and then the Boche put down their barrage for the infantry to come over.

"The frontline trenches had been one mass of flares, but when the Boche dropped their barrage, every one rushed out, everyone firing a different colored rocket. As the barrage rockets burst in the air, it was like a beast with millions of tongues of fire, our guns opened up and it was beautiful.

"Every nerve was set a tingle watching to see our own battalion commence fire, so we could notify the proper authority.

"At 4:30 it was daylight, and the air was filled with aeroplanes and soon the wireless told us the Boche were coming through 'No Man's Land.'

"As the enemy reached the different points, the aviators advised us. At 12, the Boche had reached our line of resistance and had a few footholds.

"Things looked bad for us, but we just kept pouring more shells into them, and soon the report came back that we had counterattacked and driven them out of our lines at great losses.

"The Boche then dropped back to form a new line of attack. The firing continued all that day and night. At 5 in the morning, they tried again to break into our line, only to be driven back.

"Five times they tried, with renewed effort every time, assisted by tanks, but not a single time did they get past our artillery fire. We know they were beaten bad with severe losses.

"On our right, they made out a little better, for they got into our lines in several places, and the Americans, to the honor of the Alpine Chausseurs (a French division in the line next to the 42nd), stripped themselves to the waist and without firing a shot, with a yell, jumped from their trenches and with the cold steel glistening in the sun, killed the Boche right and left and drove them back.

"The infantry action was over, but the artillery continued for two more days.

"We were with the Alpine Chausseurs and the best artillery in France, so we showed up well, so you know we have a good division.

"The Chausseurs say the American infantry is wonderful, and they consider themselves the best infantry in the world. The French artillery officers say our fire was perfect, and that they never expected Americans to do.

"We took many, many prisoners, and the morale of all is low. Twenty-one Boche slipped in between one of our American battalions and an Alpine battalion. Before the Alpines could form a party to dig them out, the Americans rushed in and killed the whole lot with the bayonet, not one American being scratched.

"I am writing these things to show that Americans are making good. Don't think the Boche aren't fighters. They fight like the devil, and this war isn't over by any means. But no man could give his life for a greater cause or die a more worthy death."

Already when Huske wrote home, the French had launched the first of the Allied offensives lasting until the armistice of Nov. 11, 1918, ended a war that had raged since August of 1914.

In later action as a forward artillery observer, Huske would be wounded by shell fragments and suffer lung damage from a German gas attack.

Survived the war

He survived World War I, returned to Fayetteville and the family hardware business.

A footnote to his letter is that Huske would have known that the "Alpine Chausseurs" he so admired called themselves "blue devils," from the color of their uniforms and the ferocity of their fighting.

And Huske, who was a football star at the University of North Carolina at Chapel Hill before the war (he graduated in 1915), may have known later that young veterans who came back from France and attended arch-rival Duke University appropriated the fierce name of the French unit and the "French blue" color of their uniforms for Duke athletic teams.

Col. Pike Was 82nd's 1st Hero

Originally published September 3, 1998

In early September 80 years ago, the 82nd Division was a year old. And it was nearing its baptism of fire in World War I.

For nearly two months in that September 1918, the 22,000 men of the "All American" division had been learning the bloody realities of trench warfare under the watchful eye of headquarters of the American Expeditionary Force, the high command for the hundreds of thousands of U.S. Army "doughboys" crowding into France that spring and summer.

The 82nd arrived in mid-May. Within a few weeks, units of the division were committed to so-called "quiet sector" front-line training, holding defensive positions not far from the AEF headquarters.

The front-line units had taken their first casualties from artillery shelling and sniper fire.

Meanwhile, behind the lines, intense training continued for an offensive operation that would fit the name of the division.

A campaign to reduce the bulging St. Mihiel salient had been planned for months to be the first all-American operation of the American Expeditionary Force.

The years-old salient was like a 60-mile bubble in the Allied lines. It was close to the main training area for U.S. divisions.

The Marbacke sector of the trenches where the 82nd was taking its "active sector" training was the western hinge of the St. Mihiel salient.

But in midsummer, the operation was taking a back seat to more urgent demands.

In this fourth springtime of the war, the German army had launched a final all-out offensive, hoping to finally overwhelm the weary British and French before the huge American presence could be felt.

In the desperate fighting for survival, U.S. troops were committed by division and even regiments to fight alongside the Allies.

By June, however, the German advance was being blunted and the Allies were beginning their own series of all-out offensives. The St. Mihiel plan was on the table again.

As the day for the offensive approached, the 82nd faced a problem chronic to many a division of the American Expeditionary Force. It was a problem of manpower.

Not so much of numbers. The division finally had all the 22,000 men that a World War War I division required.

It was of experience.

Nearly half the 15,000 soldiers in the division's four infantry regiments and four machine gun battalions were virtual newcomers to military service. Some had been drafted as late as March and had sailed for France with barely six weeks of training with the division.

The problem could be traced to the previous March.

Before then, the 82nd was a relatively cohesive gathering of draftees from the South, mainly from Tennessee, Alabama and Georgia.

These new soldiers learned together at Camp Gordon, Ga., where the division was first organized in August 1917.

Just as they ended their training cycle, however, the terrible manpower demands of the war intruded. More than 5,000 men of the division, mostly from infantry and machine gun units, were ordered away to seed units of newer divisions still in training.

The replacements who arrived that spring were draftees from throughout the country, but primarily from Northeastern states such as New York, New Jersey and Pennsylvania.

This unexpected diversity in its ranks gave the All American division its nickname.

But for veteran officers such as Lt. Col. Emory Jenison Pike, the division's machine gun officer, the inexperience caused worry as D-day for St. Mihiel was announced for Sept. 12, 1918. Pike, a 1901 graduate of West Point and a former cavalryman, would have the task of moving the division's machine gun line forward after infantry had advanced.

Attack successful

The initial attack on the salient went well, however.

The men of the 82nd attacked out of familiar trenches, toward terrain they had studied. Their mission called for an advance of only a few miles to "straighten out" a portion of the western corner of the salient.

The main effort was on the point of the salient, which was pinched off in a single day by converging attacks by the veteran 1st and 26th Divisions.

But then came three days of what would become familiar fighting for doughboys in World War I.

The Germans hunkered down in defensive positions, using artillery and machine guns with deadly effect on the green American formations that often went forward in waves like an old Civil War battle scenario.

In this maelstrom 80 years ago, Pike became the 82nd's first enduring hero.

The posthumous citation for the Medal of Honor tells the story of Pike's role and gives a vivid picture of the 82nd's baptism at St. Mihiel. It reads: "For conspicuous gallantry and intrepidity above and beyond the call of duty in action with the enemy near Vandieres, France, on Sept. 15, 1918.

"Having gone forward to reconnoiter new machine-gun positions, Col. Pike offered his assistance in reorganizing advance infantry units, which had become badly disorganized during heavy artillery shelling.

"He succeeded in locating only about 20 men, but with these he advanced, and when later joined by several infantry platoons, rendered inestimable assistance in establishing outposts, encouraging all by his cheerfulness, in spite of the extreme danger of the situation.

"When a shell wounded one of the men in the outpost, Col. Pike immediately went to his aid, and was himself severely wounded by a shell burst in the same place.

"While waiting to be brought to the rear, Col. Pike continued his command, still retaining his jovial manner of encouragement, directing the reorganization until the position could be held.

"The entire operation was carried on under terrific bombardment, and the example of courageous devotion to duty as set by Col. Pike established the highest standard of morale and confidence to all under his charge.

"The wounds he received were the cause of his death.

"Mrs. Martha Agnes Pike, daughter, Des Moines. Residence at appointment: 806 Des Moines, Ohio."

The machine gun officer was among 816 casualties of the 82nd Division's first offensive.

Pike Field at Fort Bragg memorializes the heroism of Lt. Col. Emory Jenison Pike.

Doughboys Storm Hun Positions

Originally published November 5, 1998

Pvt. Kenneth McNeill was full of all the exuberance of his 18 years when he arrived in France in the summer of 1918.

He wrote home to Fayetteville: "We are in fine feather. We had a little fun with a submarine. I got two sweet kisses when we came through the streets. I won't be back until I get the Kaiser's helmet!"

Now, it was a gloomy September morning three months later. McNeill was a 19-year-old platoon sergeant in Company F, 119th Infantry, 30th Division.

And now, the young physician's son was about to test his exuberant courage in the bloody conflict of World War I.

At dawn on Sept. 23, 1918, the 30th Division was on the jump-off line to storm the German army's last formidable defensive position in France, the Hindenburg Line.

The division's objective was to push across the ground over an underground portion of the St. Quentin Canal and seize the village of Bellicourt.

This is McNeill's narrative of his baptism of fire, a letter to his father 11 days after that dawn:

"Late one afternoon, the captain called the sergeants together and said, 'Well, boys, we are going to break the Hindenburg Line.' He gave us a few maps and told us to go back and tell the men about it. We went back and told them and got things ready.

"The next morning at zero hour we were all ready. Then all of a sudden 'hell' broke loose just in front of us.

"That was our barrage. It stayed there for a little while, and then began to move forward, and we began to follow.

"Oh! That barrage was wonderful!

"I looked down my line of men and didn't see a single one who seemed to be afraid. We advanced on and crossed one or two of Jerry's lines.

"We hit the Hindenburg Line with every man cheering and went right on through.

"We went on a little further and then when two of my fellows were wounded, I stopped a minute to bandage one who was hit in the arm. When I got through, I looked around and my company had left me.

"I started off to find them, but got lost out there in No Man's Land. I had one corporal with me. We moved around a little bit, and found out that we were surrounded by machine guns.

"He said: 'What will we do, sergeant?' I said: 'We had better try to get to battalion headquarters and see if they have heard any more from the company.' We started off down a trench, and after a while looking in a dugout we found out that someone was in there.

A challenge issue

"We got up to the hole with our bayonets, and hollered for them to come out.

"Out came a bunch of Jerries with their hands stuck up in the air and trying to say something, but I couldn't understand that frog language. I pointed out the direction I wanted them to go and they looked like they were tickled to death to go.

"I got a little way and counted them and guess how many? Forty-one.

"I carried them on to battalion headquarters and then set out for the company again, and came across Captain Lamb and found out that we had come about two miles taking our objective and the Australians had gone on over us.

"We have the bravest bunch of men you ever saw. They are true REDBLOODED AMERICANS. I have a souvenir captured from a German. Tell all the boys at Donaldson (School) that Bill Hines and I came out all right."

Capt. Robert Lamb of Fayetteville was commander of Company F, which included several dozen men from Fayetteville, Hope Mills, Gray's Creek and from Cumberland County farms.

What did Lamb do on Sept. 29? This is the citation for the Distinguished Service Cross, awarded to him soon after the war: "At Bellicourt, in command of a company, he, with two other men, rushed a machine-gun post which was holding up the advance, killing the German crew. Later, separated from part of his command, owing to a dense smoke screen, he found himself with a few men in front of a German machine-gun nest. Leading the attack, he captured the positions with 25 prisoners."

Letter to comrade

Lamb described Sept. 29 in a letter to Sgt. William Hall, who was wounded that day:

"Sgts. Gordon Rhodes and Gilbert White were killed on the field.

"Sgt. (Lonnie) Tew from Sampson County was captured, as were Lts. Nelson and Hawkins and 11 men. All the prisoners are back with us.

"Company F entered the fight with 181 men and two lieutenants. Lost 109 men and two lieutenants.

"But we went through our first and second objectives and to the La Catelet switch trench, which as you know was our objective of exploitation.

"No other company went quite so far, and I finally had to retire a distance of 1,000 yards as the 27th Division which was on our immediate right at the tunnel and left my flank in the air."

In another letter to his mother, Lamb painted the overall picture:

"I can tell you that the 119th with Company F and G Company (Raeford) smashed through to a depth of 3,000 yards in three hours, and the regiment got about 2,000 prisoners.

"I never enjoyed a hunt more, but it was costly, for they fought like demons for the front lines.

"I with my company captured or killed the crews of two field positions, and numerous machine guns.

"I am glad that none of the dead are Fayetteville boys, and no Fayetteville boys have been seriously hurt."

Sgt. Herbert Henderson of Fayetteville, 22, who like Hall was wounded in the first minutes of the attack, described the role of the sergeants:

"Sergeant Kenneth McNeill was in command of 1st Platoon with Capt. Lamb in between where he could command the entire company.

"Sgt. Hall in command of the 2nd Platoon.

"I led the attack in charge of eight Lewis gun gunners.

"It was some exciting time. Every man was on the alert and every nerve was tingling.

"I was wounded in the first hour's fighting and was carried back to the First-Aid Station."

To the right of Company F and Company G of the 119th Infantry also struggled through the maze of German machine-gun positions.

Local men killed

While Lamb counted no Fayetteville dead in his company on Sept. 29, there were actually eight Cumberland County men killed on that day or who later died of wounds suffered there.

McNeill, Lamb, Hall and Henderson all were alive when the armistice ended World War I on Nov. 11, 1918.

They all returned to Fayetteville to attend the observance of the first Armistice Day (now Veterans Day) in 1919, and to spend part or all their postwar lives there.

Eighty years later, on the 80th anniversary of the end of the "war to end all wars," the exploits of the men of the 119th Infantry at the Hindenburg Line are cherished reminders of the service of all veterans in that war and in all wars.

Area Veteran Was Among First Medics

Originally published November 10, 1994

Edgar Draughon grew up in the peaceful turn-of-century rural community known as "Flea Hill," today's Eastover in Cumberland County.

When World War I came, the 23-year-old Draughon earned enduring military distinction in the maelstrom of no man's land, the deadly ground between the trenches in France.

Among the Cape Fear region's several farm-boy heroes of the war that ended 76 years ago on Nov. 11, 1918, Draughon's record is unique because he earned his distinction not by taking lives. He saved them.

Draughon was a new type of soldier then.

He was what today is called a "medic." During 1917-18, however, it was an innovation to have enlisted men whose job was to treat and rescue the wounded on the battlefield.

Draughon was a member of a newfangled outfit, officially known as "the Sanitary Detachment" of the 120th Regiment of the 30th Division.

His performance in that new service is simply described in the citation that accompanied the awarding of the Distinguished Service Cross, the Army's new medal for battlefield service that ranked just below the Medal of Honor.

Of Draughon, the citation said:

"In the attack between St. Quentin and Cambrai and subsequently from Sept. 29 to Oct. 20, 1918, Pvt. Edgar S.W. Draughon was conspicuous for his constant devotion to duty and indefatigable energy in evacuating wounded from the front lines to the Battalion aid station. On Oct. 19, he with complete disregard for his personal safety advanced under heavy shell and machine-gun fire beyond the front line, rendering first aid to a wounded officer and assisting him to the rear. It is recommended that steps be taken to cause him to be awarded the Distinguished Service Cross for extraordinary heroism in action."

The citation was signed by William B. Hunter, Surgeon, 1st Battalion, 120th Infantry.

The young lifesaver from Flea Hill was among about a dozen Cape Fear region soldiers who would get the Distinguished Service Cross for World War I heroism.

Most of them were in the 30th Division, the "Old Hickory" outfit made up of former National Guard units from the Carolinas and Tennessee.

The 120th Infantry alone produced more than 100 of the division's 305 winners of the award.

The many medals, a record for a division in the American Expeditionary Force in France, reflected the 30th Division's extended service in combat.

That service involved early experience beside British divisions in the summer of 1918.

Then, in early autumn, the 30th made World War I history with the famed "breaking of the Hindenburg Line," the action that began Sept. 29, 1918.

In that fighting, Draughon's work was heavy. The division suffered heavy casualties.

More than 8,000 men were killed and wounded during the period covered by Pvt. Draughon's citation. Dozens of men from the Cape Fear region were in the casualty lists because most of them were members of the 120th Infantry or its companion North Carolina outfit, the 119th Infantry. Those were the units that cracked the German defenses.

Draughon described the opening of that drive in a letter to The Fayetteville Observer. The letter revealed him as a writer of good detail and with a poetic bent.

He recalled having "a good breakfast of hot coffee, cheese, butter," before unloading near the tapes that marked the jump-off.

On the day before the attack, "it rained all day a cold mist on us."

Then, on the starlit morning of Sept. 29, at 2:15 a.m., "a deathly still prevailed" just before the regiments went over the top.

Draughon waxed eloquent, but with few details, in describing the fighting, saying only that "our losses were great, many a boy met eternity that day."

The dead, he wrote, were "heroes who died to establish democracy and for the advancement of civilization and to liberate poor, suffering humanity."

The serious-minded attitude in Draughon's letter was reflected, too, in an experience that he had after the armistice ended the fighting Nov. 11, 1918.

In a letter to his mother, Mrs. George Draughon, he told first of being sent to the French city of Le Mans, where he and dozens of other Distinguished Service Cross winners were photographed wearing their medals. He asked that a copy be sent to his family, but they were being saved for some official publication.

The officer then told the men that they could "go where we pleased and enjoy ourselves."

Instead of heading for the nearest cafe, the young man from Flea Hill opted to taste some of the culture of the Old World.

He "visited the museum of the Le Mans cathedral."

He wrote, "Every age of the human race is well illustrated in there, sculpture, painting, drawing, etc. It was quite amusing to go through and look at the real thing which I spent so much time studying."

In the spring of 1919, just before Draughon returned to the United States with the 30th Division, his hometown newspaper got a copy of the picture taken in Le Mans.

It depicts a serious-faced young soldier in his high-collared tunic and wearing a crisp campaign hat. He stares boldly into the camera.

With that, Draughon came home and took off the uniform. War was put behind him.

For the next half-century, he and his wife, Lois Edwards Draughon, lived together in Flea Hill, where they attended Salem Methodist Church.

He became a rural postman and a teacher. He died at 73, just after Christmas in 1968. He was buried in the Culbreth cemetery in Falcon. Lois died a few years later. They never had children.

Cape Fear Soldiers Prevailed At Bellicourt

Originally published February 13, 2003

The fog-shrouded plain in front of the men of the 120th Infantry Regiment looked deceptively benign on that morning almost 85 years ago.

Only a mile away across the rolling French farm fields, the chunky spires of the village of Bellicourt were dimly visible.

That was the objective on the morning of Sept. 29, 1918.

So, when the regiments of the U.S. 30th Division "jumped off" behind a short artillery barrage, thousands of young men from the farms and small towns of the Carolinas and Tennessee, rushing forward behind two dozen lumbering tanks, could almost believe this would be easy.

Instead, the desperate German defenders in the bunkers and trenches of the Hindenburg Line were fighting what they knew was a last-ditch battle of World War I, the Great War, to stave off an Allied offensive that might end the fighting that had raged for four years in northern France.

Before the day was half over, hundreds of the attackers would be killed and wounded in a confused maelstrom of hand-to-hand encounters between attackers and defenders.

The benign look of the front confused the Americans because they didn't see the carefully concealed spider's web of defenses in which many Germans simply stayed low in their positions as the attackers rushed by in the fog and smoke.

Then they turned their machine guns on the rear of the front waves as well as the front of the follow-up units.

The mission of the 120th Infantry in the confused struggle was to make its way to Bellicourt and to the southern mouth of the underground tunnel of the St. Quentin Canal that traversed the U.S. front.

By noon, the survivors of the struggles on the plain in front of Bellicourt had reached the canal mouth and were able to capture defenders using the tunnel. A few hours later, the remnant of the regiment was relieved by Australian troops who had managed to bridge the canal itself.

The battle for Bellicourt, known also as the Battle of St. Quentin Canal, goes down in the annals of the 30th Division, which is today's National Guard division for North Carolina, as the day when the men of the "Old Hickory" division "broke the Hindenburg Line."

Battle statistics of the 120th Infantry show that 202 officers and enlisted men of the regiment were killed in action that day, and another 759 were wounded, a casualty rate of more than 40 percent.

That is the big picture.

There are dozens of individual stories of Cape Fear men who fought on Sept. 29. Uncommon valor seems to have been the common experience.

I have written of two sergeants from the area, "Dunk" DeVane and Edgar Draughon, who both won the Distinguished Service Cross for, in peril of their own lives, bringing back wounded men from the fields before Bellicourt. They were members of units in the same attack.

Edgar Blanchard of Fayetteville was the area's hero in the 120th Infantry.

His citation for the Distinguished Service Cross reads:

"Pvt Company G, 120 Infantry. For heroism at Bellicourt, he displayed marked personal bravery capturing singlehandedly seven Germans whom he came upon in a trench and dugout, while taking a wounded soldier back to the American lines. He preferred to return to the firing line, turned the prisoners over to the wounded man and rejoined his squad. Mother is Mrs Addie Blanchard of Fayetteville."

The 23-year-old Fayetteville native wrote home two months after the war ended on Nov. 11, 1918, to tell his mother about what happened.

He wrote:

"We are having a good time since the war is over. I have been in several battles and in all the Hindenburg drive.

"I didn't think I had a chance but I came out without a scratch, but I had some close calls. I got a hole shot through my helmet.

"I guess you heard about Herbert Henderson being wounded. I was in the battle of Bellicourt and it was some battle. We went into the drive with 190 in my company and when we came out we had only 53.

"The last letter from you came to me in the front line, and it certainly did make me feel better.

"We go to church every Sunday and I hope I am a better boy than I used to be. Write soon for I surely do love to hear from home."

A few weeks after the armistice, he had written to report that his brother, Leon Blanchard, who was also in Company G and had been reported killed in action, was instead severely wounded, but had returned to the company and "is doing all right."

And then, his own first report on his experience of Sept. 29, 1918, "we have been in one of the biggest battles ever fought."

Just like in the stories of Dunk DeVane and Edgar Draughon, Edgar Blanchard came home from his experiences in France and seldom mentioned them again.

He died at 53 in 1948 at his home on Canal Street. He was a "well known painter and contractor who had been in declining health for several years."

He left a large family — three sons and four daughters — and descendants remain in the area today.

Like so many sturdy veterans of the "war to end all wars," Edgar Blanchard's military service in the blood and mud of France lasted only a few weeks. Yet, like so many others he left a legacy of heroism worthy of warriors of long tenure.

His earned his particular niche in military annals fighting before Bellicourt on that misty day 85 years ago.

Despite His Youth, George Ward Was A Hero

Originally published April 1, 2004

"He is modest to a fault."

That was the hometown assessment of 18-year-old George Ward of Fayetteville, who may have been the youngest soldier to win the Distinguished Service Cross for bravery in action during the Great War of 1914-18, now known as World War I.

Ward fought with the other Cumberland County men of the 119th Infantry Regiment of the 30th Division.

Like so many of them, he was there Sept. 29, 1918, when the regiment dashed across rolling French farmland to attack German positions around the village of Bellicourt, positions known as the Hindenburg Line.

Ward's citation read: "When his company was halted by enemy machine gun fire, Pvt. Ward rushed the hostile position and killed one gunner with his bayonet. He bayoneted three of these and took the other prisoner. Pvt. Ward was severely wounded in this action."

A year later, in a conversation with Mrs. John Anderson of The Fayetteville Observer, Ward told his version of the day:

"We were two weeks drilling, and on Sept. 27 I was ordered to give up my Lewis machine gun and be prepared to go over the top with a rifle.

"Very early in the morning of Sunday, Sept. 29, the tanks went through the barbed wire entanglements to lead the way for the infantry.

"The German machine guns opened up on us. When we reached the 'nest,' my companion held the belt of one of the machine guns while I fired it on the gunners, killing them.

"I was then hit on the knee with a bomb and fell on the ground with weakness. But we had set out to stop those guns, and it wasn't anything to be cited for.

"My companion then left me to send back some stretcher bearers, and I haven't seen him since. He was a fine fellow.

"Our infantry had been reinforced by the Australians (the finest fighters next to our own Americans I ever saw), and just then an Australian sergeant came up to me.

"He quickly secured four 'Tommies' with a stretcher to pick me up.

"The Germans were shelling the sunken road to prevent communication from our men and a shell hit my four stretcher-bearers, killing them instantly.

"One shell struck my foot, splitting it open. This third wound was too much for me, so I collapsed and didn't know when I reached the aid station, where some of the German prisoners were made to carry me.

"When I came to, I was in a French hospital being cared for by the Red Cross, who are everything 'over there.'

"I was then sent to a hospital in Bristol, England, and after some weeks to Winchester.

"After being off my feet for four months, I was able to embark for the states and was thankful to get back to God's country.

"Some people wouldn't believe all we could tell about our experiences, but we could each write enough to fill a book.

"It's awful to remember those human beings that we had to kill, but when we think of the brutal treatment of women and children we are glad for every German we put an end to. Every American would go through it again if necessary."

When Ward referred to his "third wound," he was counting from the day nearly three months earlier when, near the Belgian town of Ypres, as he described it: "We stayed in the trenches while the Germans threw a barrage for 12 hours, mixed with mustard, chlorine and every kind of gas on earth."

He was gassed and sent to a hospital, where he stayed until early September, then rejoined his unit in southern France.

"We were glad to get there because the dugouts were way under the ground, whereas in Belgium they were on top. Belgium is a beautiful country.

"I got back just in time to prepare for the big Hindenburg Line drive."

15-year-old enlistee

Ward's military career started when he was 15 years old in 1916, when the local militia outfit known as the Fayetteville Independent Light Infantry was called into federal service to go to the Mexican border, where American troops were chasing the bandit Pancho Villa.

To fill its ranks, the company eagerly enlisted any young man who could get parental permission.

For instance, among Ward's comrades on the Mexican border was Robert Porcelli, only 16, the son of an Italian immigrant, who also went to France with the 119th Infantry. At about the time Ward suffered in the gas attack, Porcelli was killed by a shellburst, becoming the first soldier from Fayetteville to die in action in World War I.

When the 30th Division was formed and trained for the war in late 1918, Ward wound up in Company I of the 119th, originally a militia company from Edenton. When the division went overseas, he was assigned to Company D, known as the Goldsboro Company.

The young hero of the Bellicourt attack came home in January 1919, still limping from the wound in his thigh and, like a half-dozen other Fayetteville men, still feeling the effects of the gas attacks in Belgium.

He was greeted by his father, Thomas Ward, and a circle of friends. He went on to a civilian life as an interior carpenter, living with his wife on Branson Street.

His story was told, sometime with embellishments, in at least two other articles in the Observer written by admiring comrades.

But in all of them, Ward maintained his modest, claim: "I only did what we were sent over there to do."

Local Soldiers Distinguished In WWI Fight

Originally published September 29, 2005

Sept. 29 is a day for local memories of the Great War of 1914-1918, now known as World War I.

Eighty-seven years ago today, Sept. 29, 1918, dozens of young soldiers from the Cape Fear region stormed across the rolling farm fields of France toward a village named Bellicourt on their way to "break the Hindenburg Line."

Before the day was over, seven men from Cumberland County were dead or mortally wounded.

Scores survived, however, and lived to tell stories of the confused day of combat when the 30th Division of the American Expeditionary Force took on the German defenders in strong positions, the Hindenburg Line, at Bellicourt and along the St. Quentin Canal.

Many were in Company F of the 119th Infantry Regiment, commanded by Capt. Robert Lamb of Fayetteville.

They were from Fayetteville, Hope Mills and Gray's Creek and from Cumberland County farms. Some had been soldiers before the war in the old volunteer company known as the Fayetteville Independent Light Infantry, which when "federalized" became Company F.

Several told their stories in letters published soon after the war in The Fayetteville Observer.

I have quoted from some of them, such as 19-year-old Sgt. Kenneth McNeill of Fayetteville, a platoon sergeant in Company F, and 22-year-old Sgt. Herbert Henderson, who led the company machine gun platoon, shoulder to shoulder with Lamb in the charge toward Bellicourt.

Here are some other stories of Sept. 29:

An article by a Fayetteville native, A.M. Myrover, in the Greensboro Daily News told of several. Myrover wrote:

"The modesty of heroes is proverbial.

"Notable is the case of Cpl. C.M. Huggins of Company H, 119th Infantry, who has, or at least had, a British military cross of honor, which was pinned on his breast by the king of England, and hides it even from his comrades in arms. Huggins lived on a Cumberland County farm before enlisting."

On his way home to Fayetteville from Camp Jackson, S.C., after being discharged in 1919, Huggins was asked: "Where is the medal King George gave you? Why don't you get it out and let somebody see it?"

Huggins replied: "Oh, I don't know where it is. I reckon it is in that old grip," kicking a grip (suitcase) he had bought before leaving Fort Jackson.

His traveling companion continued: "But he refused to open the bag. Finally the grip got busted in two. I don't know where the medal was lost or not, and I don't think he did either. It didn't give hang whether it was or not."

Erasmus McMillan

Myrover's article continued: "There was Erasmus McMillan, also of Company H, born and raised in Gray's Creek Township, Cumberland County.

"McMillan was a regular, dyed-in-the-wool, blue-stocking predestined Presbyterian. Nobody ever heard him utter a 'cuss' word, until he hit that barbed wire on the Hindenburg Line.

"From then on, he was a demon! He seemed to bear a charmed life, or was predestined to come through, or something like that. He was everywhere when the machine guns were thickest.

"He was one of the nine from Company H who captured a French town from the Boches, under the leadership of Capt. Theodore Fry of Fayetteville.

"Then on October 30, 1918, the Old Hickory (30th) Division was withdrawn from the front line, and McMillan became the most devout unoffending Calvinist you ever saw."

(McMillan's story of capturing a French town in an action led by Capt. Ted Fry is confirmed by the citation of the Distinguished Service Cross awarded to Fry:

"At Ribeauville, disregarded danger and with great bravery led the remainder of his company, 15 in number, unassisted on either flank, through an enemy barrage, captured the town, and successfully repulsed the enemy. Upon arrival he captured two prisoners and nine machine guns and was not until ordered by superior authority that he withdrew from the town."

Fry went on to be a Regular Army officer. In April of 1942, Col. Fry and his regiment of Filipino scouts surrendered to Japanese troops on Bataan in the Philippine Islands.)

Among the most startling recollections of Sept. 29 are from Sgt. Charles A. Kelly, a Fayetteville soldier who was "gas officer" in headquarters of the 2nd Battalion of the 119th Infantry.

Kelly wasn't in the attack of the 119th toward Bellicourt, but he was shortly out on the battlefield and touring the captured German positions, including the tunnel of the St. Quentin Canal.

When he arrived back in Fayetteville in April 1919, he had "stacks of pictures, pieces of money and other souvenirs he took from the bodies of dead Germans as they lay on the battlefield."

The Observer article noted: "His description of the 'great vat' in the Hindenburg tunnel is graphic.

"He says that around this vat lay the bodies of men of all nationalities, including Germans and Americans, from which had been cut squares of fat that were thrown into the vat to make oil, a quantity of which was in the vat at the time he saw it."

Is Kelly's macabre recollection to be believed? Or is someone pulling his leg.

He seems to mean it. He went on: "The American boys went through experiences and saw things which they are not telling, lest they be deemed exaggerations. But the pace was awful and the happenings unbelievable."

The writer concluded. "Kelly talks none about himself but only of the dreadful conflict in general."

Sgt. Kelly wasn't modest about the war in general: "The Americans won the war, and they didn't get to France any too soon."

Bloody Battle Fatal To Thousands

Originally published October 8, 1998

By Oct. 6, 1918, the Argonne Forest had become the slaughterhouse of the U.S. 1st Army's hopes for a breakthrough that would end World War I.

A thousand men were dying each day, and thousands more were wounded, in an attack that opened Sept. 26.

The German enemy was fighting from hundreds of machine gun nests firing from wooded hills and narrow ravines of small rivers that ran down to the larger Meuse River.

Despite the casualties, a dozen U.S. divisions had failed to clear the way for an assault on the main German defense line five miles away.

For two days, the Army's most experienced big outfit, the 1st Division, known as the "Big Red One," and the 77th Division, called the "ghetto division" because so many of its soldiers came from the sidewalks of New York, were ground down in frontal attacks along the eastern edge of the Argonne Forest plateau where it dipped down into the valley of the Aire River.

Early on Sunday, Oct. 6, a frustrated Col. Ralph T. Ward of the operations section of Gen. Hunter Liggett's 1st Army looked at his map and hatched a bold scheme that if successful could break the gridlock.

He took it to his boss and to the commander in chief of the American Expeditionary Force in France, Gen. John J. Pershing.

They gave the go-ahead.

On that October morning 80 years ago, the 82nd "All American" Division was about to make its mark in World War I.

Ward's scheme called for sending the 82nd straight up the narrow Aire valley beside the 1st Division on its left. It would then pivot sharply to its left, striking across the river and up the Argonne escarpment at the tiny village of Chatel-Chehery and two nearby adjoining hills, known simply by their heights in meters as Hill 244 and Hill 223.

Such a flanking blow just might dislodge enough of the German machine gun line to get the adjoining divisions moving again through the forest.

When 1st Army headquarters alerted Gen. George B. Duncan, who had been the 82nd's commander for only two days after replacing Gen. William P. Burnham, the 24,000 men of the division were in a reserve area eight miles behind the front.

Without an attack order in hand, Duncan sent a long column marching off in late afternoon, threading its way along a road jam-packed with the rear-area traffic of the battered divisions on the front line.

Brig. Gen. Julian R. Lindsay of the 164th Brigade rushed ahead to reconnoiter to find a firing line for the division artillery.

Even as he moved, out the order came from 1st Army, outlining the mission of the dawn attack the next day.

No artillery help

The order directed the two regiments of the 164th Brigade to attack at 5 a.m. on the morning of Oct. 7. There would be no advance artillery barrage.

Throughout the night, the division column struggled along the road, lashed by a heavy rain that began after dark.

At jump-off time, neither regiment was at the line of attack. By 6 o'clock, however, the weary marchers wheeled into position.

Fewer than 2,000 men of the division launched the first attack, a battalion of the 327th Infantry Regiment.

But within hours, they had stormed the height of Hill 244, were half way up Hill 223 and were fighting in the buildings of the crossroads of Chatel-Chehery. The other regiments were getting in position.

By nightfall, 1st Army was preparing to send the 77th and 1st divisions charging again against the positions in their front, to test whether the 82nd's attack was causing them to crumble.

Early the next morning, the 328th Infantry Regiment was in position to extend the All American attack.

A flanking movement

A platoon of the regiment slipped west of Chatel-Chehery to flank German machine gun nests holding up the advance on the village and on Hill 223.

Company G of the regiment charged through barbed wire only to be raked by fire from the German positions.

Acting Sgt. Early slipped his platoon again, getting behind the German line and gathering up two dozen prisoners who huddled by a stream while Early sought a way to get back to the battalion's lines.

Another German machine gun opened up.

Six men of the platoon were killed immediately, others wounded, including the three ranking non commissioned officers.

Cpl. Alvin C. York, late an anti-war member of the Church of Christ and Christian Union of Pall Mall, Tenn., was left in command.

Then followed what would become the single most celebrated exploit of U.S. soldiers in World War I.

The citation for the Medal of Honor, later bestowed on York by Pershing himself, read:

"After his platoon had suffered heavy casualties and three other noncommissioned officers had become casualties, Corp. York assumed command. Fearlessly leading seven men, he charged, with great daring, a machine-gun nest which was pouring deadly and incessant fire upon his platoon. In this heroic feat, the machine gun nest was taken, together with four officers and 128 men and several guns."

The citation declined to mention that York, who had resisted the draft for several months because of his religious convictions against killing, also picked off at least 28 German gunners, firing a pistol taken from a captured German officer.

York had learned marksmanship hunting turkeys in his native Tennessee mountains. He later told an investigator that he "couldn't miss" at the 30 yard range on Oct. 8, 1918.

York's feat was virtually forgotten for the next three weeks as the 82nd Division joined in the final push of the war that ended with the armistice on Nov. 11, 1918.

In that fighting, scores of other 82nd men also became combat heroes, more than 100 winning the Army's new Distinguished Service Cross, including 14 from York's regiment.

Hundreds more paid the supreme sacrifice.

The 82nd went into the Argonne with 15,000 infantrymen and machine gunners.

Nearly 1,800 were killed and wounded on those first two days.

By Oct. 31, when the division was pulled from the line and ended its World War I combat service, the 15,000 was reduced to 9,200.

Postwar reckoning listed 8,077 total casualties, including 995 killed in action, for the 82nd in World War I.

Only six other divisions of the 27 that participated in the war's campaigns had a higher cost in killed and wounded.

The 82nd came home to victory parades and demobilization.

Alvin York, promoted to sergeant, was feted in New York and Washington, and then went back to Pell Mell to marry his mountain sweetheart and settle on a farm presented to him by the Tennessee legislature.

Twenty-six years later, a reconstituted All American, the 82nd Airborne Division, would also receive a morning call to undertake an unexpected, desperate mission in a forest and villages only 70 miles north of Chatel-Chehery.

In December 1944, the division was in camp at Rheims, France, when the call came to board trucks and head north into Belgium, to the Ardennes Forest.

For three weeks, the paratroopers of the division would fight around the villages of Trois Ponts and St. Vith, anchoring the "northern shoulder" of the U.S. defense against the German army's last-gasp offensive of World War II, the Battle of the Bulge.

Early Pilot Claimed By Cape Fear

Originally published February 10, 1994

It was a chilly winter evening in early 1919, and W.A. Holmes had just finished chores on his lonely farm a few hundred yards from the Cape Fear River, a mile north of Fayetteville.

Then he heard it. In the dark, Holmes said, it sounded as if something was crashing into a tree.

Five minutes later, he recalled, there were "cries for help."

Pulling on a coat, Holmes ran to the river bank, just in time to see "an airship" as it "plunged into the water on the west side of the stream."

Holmes recalled that "as the machine struck the water, there was an explosion like the report of a pistol."

Then, he said, "all was silent."

Tangled wreckage

As he watched, the tangled wreckage of a wood-and-canvas airplane "settled down in the water and floated across the stream to the east side."

Holmes saw broken tree limbs scattered on the river bank. Looking up, he saw shreds of canvas dangling from unbroken branches.

The date was Jan. 7, 1919.

That day 75 years ago was the genesis date of the place that was to be known as Pope Air Force Base.

For when that two-seater Curtis "Jenny" airplane sheared off the top of the tree and crashed into the Cape Fear, it took the lives of two men of the Army Air Service

— the pilot, 1st Lt. Harley H. Pope, and his "mechanician-observer," Sgt. W.W. Flemming.

In March 1919, when the Army formally chose a name for the 5-month-old landing field on the north corner of newly built "Camp Bragg," it honored the 39-year-old pilot who had been among the earliest flyers to use the dirt strip.

For Harley Halbert Pope, death in a crashed airplane ended a life that had been filled with adventure since his boyhood in the Midwest.

A native of Bedford, Ind., he had taken camping trips far into Canada's Mackenzie River wilderness as a teenager.

Pope was in his late 30s when World War I began. He joined the air service soon after the hostilities began. He spent a year in France as a flyer with the U.S. Army's 276th Aero Squadron.

"Fancy aviation"

A flattering newspaper article said that the flyers in the squadron were "well known for their numerous exhibitions of 'stunts' and fancy aviation."

Pope, accompanied by Sgt. Flemming, had been sent to the Camp Bragg field in December 1918 as an advance party from the Army airfield at Fort Jackson in Columbia, S.C., with orders to prepare the Bragg strip for his unit.

The landing field was part of a big military reservation being carved out of the pine woods of North Carolina. It was still a wild place, so much so that for several years, standard landing instructions called for first buzzing the field to scare off deer that might be grazing.

"Fine exhibitions"

After working for several days during the holiday season, Pope and Flemming had returned to Emerson Field at Fort Jackson, "giving some very fine exhibitions of flying before their departure."

Then they escorted another plane of the squadron back to the Cape Fear country.

Using railroad tracks as navigational markers, they took a wrong turn and wound up at Raleigh.

In the "approaching dusk" of Jan. 7, 1919, they took off again, heading for the new field in the pine woods only an hour's flying time away.

In describing the last moments of "Curtiss bi-plane No. 38199," an Army historian echoed the conclusions of the official investigation:

"Several times Lt. Pope circled over Fayetteville trying to find his bearings. With the fuel in the 'Jenny' exhausted, he attempted a forced landing in the Cape Fear River. Breaking through the trees near the Clarendon Bridge, the airplane crashed headlong into a railroad bridge."

Overly dramatic

The official version was overly dramatic about the final crash.

The plane never got to either the Clarendon (now Person Street) bridge, or to the superstructure of the railroad bridge just downstream.

Instead, farmer Holmes had been right. The plane sank into the river short of the bridges.

It was recovered two days after the crash by troops under Col. Maxwell Murray, commanding officer at Camp Bragg.

However, when the bodies of the two men were recovered — Flemming's in March and Pope's in April — it was from the river at a point between the highway bridge and the railroad bridge.

Medical authorities concluded that the men drowned when the swollen river pulled the wreckage under. A search in the river near the crashed plane failed to find the bodies. Instead, the bodies gradually drifted under water from the crash site. When they were found, it was "in a very conspicuous place under some willows where there are several eddies."

Fisherman's discovery

Ironically, the bodies were discovered by the same fisherman, Orrie Johnson, within a few yards of each other, but more than a month apart.

When Flemming's body was recovered, it was quickly identified by its uniform, although his military ID card had been found in the wreckage two months earlier.

Accompanied by Sgt. Leonard of the "Aerial Photography Section, Section 84," the body of the young serviceman was shipped to his hometown of Providence, R.I., riding on "ACL train 82."

Watch recovered

Several weeks later, on the same night train, the body of Harley Pope was shipped to Indiana. In his hometown, his casket was met by his mother, Mrs. Lida Pope, accompanied by her priest.

When his body had been found in the river, Pope's pockets had contained "one hundred and six dollars ... and a fine gold watch in a perfect state of preservation."

His aviator's cap, with the initials "H.P.," had been recovered with the wreckage of his plane.

As his funeral train pulled out of the Fayetteville depot, Pope received a final salute from two of his young colleagues of the 276th Aero Squadron, Lts. L.C. Mallory and E.A. Wine.

Pope's remains were accompanied on the journey to Indiana by Maj. Norman W. Peek, his commanding officer with the 276th Squadron.

On the journey, Peek had a brand new title. He was "commander of aerial activities at Pope Field."

The official designation of the installation had been announced only days before the body of its namesake was discovered.

In a memorandum of March 27, 1919, the War Department announced:

"The Secretary of War directs that all concerned be notified that the flying field to be established in N.C., is named Pope Field in honor of 1st Lieut. Harley Halbert Pope, A.S.A., R.M.A., who was killed in an airplane accident January 7, 1919."

War Department Mobilizes To Build Bragg

Originally published February 29, 1996

Fort Bragg is now in its first months of its 77th year as a military place.

What was it like in the first year?

That would be the first few months of the year 1919, when "Camp Bragg" had been carved out of the Carolina Sandhills and the first troops arrived.

To go back a bit, the site for the artillery range and cantonment was personally chosen in August 1918, by Col. Edward P. King of the Artillery Branch.

He reconnoitered south from Washington riding in a big white touring car, under orders from his chief, Maj. Gen. C.P. Snow, to find a place suitable for training with the big guns then being fired on the battlefields of France.

King returned and reported he had found an artilleryman's dream, with miles of room for shooting.

The Army drew up plans for a 135,000-acre range and a cantonment housing 16,000 troops, with an airfield and a 500-bed hospital.

Construction began in September 1918. Thousands of Puerto Rican, Haitian and Cuban laborers recruited from the streets of New York and directly from the Caribbean found themselves in the Carolina Sandhills.

By the middle of December 1918, scores of buildings were completed, but scores of workers were dying or sick, victims of the influenza epidemic in the southern United States.

And by then, of course, World War I was over, the armistice signed on Nov. 11.

Would Camp Bragg continue?

In early January 1919, the War Department vowed that it would. It pressed Congress to continue appropriations for acquiring land.

Contracts totaling $17 million already had been issued for the construction program.

Seven days into the new year came a spectacular tragedy when a "Flying Jenny" two-seater airplane flown by Lt. Harley Pope crashed into the railroad bridge on the Cape Fear River. Pope and his observer, Sgt. W.W. Flemming, were killed. Pope's body would not be recovered until three months later.

Col. Maxwell Murray, who had been in charge of the construction job since October, was formally named post commander Jan. 9, two days after the Pope-Flemming crash.

And then the first combat-arms troops began arriving.

They joined the detachment of officers and enlisted men of the Quartermaster Corps who had been assisting the first construction engineer, Maj. John K. Thompson, for several weeks.

Understrength companies of the 46th Infantry Regiment arrived from Alabama, where the War Department was closing down World War I training camps.

The infantrymen set up quarters in the new barracks, often having to sleep in their field gear or to build their own bunks.

They quickly discovered the amenities of the courthouse town 10 miles away. By late January, the soldiers had commandeered the small armory in downtown Fayetteville for a weekend social event, a dance attended by enthusiastic civilians and featuring a demonstration of squad drill by men of Company D.

By then, the unfinished camp was facing a shortage of construction workers, as most of the 1918 work gangs had returned north.

Workers lured here

A new construction engineer, Col. H.D. Sawyer, met the problem with a recruiting effort in far-off Cincinnati, Ohio. He advertised for 800 workers. Pay would range from 60 cents an hour for carpenters to 30 cents for laborers for six weeks of guaranteed work, with board and lodging provided at $5 per week.

Sawyer knew Cincinnati. In 1918, he had been in charge of construction of a $15 million gunpowder plant there. He knew the city was in a recession, with "20,000 idle workers."

As workers arrived by the trainload, there were other signs of permanence on the sandy post.

In early February, the first artillerymen arrived, small cadres from Fort McClellan, Ala.

With the gunners came a detachment of the 32nd Balloon Company, a small party of radiomen in the 25th Radio Detachment, and another group known as the 84th Photographic Section.

Along with the ground soldiers came the airmen. The former cotton patch runway of the as-yet-unnamed landing field was being used by a half-dozen of the ubiquitous Curtiss "Flying Jenny" aircraft, flown by young men of the 276th Aero Squadron. Several of them had been flown over the skies of France in 1918.

In March 1919, other small units from World War I artillery outfits drifted in as the Artillery Branch in the War Department scraped its dwindling manpower pool to give its new range a unit of gunners.

To hurry along the construction and land-buying, King showed up in early March to meet with Sawyer and then call on the Cumberland County Board of Commissioners.

Out of the meeting came an agreement that the county would split the cost of putting a gravel bed on the "Yadkin Road" between the post and Fayetteville. The colonial wagon path would later become "Fort Bragg Road."

Meanwhile, the Army completed work on the post's water plant on Lower Little River. The most up-to-date system in North Carolina, it included a 75-foot dam, pumping station, and two cedar tanks each holding 125,000 gallons dropping water to a filtering plant. The sluice that dropped the water from tank to plant turned a dynamo so that the facility could "supply water and light for 150,000."

The system was built by a 300-man crew under G.E. Wycoff from the Moreno Construction Co. of St. Louis.

As the War Department again announced that Camp Bragg and 15 other newly built posts would be "permanent," there erupted the first signs of an enduring tension between post and civilian community.

It had to do with what The Fayetteville Observer defined as "morals." Civilians were complaining about the sometimes-rollicking off-duty conduct of the men in uniform.

The newspaper editor, himself a veteran of the fighting in France, was having none of it. He assured readers: "We can't see any problem."

From his own observation, before and after, he concluded, "Fayetteville is no worse morally than it was before camp activities were begun." Chiding his fellow citizens about their own morals, he reminded them that local merchants learned quickly to charge "camp prices," a reference to price-gouging of soldiers.

Health care established

As Camp Bragg's first spring arrived, another big step toward permanence was taken when the post's Medical Department was formally established.

It consisted of doctors and nurses who were attached to the 46th Infantry at Camp McClellan.

The party was overwhelmed by the sheer size of the hospital complex, with space for

500 beds. They managed to set up shop in a few of the two dozen buildings.

The new season brought the sad discovery of Harley Pope's corpse. A fisherman on the Cape Fear recovered it only a few hundred yards from where his plane crashed in January.

The Camp Bragg landing field, with its four small hangars, had been formally named Pope Field a few days before the discovery.

In the first days of Camp Bragg's first April, post and civilian community joined in ceremonies as the coffin with the young airman's remains was loaded on a train to be sent back to his native Indiana.

That solemn event marked the end of Camp Bragg's first season, and the beginning of a new chapter.

A 1919 Snapshot Of Camp Bragg

Originally published August 28, 1997

What did Camp Bragg look like when it opened for business in the early winter of 1919?

Some quick things to remember:

One, it was a "wood city." If you want to get a feel for it, today's "old division area," constructed in World War II, is pretty much like it. Except paint. The original buildings were unpainted.

Two, Camp Bragg wasn't "airborne" then. It was artillery. When built, it could accommodate as many as 16,000 soldiers and 4,000 animals. But the complement actually seldom exceeded 2,500 during the 1920s and 1930s.

Three, while much of the original street layout remains today, even with their original names, you must imagine the buildings. Fewer than a half-dozen structures built in 1918-19 remain on the post. You'll read about them as you go along.

Where to start to get a feel for the original Camp Bragg?

Stand at the most abiding place of all, the point where Randolph Street ends at the parade ground flagpole and the two 45-degree streets — Hunt and Alexander — converge.

With Pelham Street on its south and Sedgewick to the north, the large space beyond the flagpole, today's parade ground, contained the big post theater and the officers' club, and within a year or so was also site of the post's first polo ground.

Standing at the flagpole on a January day in 1919, looking north and south, you could see practically the whole place, except for the "aviation field," two miles to the north (toward the present-day chapel).

The blueprint of the post looked like a north-south ladder. Anchored on two north-south roads, Reilly Road on the west and Armistead Street to the east, a latticework of connector streets began on the south about where the Officers' Club parking lot is today.

Reilly Road in that direction extended no farther than where Community Access Road peels off today.

Looking north from the flagpole, the latticework of cross streets extended to present-day Letterman Street.

Reilly Road ran north for two miles to the aviation field, soon to be named Pope Field, with its collection of two hangers, barracks and office building.

The main troop housing areas of the cantonment were grouped between Reilly and

Armistead, in regimental-sized layouts of rough wooden barracks (referred to in photo captions as "quarters for the boys"), mess halls, officers' quarters, "lavatory" huts, classroom buildings and headquarters buildings.

Also in the layout were Knox, Scott, Macomb, Woodruff and Hamilton streets.

The post's railroad siding and quartermaster warehouses were at Macomb and Knox, where they are today. Quartermaster and ordnance personnel had barracks along Scott between Armistead and Hamilton.

And in the entire block between Hamilton to Knox south of Letterman, was the post hospital, a collection of 22 concrete-brick, stucco-covered wards, barracks, surgical units and office buildings. Fully equipped, they could accommodate 500 patients. They were the most substantial structures on the post.

In 1919, access to the original post was along two axis that still serve today.

Randolph Street was the long avenue leading from the first railroad siding to the center of the camp.

The dirt road leading from the town of Fayetteville followed today's Knox Street. At the time, Fayetteville was a courthouse-and-textile town of fewer than 10,000 residents.

The original road curved into Randolph opposite Hamilton Street, cutting right across where today's brand-new post library is located.

The first post headquarters and administration building, a sprawling single-story structure, was located beyond this intersection in the southwest corner of Hamilton and Scott.

South of this "road to Fayetteville," in the corner of Randolph and Knox, was the angular 40-foot-high shed of the balloon detachment, where big sausage-shaped artillery observation balloon were housed.

Today's Fort Bragg is a place of streets lined with handsome hardwood trees and with large tracts of mature pine trees throughout the 185-square-mile reservation.

However, 1918 photographs of Camp Bragg under construction depict a landscape practically bare of trees, resembling today's airborne drop zones.

The Carolina Sandhills originally were a vast forest of longleaf pine trees, with a thick ground cover of wiregrass.

However, by August 23, 1918, when the contract for Camp Bragg's construction was awarded, much of the forest was gone, first logged and then denuded by the turpentine industry.

The area was abundantly supplied with good water, coursing through the sandy soil in smaller creeks flowing north into the Lower Little River and south into Rockfish Creek.

The post's state-of-the-art water system used Little River water as well as water pumped from McFadyen Pond, a natural lake just west of Reilly Road between the post and the aviation field.

In 1919, of course, there was no Bragg Boulevard, no All American Freeway. Not a single paved road in Cumberland County. The Yadkin Road was an ancient wagon track through the Sandhills from Fayetteville toward Carthage, the county seat of Moore County. It was among several east-west tracks that rambled through the deserted western expanse of the reservation.

Another dirt track angled off the Yadkin Road toward the site chosen for the cantonment area.

It was the "road to Fayetteville" that came into the post as present-day Knox Street, mentioned above.

The main access to the site was from the east, from a railroad siding on the line of the Atlantic Coastline Railroad, which paralleled today's Murchison Road, then often referred to as Manchester Road.

From that siding, the post construction crew quickly built an arrow-straight, two-mile-long road, named Randolph Street, to the post site.

The earliest construction also included a spur railroad line from the siding.

Paralleling Randolph Street, the railroad ran to the post's supply head, the series of warehouses at Knox and Macomb streets. The essential character of Camp Bragg's layout was vividly described in an early 1919 brochure:

"The camp is erected in groups with considerable space of forest or openfield between. The aviation field is two and a half miles from ther balloon station. The remount station (home of the horses and mules) one and a half miles from the camp center, and so on.

"Instead of one huge concentration of barracks and mess buildings spotted at random over the camp ground, the scheme now reminds one of a number of small villages placed in a natural park and each within easy access of the other one in the cantonment."

Pope's Fledgling Years A Time Of Winging It

Originally published July 11, 2002

As Fayetteville begins the yearlong countdown to May's stupendous observance of the Festival of Flight, when the air will be filled with armadas of high-tech flying machines, it is fun to look back at the days when aviation was in its infancy here.

The first five years at Pope Field, beginning in the winter of 1919, were times of the "Flying Jenny" two-seater and the De Havilland, "the flying coffin."

It was a time when intrepid young men in tight-fitting leather skullcaps and goggles pushed the limits of their planes and themselves, enthusiastically demonstrating the possibilities of flying.

Testing the limits

In early 1919, the commanding officer of the Army post's airfield in the Carolina Sandhills set out to make a long-distance flight. The Fayetteville Observer reported it:

"Although there was a steady headwind blowing, Captain B.J. Saunders climbed into his Curtiss JN 4H and in a short time was on his way to Salisbury, where he is to select a site for a permanent landing field for the mail route and also for commercial use. Sgt. Lea went with Capt. Saunders as mechanic.

"Capt. Saunders took a direct compass course but owing to the increasing of the headwind ran out of fuel when only two miles short of the race-track where he was to land.

"While landing in a soft plowed field, the end of the propeller was split, damaging it to a very small extent, although enough to prevent flying without replacing it. Lt. H.E. Sherman and Lt. H. Bigalow left via auto with a new propeller."

In the summer of the same year, another Pope aviator found a new use for aviation:

"Lt. Kenneth P. Behr, on Saturday flying low with defective compass on flight from Pope to Georgetown (had been headed for Wrightsville) spotted three moonshiners near Fort Fisher."

By the summer of 1922, Pope flyers were attempting even more spectacular feats.

The Observer reported that Lts. Leroy Walthal and E.P. Gaines were on their way to Louisville, Ky., with Walthal flying and Gaines as an observer.

Their 1,600-mile flight and return took a week. It took them first to Asheville, then to Knoxville; Danville, Va.; and into Dayton, Ohio, where Wright Field was the Army's major testing and development center.

From Dayton, the Pope pair returned by way of Moundsville, W.Va., and Washington.

The Observer said the flight was "part of a War Department policy to simulate commercial routes."

The Observer editorialized:

"Both men have done yeoman work in this regard. Gaines is just back from Onslow County. Walthal went to Charlotte and Rock Hill. Gaines to Columbia, Augusta, and Macon." The flyers who had accomplished such a long-distance flight must have been chagrined a month later when they tried a Fourth of July flight over the Blue Ridge Mountains into Boone and north Georgia.

Gaines and Walthal got only as far as Asheboro.

Transcontinental flight

Then in 1923, the most historic of the early flying exploits of Pope airmen occurred. Lt. Gaines, with Sgt. A.J. Hilton, took off for a transcontinental flight.

For three weeks and 6,000 miles, Gaines and Hilton hopped the country to San Diego and back again, with a total time in the air of 25 hours and 10 minutes to the West Coast.

When they returned to Pope on June 15, 1923, a large crowd was on hand to cheer, and the post commanding officer, Brig. Gen. A.J. Bowley, was one of the first to grab their hands.

Trying for long-distance flight was one thing, but Pope flyers also enjoyed buzzing the town.

In 1920, Lt. R.E. Davis "made a typical flight over Fayetteville." The Observer reported:

"The giant monster swooped within 100 feet and circled the old Market House. Then headed straight for Langley. The plane left Pope at 1:30 after spending several days recruiting."

Two days before Christmas in 1920, the flyers at Pope flew low again over the Market House and dropped a note. It read:

"The officers and enlisted men at Pope Field wish the people of Fayetteville a merry, merry, Xmas and a happy and prosperous new year. Signed: Air Service."

Romance in the air

In those halcyon first years of aviation at Pope, all the news was not about flights and feats of endurance. Romance was also in the air.

In June 1920, the Observer's social reporter wrote:

"Lt. Frederick Hopkins Jr. entertains at Ardlussa for Miss Bessie Marsh and Dr.

Frank Newby. Covers were set for six."

Ardlussa was a country estate used frequently for Camp Bragg social events.

Lt. Hopkins was mess and supply officer of Pope Field. Bessie Marsh was a sister of Henrietta Marsh, another beautiful daughter of Mr. and Mrs. Henry E. Marsh, a prominent Gray's Creek family.

You can bet that one of the "covers" at the meal was set for Henrietta.

Years later, Maj. Gen. Fred Hopkins wrote to a relative:

"On 1 September, 1920, I received my commission in the Regular Army in the Air Service.

"On the 9 of September I married Mer (Henrietta) Marsh in the parlor of the farm house at Gray's Creek you know so well. We left immediately for Washington. Sailed on 20th. Mer came shortly."

A daughter of that early local military marriage between a Yonkers, N.Y., officer and a southern belle from Cumberland County is Henrietta (Mrs. Rupert) Jernigan, who lives in Fayetteville today.

Early Airmen Searched To Find Their Niche

Originally published March 21, 1996

The men and women of today's Pope Air Force Base know their mission.

But in its first springtime, the men of Pope Field were trying to find a niche in Army aviation.

Their quest led one of them to pioneer the use of airplanes pursuing that ancient scourge of Southern swamps and woods — the bootlegger.

In April 1919, when it was formally named Pope Field, the new air installation in the Carolina Sandhills consisted of four angular hangars and several office and barracks buildings.

Part of spanking-new Camp Bragg, which was meant to be the Army's biggest artillery range, the airplanes at Pope Field were expected to provide air observation for the big guns.

Along with a detachment of observation balloons, the two-seater planes of the 276th Aero Squadron were supposed to seek out targets and check on the accuracy of the guns when they fired across the long reaches of the reservation.

But in the spring of 1919, there was a problem.

There weren't enough guns to get up a good shoot.

Demobilizing after World War I, the Army was desperately trying to scrape up enough artillery units for Camp Bragg.

The post was completed just as the war ended in November 1918. It was designed to be home to as many as 16,000 artillerymen and supporting troops.

In April 1919, however, the total post population was 101 officers and 977 enlisted men.

There was a single battalion of artillery, an understrength unit from the 81st Field Artillery Regiment. It arrived in early March from Camp McClellan, Ala.

At Pope Field, young men of the 276th Aero eagerly yearned to do their stuff. Their Curtiss "Flying Jenny" two-seaters became familiar sights in the skies over Fayetteville and other nearby towns as they roamed the area to familiarize themselves with landmarks.

But it wasn't the exciting stuff some of them had experienced in the skies over the trenches of France in 1918.

The men of the ground crews were also strapped for parts and gasoline as Army budgets shrank.

Maj. Norman Peek, "commander of aerial activities" at Pope, took off in early April for the sad duty of accompanying the body of Lt. Harley Pope back to his home in Indiana.

The post had been named that week for Pope, who with Sgt. W.W. Flemming died when their Jenny crashed into the Cape Fear River at Fayetteville in January 1919.

Among the pilots in Peek's command were Capt. Bradley J. Saunders and Lts. Harris Bigalow, Norman W. Gibson, L.C. Mallory, H.P. Haley, Edward Wine, Fred Hopkins Jr., Charles C. Green and Kenneth P. Behr.

Lt. Walter Black was the Pope Field medical officer.

Lt. William L. Newmeyer was an early ground crew boss, known as the engineering officer.

Sgt. S.G. Lee was a master mechanic among the several dozen enlisted men assigned to the squadron.

As spring lengthened and the post awaited arrival of more artillery units, the airmen flew farther afield.

The Army Air Service had concluded that Pope Field was an excellent stopover point between the service's headquarters base at Langley Field, Va., and other fields in South Carolina and Florida.

So pilots at Pope were encouraged to make reconnaissance flights along likely routes for the commuters.

This usually involved low-level flying along railroad tracks, which in those days were the principle landmarks of air travel, while noting mileages and important ground markers.

Behr's legacy

And so it came to pass that in June 1919, Lt. Kenneth P. Behr, with Lt. Cox in the other seat, headed out over the railroad track toward Wilmington, intending a leisurely afternoon flight. He meant to land on the beach at Wrightsville, refuel and return to Pope by nightfall.

A "defective compass" led them to another adventure, an early incident in what would become an air war against bootleggers.

Late in the afternoon, Behr and Cox found themselves on the beach at Georgetown, S.C., a hundred miles south of Wrightsville.

As the sun began to set, they scrounged gasoline and took off, flying the coast toward their original destination.

What happened next was described in an account in a Wilmington newspaper:

"Flying low and keeping near to the coast, (Behr) spotted three separate moonshine factories between Fort Fisher and Wrightsville, where he landed at 9:30 in the evening. Behr said the factories were "all being operated at full blast."

"The moon was not yet risen Saturday night when Lt. Behr was making his way up the coast, but the stills were easily distinguishable from the reflection of the fires whose heat was turning molasses into 'monkey rum.'

"All were attended by a number of men who dropped their work to search in the darkness of the sky for the sight of the airplane, which was distinguishable only by the exhaust of its engine.

"On landing on the beach, Lt. Behr made every effort to reach the Wilmington Star by telephone. But the line was in use at every attempt."

Behr nonetheless knew he was on to something.

Hello, newspaper?

Overnighting, he got in touch with the newspaper the next day before taking off for Pope Field.

And he proved himself a visionary of a new role for the airplane. The newspaper account continues:

"(Behr) was very much interested yesterday in the possibility in the detection of the blockaders by use of airplanes.

"He believed that the airplane method of detection will be generally adopted since it is the only means of access to the dense jungles where the moonshiner has his habitat."

Sheriff Jackson of New Hanover told the newspaper he was "keenly interested" in the young airman's information. He knew bootlegging was going on in that vicinity but had only been able to nab a "cache of molasses nearby."

The pioneering vision of Lt. Behr would take a back seat to the military mission of Pope when Camp Bragg's artillery complement was swelled in the summer of 1919 by two regiments of big guns. Pope's planes began their observation task.

In October, Behr and several other pilots, as well as all but 30 of the enlisted men of the 276th Aero, would be transferred to San Antonio.

Pope Air Force Base: Field Of Dreams

Originally published June 2, 2005

The daring men who came to support Camp Bragg in their Flying Jennies established ties to the Army and the community that will endure through any realignment plan.

If the Air Force does return Pope Air Force Base to the Army, why, it will be just like the day it began in 1919 when Pope Field was the aviation field for the new Camp Bragg artillery post in the Carolina Sandhills.

That will be back to the days of "those magnificent men in their flying machines," aviators of the fledgling Army Air Service who piloted their canvas-and-wood Flying Jennies and DeHavilands off the dirt runways of what had been pine woods and cotton fields.

I have written about 20 reports here over the years about such "birdmen of the air" as Lt. Kenneth Behr, Capt. Bradley Saunders, Fred Hopkins, Harris Bigalow and, of course, Lt. Harley Pope and Sgt. W.W. Flemming.

Pope and Flemming, his sergeant companion, died Jan. 7, 1919, when their Flying Jenny clipped a pine tree and crashed into the railroad bridge across the Cape Fear River at Fayetteville.

Their bodies were recovered months later when the river went down.

Also later, the War Department ordered the Camp Bragg Aviation Field named

Pope Field in honor of the World War I flier who was among the first of the 276th Aero Squadron aviators flying their Jennies from their base in Columbia, S.C., to the field at the new Camp Bragg.

Pope's remains were given a military sendoff and then shipped by train to his native Bedford, Ind., "accompanied by Maj. Norman Peek, commander of aerial activities at Camp Bragg."

After they gave the field its name, the aviators, many veterans of World War I, took up a busy schedule of activities, some highly inventive.

They were waiting for the Army to bring in big artillery units so the fliers could take up their official duties flying as observers for the big guns.

Lt. Kenneth Behr was among the early arrivals with the 276th Squadron detachment.

He made history by inadvertently discovering how useful an airplane could be as bootleg busters.

On a June evening in 1919, flying low with a defective compass and off course from a flight from Pope to Wilmington, he was "going full blast" over the beach at Fort Fisher when he spotted three moonshiners at work on the ground.

He landed on Wrightsville Beach at 9:30 and later called the New Hanover County Sheriff's Office.

Behr and Capt. Saunders, who ranked as commanding officer of the 1919 Pope Aviation Field detachment, were pioneers in laying out flying routes and helping North Carolina towns locate landing fields. Among their 1919 choices was Audubon Field near Wilmington.

"Flying parson"

A notable North Carolinian among the early magnificent men was Lt. Belvin Maynard, known as "the flying parson" because he mixed evangelical witnessing with his sensational stunt flying and his fame as a cross-country pioneer.

In November 1919, he and Lt. Edward L. Wine, accompanied by their police dog, Trixie, flew from Clinton to Pope and then on to Audubon Field.

Maynard was to die three years later while stunt flying in New England.

When he was buried near his hometown of Wallace, fliers from 1922 Pope Field flew a saluting formation and dropped flowers.

Other 276th fliers whose names appear in accounts of Pope activities in 1919 are Lts. Norman Gibson, H.P Haley, L.E. Sherman and (no first name) Cox, and William Newmeyer. Lt. Walter Black was listed as "Pope surgeon."

Enlisted men, few identified by their first names, included Sgt. Murray and Sgt. S.G. Lee, "a mechanic."

Jack Clark was listed as the field meteorologist.

Lt. L.C. Mallory was among the first Army fliers stationed at Pope Field. He quickly made a name across the area with his daredevil exploits in his Flying Jenny two-seater. He quit the Army in mid-year to pursue full-time exhibition flying.

A report from the Welcome Home celebration in Raeford in September extolled an exhibition by Pope fliers and by "ex-Lt. L.C. Mallory, who was the famous stunt flier of the field."

Reflecting the social ties already developing between post and town, at a Christmas dance in Fayetteville in December 1919, "Ex-Lieutenant L.C. Mallory of Milwaukee" was matching his dazzling flying technique with an equally dazzling technique on the dance floor as he and Miss Bessie Cotton joined the younger set of Fayetteville and Camp Bragg welcoming in a new decade, the Roaring '20s.

A wedding

The ties got even stronger in early 1920, when Lt. Fred Hopkins Jr., who had arrived in the autumn of 1919, found his wife among Cumberland County's belles. He and Henrietta Marsh were married at the Marsh family country home in Gray's Creek and immediately left for a new assignment. Hopkins would return for a second stint at the Bragg post's flying field.

In September 1919, Pope's fliers were all over the area. Lt. Charles C. Greene, "one of the air kings of Pope Field," and Mallory were "performing stunts" over a Welcome Home Celebration in Raeford.

In October for the Cumberland County Fair, it was Greene and Bigalow at the controls of the Flying Jennies, with the latter carrying along Hopkins and Sgt. Murray as observers.

There was also time for off-duty fun.

A "Saturday shoot" at the Pope Field Officers Gun Club brought together Capt. Bradley J. Saunders, Lts. Harris Bigalow, Norman W. Gibson, H.P. Haley, Edward Wine, Fred Hopkins and Charles C. Greene. Of course, Saunders won the prize.

By the time of the autumn shoot, orders from Washington were reducing the Pope detachment to a few dozen men. By spring of 1920, Hopkins at times would be the only officer on duty.

But there was one more big hurrah.

In the first week of December 1919, the not-yet-year-old Bragg and Pope Field played host to a touring visit from none other than Gen. of the Armies John J. Pershing, World War I commander in chief of the American Expeditionary Force in France and Belgium.

Pershing and a bevy of brass from Washington trooped the line of neatly parked Flying Jenny aircraft.

Those magnificent men in their flying machines thus finished their first year at the new aviation field of the Carolina Sandhills.

So today, the Pentagon can change ownership and the mission, but it can't take away the rich history of a grand old aviation place that will observe its 90th birthday four years from now.Town Suits Up In Deference To "Black Jack"

Originally published November 30, 1995

He was the most famous American soldier since Ulysses S. Grant, wearing the most exalted rank since George Washington.

But in December 1919, the Army didn't know what to do with General of the Armies John J. Pershing.

So he came to Fayetteville and to brand-new Camp Bragg.

"Black Jack" Pershing had come home from Europe in September, returning as the triumphant hero-commander of the American Expeditionary Forces in World War I.

The War Department was, at the time, bereft of a job befitting his rank as the first active-duty five-star general.

The puzzle was quickly solved. He would tour the country, visiting camps where the doughboys of the American Expeditionary Force had learned the martial business, also accepting a string of invitations to various civilian functions.

Camp Bragg, only a year old, had come along too late to be a training camp for the gunners in France. But artillerymen in the War Department had high hopes for the huge cantonment and firing range in the Carolina Sandhills. They were pressing the War Department to make it a permanent post-war post.

At the time, fewer than 2,000 soldiers were at the camp, along with two understrength artillery regiments and a small squadron of "Flying Jenny" planes.

Pershing got to the 4:40 p.m. train on time.

Several members of a congressional committee had visited earlier that fall and were deciding whether to appropriate money to finish land acquisition.

When the Pershing tour was announced, the War Department artillerymen and town officials of Fayetteville pressed for a stop at Camp Bragg and Fayetteville as Black Jack made his way from Camp Lee, Va., south toward big Fort Jackson in Columbia, S.C.

The general agreed to be there on Friday, Dec. 5, 1919, but said that most of the day would be spent inspecting Camp Bragg.

Fayetteville at the time was a town of 9,000. It had spent most of 1919 welcoming home the hundreds of local men who had gone off to war, as well as coping with the post-World War I economic slump and a deadly epidemic of flu. It had just finished its first big Thanksgiving week parade honoring veterans and the families of young men killed in the war. It had less than two weeks' notice to get ready for the Pershing visit.

Nonetheless, it put its best foot forward.

A professional decorating firm was hastily hired to bedeck the town with patriotic bunting. Professional musicians — the Icemorless Band of Monroe, 38 strong — were brought in. Dr. Vance McGougan, who had already made a reputation as host for the congressmen, laid out a barbecue at Eutaw, his country place on the Bragg Road.

Fayetteville merchants, never shy about a sales opportunity, got into the swing.

An advertisement in The Fayetteville Observer proclaimed:

"Hail General Pershing. Will you be one of the men who will escort him? Remember, the order reads that civilian clothes will be worn. No doubt you want to make an impression in civilian clothes, as you did in uniform, then dress up in a SUIT or OVERCOAT, and put on the finishing touches by wearing a Manhattan shirt and a Stetson or Schoble Hat. A pair of Edwin Clapp, Florsheim, or Bostonian shoes, all to suit particular purposes." Where to get these fashionable 1919 duds? Why, Stein Brothers at "The Big Store on the Corner."

And so, on Dec. 4, 1919, the Observer threw an eight-column headline across its front page, "General Pershing Program Is Complete,"and the story reported:

"The city presented a handsome appearance today, the red, white, and blue being conspicuous in the decorations. A decorator from Winston-Salem has been in charge of this work. The Old Market and many other buildings have been adorned with flags and streamers stretched across the principal thoroughfares added to the holiday aspect.

"The committee in charge of the barbecue has completed its work, except for a few minor details, and it is expected that this feature of the day will be a noteworthy and enjoyable one."

Pershing, with a staff of 10, would arrive at Camp Bragg on an 8 a.m. train. After breakfast, an inspection tour would be conducted by Col. E.P. King of the artillery branch, the man who had chosen the site for Camp Bragg 18 months earlier.

At 2 p.m., the party would motor to Dr. McGougan's, to be greeted by Mayor Underwood.

At 3:15, he would arrive at the "city limits," greeted by music and a throng of citizens. He would wave to the crowd from the "west veranda" of the Market House, after being introduced by Lt. Gov. O. Max Gardner.

He would then go inside the upper floor of the Market House, to the "Ladies' Rest Room" operated by the Fayetteville Civic Association, where he would "meet the ladies, and especially the mothers of young men who lost their lives in the service."

At 4:40 p.m., he would take the train to Columbia.

Now, for a historical conundrum.

Did it all happen that way?

You can't tell from the records of the Observer.

There aren't any.

Only a half-dozen times in the first 75 years of the newspaper is there a gap in the preserved editions.

This was one. There are no editions for Dec. 5 or 6, 1919.

Did they run out of historically significant copies? Somebody rip off the bound files?

Whatever, if you have ever seen a copy of these editions, they are truly valuable, and we would give eyeteeth to make copies of them.

Of course, the story appeared in many other papers, and yes, it did come off that way.

Black Jack, resplendent in his American Expeditionary Force commander's uniform and long braided coat, was in an expansive mood as he greeted old comrades from the 21st and 5th Artillery, units that had fought in France.

The tall, handsome Pershing met with local veterans who had served under him in 1916 when he commanded the U.S. Punitive Expedition in Mexico.

Pershing got to the 4:40 p.m. train on time.

Three days after his stop in the Carolina Sandhills, he was speaking to thousands at the Southern Commercial Convention in Savannah, greeted by the governors of Georgia and South Carolina.

Blue Yonder Not So Wild Over Pope In 1920

Originally published August 8, 2002

Lt. Fred Hopkins Jr. was a lonely guy at Pope Field in the early months of 1920.

Hopkins arrived at the dirt-strip airdrome at Camp Bragg in the autumn of 1919 to join a group that included a dozen or more aviators and several dozen enlisted personnel.

But then on Oct. 17, 1919, practically the entire command was ordered to move to Ellington Field in Houston, Texas. All but four officers and 30 enlisted men of the 276th Aero Squadron were sent away.

Hopkins, who had yet to earn his flying wings and was still classed as an observer, nonetheless managed to solo in unofficial flight training sessions with Lt. Lou Mallory, a veteran flyer who had seen air combat in France in the Great War, now known as World War I.

Returning after a 1919 Christmas furlough from his New York home, 25-year-old Hopkins found himself acting as both mess officer and supply officer for Pope, in effect the field's commanding officer, presiding over an enlisted cadre comprising mostly young recruits to the Army Air Service.

"Flying coffin"

Hopkins described the early aircraft at Pope:

"The squadron was initially equipped with Curtiss JN 44-H airplanes equipped with the 300 HP Hispano-Suizo engines. Later a De Havilland DH 4-A with a 400 Liberty Engine was assembled and flown. This model had an 80-gallon gasoline tank between the front and the rear cockpits and was known as the 'flying coffin.'"

While Hopkins held the fort at Pope, however, the Air Service was busy in North Carolina in that summer of 1920, barely a year after the beginnings of Camp Bragg's history as an artillery cantonment.

Flying out of the Air Service's major installation, Langley Field in Hampton, Va., men and machines were busy across all of North Carolina in May 1920.

They were flying to carry the message of the Air Service that aviation was a major new player in the military arsenal of the small peacetime Army.

They were officially on recruiting trips, trying to sign up flyers and ground crewmen. Anyone who signed up could be flown to Langley in the big De Havilland machines.

On May 12, 1920, several of these young men in their flying machines came to Pope, including Lt. R.E. Davis, "who was shot down in Germany and was a prisoner- of-war in the Great War."

The experienced Davis took off for Greensboro, where he hoped to sign up some recruits at that city's Friendship airport.

Instead, the plane, a De Havilland DH-4, was "completely wrecked" in a landing attempt.

Davis blamed the crash on a "defective axle which gave way when it hit the ground and the plane was practically demolished."

Davis and Lt. Charles Potter, flying another DH-4, "were not discouraged and continued their recruiting tour of the state to the west."

Davis vowed to "return to headquarters" and get other planes for recruiting swings in the North Carolina Piedmont.

On May 28, 1920, the veteran Davis bade farewell to Pope by flying the lumbering DH-4 over downtown Fayetteville, "the giant monster swooped within 100 feet and circled the old Market House, and then headed straight for Langley."

While Pope languished, the Air Service carried out a major training exercise from Wilmington's 8-year-old civilian airfield, known as Audubon Field.

Six planes of Flight B of the 50th Aero Squadron arrived in the city in late July for a two-week stay.

Artillery spotting

Although rainy weather curtailed much of the exercise, some flights were managed to observe firing of coast artillery guns from Fort Caswell at the mouth of the Cape Fear River.

And among the aviators at the Wilmington camp was the ubiquitous Lt. R.E. Davis, who swung around by Pope to greet Hopkins and other buddies before flying to Audubon Field.

The whole exercise was commanded by Capt. Fraser C. Hale of the 50th Aero.

A week after the aviators left Wilmington, Pope Field got new life as Flight B of the 8th Aero Squadron arrived.

Four planes flew cross country from Laredo, Texas, and others came by rail.

The 11 officers and 35 enlisted men were first tasked with providing the air component for training Reserve artillery outfits in summer encampments at Camp Bragg.

Early casualty

Most of the flyers who arrived with the 8th Aero were newly minted aviators who had earned their wings at Kelly Field in Texas. Among them were Lts. Joseph Virgin, Rex G. Stoner and Harrison Hartman, who tested their skill in the cross-country flight from Laredo.

In a look into the future for these three young men, history records that Lts. Virgin and Hartman would be killed in 1921 in Pope Field's first fatal plane crash.

Rex Stoner would briefly command the field before going on to become a pioneer test pilot who in 1923 first managed to hook an airplane onto a C-3 dirigible, demonstrating the possibility of using lighter-than-air airships as airborne docks for light aircraft.

Meanwhile, soon after the arrival of the 8th Aero, Virgin was testing Pope Field's Curtiss Flying Jenny planes by taking Lt. Hopkins for a spin over Fayetteville.

Hopkins presumably didn't tell him that he had unofficially learned to fly in the previous autumn.

By September, The Fayetteville Observer described the busy scene at Camp Bragg that season:

"Camp Bragg is gradually becoming one of the most important posts in America. Troops for training are constantly coming and the booming of cannon may be heard almost every day, while airplanes acting as scouts in getting the range may be seen in the sky every day. A few days ago a party of aviators reached Pope Field and the entire camp looks warlike at times with the airmen soaring through the air and any number of men being drilled in the use of artillery."

Hopkins meanwhile had taken other steps to cure whatever loneliness lingered.

He married a local belle, and they took the train for a new life together in the Regular Army. He would wind up as a major general, a key supply expert in Air Force headquarters during World War II.

Meeting Preserves Fort Bragg's Future

Originally published September 19, 1996

The overnight train from Washington brought a high-level delegation to the depot in the courthouse town of Fayetteville on that early Sunday morning in September 1921.

Secretary of War John W. Weeks was keeping a deal made several days earlier with Brig. Gen. Albert J. Bowley, the 45-year-old commander at Camp Bragg. That highly-regarded artilleryman had taken command of the post in late 1920 and vowed to make it the most important installation in the Army.

Fighting for life

Now he was fighting for the very life of the post.

For this was no ordinary VIP visit. There were no military ceremonies, no formal announcement that it was even taking place.

The trip was part of a do-or-die campaign Bowley was waging to save Fort Bragg.

Seven weeks earlier, on July 27, 1921, Weeks had listed Camp Bragg among several Army posts that were to be shut down, abandoned in a new economy drive for the already-tiny post-World War I Army.

At the time, the post, built in 1918-1919 to accommodate as many as 16,000 artillerymen, was home to fewer than 2,000 troops organized in three understrength artillery regiments.

Using political wiles as well as military arguments, Bowley persuaded the new member of the Cabinet of new Republican President Warren G. Harding to come down for a personal visit.

The secretary's visit

So, on Sunday, Sept. 11, 1921, the new secretary arrived.

His traveling companions were instructive.

The military side was represented by Maj. Gen. James G. Harboard, who had just been named deputy to Gen. John J. Pershing, chief of staff of the Army.

Harboard and Bowley knew each other well, having soldiered together in World War I, Harboard as commanding officer the 2nd Infantry Division and Bowley as the division artillery commander.

The civilian side was represented by John Motley Morehead II, a wealthy Charlotte businessman who was North Carolina's member of the Republican Party's national committee.

A strategic liaison

Bowley had sought out Morehead as key to his strategy for saving Camp Bragg. Knowing Morehead to be a personal and political friend of Secretary Weeks, Bowley convinced Morehead that saving Bragg would boost Weeks in 1924 when Weeks was expected to seek the Republican nomination for president.

In the week before his visitors arrived, Bowley had assembled elaborate maps and reams of information extolling the glories of Bragg, comparing it to other artillery posts such as Fort Knox, Ky., and Fort Sill, Okla., emphasizing its generous size of nearly 120,000 acres.

At Bowley's elbow was another key artillery officer, Col. E.P. King.

In 1918, King literally earned the title of "father of Fort Bragg" when he returned to Washington after a reconnaissance trip by automobile to the Carolina Sandhills singing the praises of the site he had selected along the Lower Little River.

On that bright September Sunday 75 years ago, however, it may well have been that briefings and inspections took a back seat to food.

Weeks and his traveling companions ate their way through a full menu of Carolina cuisine.

A convincing feast

Breakfast at Bowley's quarters included, according to his recollections, "honeydew melon, cereal, North Carolina ham, eggs, fried chicken ... toast and coffee."

After the briefing and a motorcade around the post, the party moved out to McKellar's Pond for a lunch with eight invited civilians from Fayetteville.

Breakfast had been prepared by the post commander's sister-in-law, Betty Bowley, wife of Maj. F.W. Bowley.

Lunch was prepared by Maj. Bowley "and his expert assistants," a team of Army cooks and local barbecue chefs.

Eaten "under the pines," it included "scuppernong grapes, barbecued pig, broiled chicken, corn pone, cold slaugh, dill pickles, iced tea and watermelon."

Polo and airplanes

Then it was on to the dirt landing strip at Pope Field, where the airplanes were parked and a spirited polo match was observed, played by officers of the post.

As the sun set, it was back to town, to an old country plantation on the Lumberton Road known as "Ardlussa," the home of Mrs. Fred Vaughan.

There, said Bowley, "we gave the secretary a good mint julep, an excellent dinner with scuppernong wine."

After dinner, according to Bowley, "civilians were dismissed" while he and Col. King spent a "solid hour" pleading Bragg's case.

In the end, they struck a deal.

Camp Bragg would be taken off the hit list.

However, Bowley would have to salvage furnishings and equipment from abandoned posts to finish the bare-bones look of Camp Bragg.

Weeks also insisted that the civilian community come up with the money for a "streetcar" line between the camp and Fayetteville.

During his visit, he said he heard many complaints about the sandy track that served as the connecting road between military post and town.

Bowley quickly agreed.

"You work fast," Weeks said to Bowley.

For his part, Bowley got promises of troops from units being deactivated. Weeks also reversed a decision to move the influential Field Artillery Officers School back to Bragg from Fort Knox.

As he boarded the night train for Washington, Secretary Weeks spoke briefly to The Fayetteville Observer's reporter. He would only say:

"I am greatly impressed with the topography of Camp Bragg as a location for a field artillery range. The great problem to overcome is one of finance."

The stage was set for a triumphant day.

How Flight Took Off At Bragg

Originally published January 15, 1998

Pope Air Field in the early 1920s was seldom home to more than a half-dozen young aviators of the Army Air Service.

They made for numbers with energetic, often daredevil flying feats.

Lt. Rex Stoner was among the earliest pilots to make his mark on Army aviation at the air field next to Camp Bragg.

And Stoner did more than add to Pope's early aviation record.

He found a wife in Fayetteville.

Stoner arrived in the late summer of 1920 with Flight B of the 8th Aero Squadron, ordered to North Carolina from Laredo, Texas.

The unit of 35 enlisted men and 11 officers was commanded by Maj. William Howard.

The "Flying Jenny"

Four of the young pilots including Stoner flew their Curtiss "Flying Jenny" two-seaters in stages from Laredo, accompanied by an observer in the rear cockpit.

Seventy-five years ago, Pope Field consisted of a spacious dirt runway surrounded by patches of pine forest.

The field was at the far end of Reilly Road, where it is today.

More than two miles away over the dirt road was the main post area of Camp Bragg.

Time-Honored Design

At the northeast end of the field were four standard wooden hangars of a design familiar in old World War I dogfight movies.

Closer to the main post beside Reilly Road was a small collection of barracks and office buildings, designated as the "military aeronautics" section of Fort Bragg.

The field and its buildings were built in the winter of 1918-19 as the air component of Camp Bragg, an artillery cantonment for 16,000 soldiers.

Airplanes at Pope were assigned the new military specialty of scouting and observing for artillery.

But when Stoner and his colleagues arrived that summer 78 years ago, there was little observation to be done.

Fewer than 50 artillery pieces were on post.

A shaky future

The whole future of Camp Bragg was in doubt as the War Department tightened budgets of the already-tiny Army of the post-World War I period.

The young flyers at Pope did more than fly.

They quickly were caught up in the social whirl of Fayetteville, the courthouse-and-textile town 10 miles down a dirt road from the field.

If there wasn't much artillery to observe, there was an abundance of eligible young ladies in the town and many opportunities to meet them.

The early 1920s were big years for socializing.

Ballroom dancing became a craze. Motoring in automobiles was a popular recreation.

And going for a spin in an airplane was a daring pastime.

In fact, the social connections between Pope flyers and the young ladies of Fayetteville were firmly forged by the time Rex Stoner arrived.

The earliest existing aerial photographs of Pope Field and Fayetteville were taken in the summer of 1919 by a young lady's box camera from the cockpit of a Flying Jenny.

A treasured album

A family photo album of the late Dr. Lucile Hutaff of Fayetteville contains aerial photos of Fayetteville, Pope Field and Camp Bragg taken in July 1919, from a Jenny.

The caption on the album says:

"From the Birds of the The Air."

Stoner did more than socialize. In early 1921, he married the daughter of Mr. and Mrs. J.R. Tolar Jr. of Fayetteville.

Shortly after the wedding, the young couple left for his new assignment at the busy Army air field at San Antonio, Texas.

Stoner returned to Pope Field again only once, two years later.

It was a spectacular visit.

By early 1923, Pope was humming.

Camp Bragg was now Fort Bragg, a permanent post.

Flying picks up

Army aviation was also humming, as pilots of the Air Service eagerly competed for headlines with Navy and civilian pilots in flying feats, attempting to set new records for distance or speed.

In mid-March 1923, a flight of six Army De Havilland two-seaters and four B-1 "tanker" planes took off from San Antonio. Destination: San Juan, Puerto Rico.

Stoner was at the controls of one of the planes.

The flight plan called for eight stops, a total distance of 5,365 miles.

A week later, it was front-page news when the planes were reported on their last over-water return leg, flying from Havana, Cuba, to Curtiss Field near Miami in three hours, 23 minutes.

While his colleagues took a couple of days rest in the Florida sun, Rex Stoner got permission to leave the next day. Destination: Pope Field.

Popular homecoming

That day, The Fayetteville Observer reported:

"Necks will be craned and all eyes cast in an upward direction."

And a front page headline proclaimed:

"Stoner Flying To Fayetteville To Be With Wife."

All of Fayetteville was buzzing because Mrs. Stoner was waiting on the ground.

She came to spend the Easter holidays with her parents as her husband attempted to make Army aviation history.

Stoner left Miami early the next morning, March 30, 1923.

He arrived just before noon at Pope, after flying three hours from Savannah. Before putting down at Pope he "buzzed" the Tolar home.

But Mrs. Tolar wasn't there.

She was waiting beside the landing field at Pope.

Fort Bragg and Fayetteville first welcomed the lone pilot, who then spent the day, Easter Day, with Mrs. Tolar and her family.

Then, on Monday, the Observer reported:

"Triumphant in their conquest of the air lines between the United States and the West Indies, officers of five of the six planes returning from the flight to Puerto Rico were entertained at Fort Bragg."

The post commander, Brig. Gen. A.J. Bowley, met the flyers as they arrived and they "were warmly congratulated."

Another celebration

With that, the flight took off for Langley Field near Washington, where there was another big celebration, led by Maj. Gen. Mason M. Patrick, chief of the Army Air Service, and Brig. Gen. Billy Mitchell, his assistant.

CHAPTER FIVE
World War II

In 1939, Post Had Yet To Feel The Steel Of War

Originally published April 28, 2005

For a 10-year-old who followed war maps like other kids read baseball scores, April-May of 1940 was an unforgettable season of headlines and horror for me.

Those weeks 65 years ago marked the real beginning of World War II, when German panzer armies swept across Holland and Belgium in a mere five days and shortly afterward crushed British and French armies on the border. By the end of June, France had surrendered to Nazi Germany.

It was also the beginning of a new Fort Bragg, when the small artillery post in the Carolina Sandhills began its World War II destiny as a huge training city that by summer of 1941 was turning out tens of thousands of newly minted soldiers for the U.S. Army.

Remembering the anniversary of that remarkable historical story, a series of columns is coming up.

But first, just to set the stage for these opening scenes, let's go back to 1939 and take one last look at the peacetime "old post."

In its physical appearance, Fort Bragg in 1939 was in the heyday of its brick-and-tile-roof look, with an array of handsome new permanent barracks, gun sheds, quarters, headquarters and hospital. About 3,000 artillerymen lived there and trained with the scarce guns and equipment of an Army still pulling itself out of the doldrums of the Great Depression.

It was still a time of horse artillery, of Sunday afternoon concerts by the band of the 17th Field Artillery directed by Warrant Officer Theo Bingert, of staff officers wearing white shirts and ties at work, of sports events, of autumn hunting trips.

When Germany invaded Poland in early September 1939 and the world braced for what seemed fated to become another world conflict, some of the senior officers and old noncoms at Bragg could still remember service in the last world war.

But there were some changes.

On Sept. 21, the entire post garrison held its usual "mass unit parade."

But instead of the customary "squads right" formation, the units marched in company blocks, the new "mass" standard for military parades.

At the time, the post complement included the understrength 17th Field Artillery Regiment, the 2nd Battalion of the 36th Field Artillery Regiment, and the headquarters of the 13th Field Artillery Brigade, which also doubled as the post commander's staff.

"Sound and Flash"

The 17th was armed with 155 mm cannon, the Army's newest. The two batteries of the 2nd Battalion of the 36th, formerly a battalion of the famous 5th Field Artillery Regiment, was "motorized" and fired the biggest of the Army's howitzers, the 155 mm and the monster 240 mm.

Another notable unit was the famous "pack artillery," the 2nd Battalion of the 83rd Field Artillery. It carried its disassembled 75 mm cannon on the backs of pack mules.

The most unusual unit on post was the "Sound and Flash" company, a pioneering artillery observation outfit equipped with the latest equipment for spotting enemy batteries.

At Pope Field, which had been modernized with hardened runways and new buildings in 1934-35, the Army Air Corps contingent consisted of a detachment of the 16th Observation Squadron, a complement of just more than 100 officers and enlisted men and a force of exactly four 0-16 aircraft.

The most colorful unit, and the outfit that got the most publicity, was the 2nd Balloon Squadron, with its 107-foot-long motorized lighter-than-air machines, a complement of 131 officers and men commanded by Maj. Neal Creighton, one of the Army's most enthusiastic devotees of the observation balloon.

As October came and German armies lashed across Poland, Fort Bragg was still a place of peacetime calm. The post's longtime cordial link with the courthouse-and-textile town of Fayetteville was epitomized on Oct. 5, the day when all Polish resistance collapsed before the whirlwind panzer invasion.

In crisp fall weather, officers of the post were guests for an evening aboard the Cape Fear River yacht "Fordell," a sleek white Florida-style vessel owned by Fayetteville businessman Oscar Breece. Among the guests were Majs. Paul Reichle and Creighton.

But soon changes more significant than "squads right" were under way.

On Oct. 13, the War Department announced the first expansion of the Army in response to the darkening situation in Europe.

Fort Bragg was to get 1,478 additional soldiers, with the 17th Artillery getting another gun battery and the 36th converted from a battalion to a two-battalion regiment. The 4th Field Artillery got another battery.

A light maintenance company was authorized for the 38th Quartermaster Regiment, as well as an enlarged medical detachment.

And presaging wartime Fort Bragg's role as a major center for black soldiers, the newly formed detachment of the 48th Quartermaster Regiment (Colored) was organized as three truck companies, the first black military unit in North Carolina since the Spanish-American War.

With these additions and others, Bragg was counting on a post complement of 140 officers and 6,200 enlisted men with a $2.5 million annual payroll.

The post employed more than 200 civilians with a $300,000 payroll, and paid $600,000 locally for forage for 300 animals and for coal and heating oil.

The expansion authorized in October 1939 would still be going on when the Germans struck across Holland and Belgium seven months later.

1940 Mobilization: Troops Flood In

Originally published August 11, 2005

For Fort Bragg in World War II, Aug. 1, 1940, was the beginning of the beginning.

Seventeen months before the shooting war started on Dec. 7, 1941, at Pearl Harbor, the small artillery post in the North Carolina Sandhills began a new role that, within a year, would make it the Army's largest place.

Officially designated as Mobilization Day, the typically steamy summer moment 65 years ago was when new units were officially constituted at Fort Bragg, principally the 9th Infantry Division, the 41st Engineers and several new anti-aircraft regiments, the latter to be composed of black enlisted men and white officers.

The mobilization was the nation's response to the gloomy news from Europe, where German armies had rolled through the Low Countries and France and left only Great Britain standing in the way of Adolf Hitler's world ambitions.

For a while, it looked like they were calling a war and nobody showed up.

A single officer arrived to open the headquarters of the 9th Infantry Division.

On that August day 65 years ago, six officers of the Regulars and 10 mobilized Reserve officers set up shop for the 76th Coast Artillery. Lt. Col. Harry R. Pierce was the commanding officer.

The first enlisted men to arrive were two veteran black noncoms, Cpls. Elias Board and John J. Moore.

They were from the most famous all-black unit in the peacetime army. They were Buffalo Soldiers of the 10th Cavalry Regiment, whose peacetime station at Fort Leavenworth, Kan., hearkened to the regiment's history as Indian fighters.

Board and Moore, however, were stationed with a small detachment of horsemen at the U.S. Military Academy at West Point.

The regimental history of the 76th Coast Artillery says they "arrived (at Fort Bragg) by private automobile on August 6, 1940."

Two officers, Maj. Logan Shutt and 1st Lt. Charles G. Young, set up headquarters for the first battalion of the 77th Coast Artillery.

The 41st Engineer General Service Regiment was activated by its first commander, Lt. Col. John E. Wood. The North Carolina native and much-respected engineer veteran was keenly aware that his new outfit would, in the words of his first anniversary letter to his troops, "occupy a unique position in the history of our regiment."

The 41st, Wood told his men, was the "first Regular Army Engineer Regiment to be composed of colored troops," and that it would be "the mother regiment of all future engineer regiments of colored soldiers."

Within the next few days, larger enlisted cadres arrived, including 101 men for the 41st Engineers, 106 for the 76th Coast Artillery and 103 for the 77th.

All were Regulars from the 9th and 10th Cavalry or from the peacetime Army's two understrength all-black infantry outfits, the 24th and 25th Infantry Regiments.

Rainy season

Mobilization Day in 1940 was memorable not only because of the military news. The weather also made headlines. A midsummer rainy season began on the very day that the Bragg buildup was to begin.

Rain fell for 14 straight days. The climax of the weather onslaught came Aug. 12, when Charleston, S.C., was battered by a hurricane that sent more torrents of water into the Sandhills.

As Fort Bragg scrambled to find room for new arrivals, the town of Fayetteville was busy in the same business.

By Aug. 14, Mayor Hector Clifton Blackwell reported that the long moribund housing industry was booming "at a furious pace."

Six single-family dwellings were under construction as well as one duplex, and 23 residents had signed up offering rental quarters. In another two weeks, the number of new houses under construction grew to 40. In three weeks, "homes have been found for 160 families," according to the Fayetteville Chamber of Commerce.

But that barely scratched the surface, because 500 more officers were on their way, and 200 of them were married and had families.

Another 400 homes would be needed for noncommissioned officers with families.

This unprecedented economic boom for the sleepy textile-and-courthouse town of Fayetteville ignited the inevitable nefarious practices.

The Fayetteville Observer was moved to send up a warning. An Aug. 2 editorial demanded "No Gouging, Please."

And a week later, the mayor issued a public statement, "a plea for reasonable rents."

School successes

Meanwhile, a Fayetteville institution established long before Fort Bragg, the "Local Negro College," reported a successful academic year.

The forerunner of Fayetteville State University said that 632 students had attended summer school, the largest number in its 75-year history.

A new building was ready for science classes, and a new dormitory for male students was occupied.

Dr. J.W. Seabrook, the college's longtime president, praised his faculty and listed his needs. The school needed five faculty cottages and a new kitchen. The state legislature had been asked for an appropriation of $123,000 for new faculty positions.

Finally, Seabrook pointed out some practical needs. He very much wanted money to hire a night watchman. And the college badly needed a pickup truck.

On the darker side of civilian life, Fayetteville police reported the breakup of a "gambling den over the New York Cafe on Hay Street."

The host and six "poker patrons" were nabbed in the raid and bound over to Mayor's Court.

Three weeks after Aug. 1, the Army announced that more than 9,000 soldiers had arrived at Fort Bragg for the new units.

There were 1,800 recruits of the 9th Division living in big, pyramidal Army tents, a vast canvas tent city sprawling for a mile along the road from the Army post to Fayetteville.

Black soldiers were in the old Civilian Military Training Camp area, also a tent community, on the north side of the post.

The beginning of the beginning had begun.

Bragg Transformed Itself In Just 9 Months

Originally published December 15, 2005

They weren't building the pyramids, but by any measure, the achievement was gargantuan.

In little more than three months in the fall of 1940, an army of 18,000 construction workers raised more than 2,000 structures, transforming Fort Bragg, the small artillery post in the Carolina Sandhills, into the nucleus of the Army's largest training center.

By mid-December, 12,000 soldiers who had been sleeping in tents were moving into wooden barracks that had not existed 30 days earlier.

A 1,600-bed medical complex — an array of 112 wards and other buildings connected by covered breezeways — was due for completion in a month.

Engineers were laying out the right of way for a road linking the post with Fayetteville.

The goal was to have 67,000 soldiers on the Army post by the spring of 1941, and by December the number already exceeded 15,000 — three times that of only the previous August. The population of Fort Bragg was officially larger than that of Fayetteville.

The figures were awesome, a story since told many times.

Yet to be explored is an adequate account of the human dimensions of the 18,000 workers, carpenters, plumbers and pipe fitters, whose round-the-clock labor achieved every goal.

Where did they come from? How did they live? What were their days like?

One description comes from the report of a government photographer assigned to capture images of the work force.

A caption accompanying one image reads: "Part of a trailer camp for migratory workers, 12 miles from Fayetteville, North Carolina, employed at Fort Bragg, North Carolina.

"Trailer space is one dollar a week, and small bunk houses five dollars a week.

"One water spigot near the general store serves the whole camp until 6 p.m.

"On the left is an old trolley car used as living quarters by a small family who came one hundred miles to work at the Fort.

"Some of the families here came from Texas, Idaho, Georgia, South Carolina and other parts of North Carolina."

Another measure of the living conditions was a list prepared by the embattled Fayetteville Chamber of Commerce for distribution to workers. It pointed out accommodations available in no less than 40 towns surrounding the post, some as far as 100 miles away.

The swarm of workers created massive traffic on the mostly rural highway network of Cumberland County.

Another photograph from the 1940 government album shows Gillespie Street in downtown Fayetteville lined bumper to bumper with 1930s flivvers and sedans as workers took to the roads.

The traffic was not only overwhelming, it was often deadly. In December 1940, construction workers were killed in crashes at unattended rural intersections. Four died in traffic accidents in October, including two workers.

Alongside the construction program, the military work went on at an ever-increasing tempo.

The post was the training site for the 9th Infantry Division, field artillery recruits, anti-aircraft artillery regiments, engineers and cooks and quartermaster troops and the recent civilians who arrived as draftees through the nation's first peacetime selective service program.

Cumberland County's draft board was appointed in early December, with Lacy McBryde as chairman.

The first home of the draftees on Fort Bragg was the Recruit Reception Center.

A December description by the post public relations office reported: "Completed in just seventy-five days, the Reception Center, commanded by Lieutenant Colonel Earle C. Ewert, was ready to process 1,000 men daily. This rapid rate construction would turn Fort Bragg into one of the largest military installations in the United States within nine months."

In the first weeks of December, the inductees showed up. Gov. Clyde Hoey of North Carolina came down from Raleigh to greet the first new soldiers from his state. As camera bulbs flashed, the governor met the 27 at the courthouse, where they were sworn in before heading for their new home at Fort Bragg.

Dallas McQueen Campbell of Elizabethtown was the first to arrive. Steve Little of Clarkton came next, followed by his brother Max.

A week later, the post reported that the Reception Center had welcomed 395 white and 130 black inductees, and that there was "no bottleneck" in the operation.

And while Fort Bragg was transformed, what was going on in Fayetteville?

Caught up in its new and unexpected role as the nation's premier Army town, the old courthouse-and-textile town argued over Sunday movies.

Pressed by military authorities to provide something besides pool halls for the weekend off-duty recreation of thousands of soldiers, the owners of three of the town's motion picture houses took the plunge and announced they would show movies on Sunday afternoons.

The doors opened at the Carolina, Broadway and State theaters, only to be caught up in a political tug of war between Mayor W.C. Blackwell and the city board of aldermen.

A police officer showed up to serve arrest warrants on the hired hands who operated the theaters.

The town ordinance seemed clear enough that Sunday movies were illegal. Blackwell, however, preferred to make a case of it, and because he was also the judge at the weekly Mayor's Court, he put off the cases of the arrested ticket-takers.

On the next Sunday, the scenario was repeated.

Meanwhile, mayor and aldermen negotiated in an effort to get out from between the rock of the military and the hard place of preachers and many churchgoers who were the bulwark of the town's blue law ordinances.

By the holiday season, the impasse was over. And as The Fayetteville Observer noted, "history was made" as motion pictures began rolling on Sunday afternoons.

"Tent City" Launched New Era

Originally published August 4, 1994

On Oct. 25, 1940, when 23-year-old James C. Smith left his Robeson County farm and reported for Army duty with the 9th Infantry Division at Fort Bragg, he checked in to "tent city."

For soldiers in the first big unit that trained at Fort Bragg and later went to fight in World War II, tent city was home during the fall and winter of 1940-41.

It was a sprawling matrix of big, pyramidal Army tents, a canvas camp that stretched for more than a mile along both sides of the two-lane road that led from the Army post to the town of Fayetteville.

Tent city had started to go up within days of the Aug. 1, 1940, order from the War Department activating the 9th Division.

The department had announced its plans on July 13, 1940.

The Associated Press reported that day:

"Fort Bragg, already the nation's largest artillery post, will become the base of a new 'streamlined' division under plans announced by the War Department."

Originally, the new 9th (a division of that number had fought in World War I) was to contain about 8,000 officers and men in infantry, artillery, engineer, signal, and medical units. By World War II, the division had 15,000 soldiers.

In the same 1940 list of new units planned for the rapidly expanding Army, the department said that Fort Bragg would also be the home for the 41st Engineer Regiment and the 1st Battalion of the 76th Coast Artillery Regiment, the latter an anti-aircraft outfit.

In the racially segregated Army of 1940, the 9th Infantry Division was to be all-white. All the enlisted men of the 41st Engineers and 76th Coast Artillery would be black. The officers were white.

The 1940 activation of the 9th Infantry Division at Fort Bragg was the opening chapter in the mobilization of the U.S. Army that was prompted by the war news from Europe, where German forces had defeated Holland, Belgium and France in a lightning summer campaign.

For Fort Bragg and the surrounding civilian community, it marked the entry into a new era.

A major buildup turned the former small artillery post into one of the world's largest military cantonments, with tens of thousands of men in uniform.

Down the road, the quiet courthouse-and-textile town of Fayetteville suddenly became the quintessential "Army town," the inadvertent weekend host to hoards of off-duty soldiers.

In early August 1940, however, neither post nor town was prepared for the change.

At Fort Bragg, barracks space was sufficient to house only 6,000 soldiers, mostly in artillery outfits.

The town of Fayetteville had a population of fewer than 15,000. Recreational facilities consisted of a few motion picture theaters and a sprinkling of restaurants and pool halls.

Ready or not, soon after Aug. 1, scores of Army officers and senior noncoms began arriving to form the "cadres" of the 9th Division and the other units.

In three weeks, "homes were found for 160 families," according to the Fayetteville Chamber of Commerce. But that wasn't enough because 500 more officers were on their way and 200 of them were married and had families, and 400 homes would be needed for noncommissioned officers with families.

At the same time, The Fayetteville Observer was reporting that "approximately 1,800 men of the new 9th Division are camped in a tent city on both sides of the old road to the post in the eastern end of the reservation."

Moreover, it reported, "colored troops comprising the 76th and 77th Coast Artillery regiments and the 31st Engineers are encamped in the old Civilian Military Training Camp site in tents."

By the time James Smith arrived as a recruit in October, joining other Robeson County recruits such as Felton Byrd and Gardner Wilkinson, each tent had been "improved" with wooden floors and had been equipped with the infamous Army oil-fired heating stove.

Smith recalled that one of the first orders was to grab a mattress tick and fill it with straw, an amenity meant to relieve the hardness of an Army cot.

The Robeson County men were first assigned to the 34th Field Artillery Battalion of the division. Among the young officers with the battalion were Capt. William Westmoreland, who a quarter-century later would command U.S. forces in Vietnam.

A barracks construction program that began in September 1940 would in nine months provide quarters to accommodate more than 70,000 soldiers and make Fort Bragg the largest Army post in the country.

But Smith and others who joined the first new units at Fort Bragg would stay under canvas until late February 1941.

By then, Smith had been transferred out of the 9th to train with a newfangled military outfit, the 609th Tank Destroyer Battalion.

He would go on to fight in Europe in the 3rd Army of Gen. George Patton. His unit, equipped with big guns mounted on halftrack vehicles, would participate in the fight around Bastogne during the Battle of the Bulge in December 1944. He became a sergeant and a gun commander. Today, he lives in a nursing home in St. Pauls.

His outfit, the 609th TD, will hold its reunion in Fayetteville at the Bordeaux Inn on Sept. 15-18. He expects to be there.

Felton Byrd went on to spend his entire World War II fighting service with the 9th Division.

For the 9th, it was a long war. The division left Fort Bragg two years after its activation to take part in the invasion of North Africa in November 1942.

Thereafter, it fought in Sicily, France, the Ardennes and Germany, earning Gen. Dwight Eisenhower's assessment that it "rated second only to the 1st Division" as a combat unit in Europe. Nearly 4,500 men who fought as "Old Reliables" in World War II were killed in action or died of wounds.

Fifty-four years after the summer it began under the canvas of "tent city" at Fort Bragg, the 9th lives on only in the history books and in the many artifacts of the division museum at Fort Lewis, Wash.

For Fort Bragg and Fayetteville, the 9th, along with the 41st Engineers and the 76th and 77th Coast Artillery, represents the opening of a momentous chapter of change in the history of post and community.

Cannon-ized: Artillery Training Center Gave Post Its World War II Role

Originally published May 18, 2006

Fort Bragg started as a field artillery post and for 20 years was home to a few hundred soldiers serving a few dozen guns.

In World War II, Fort Bragg became the largest field artillery post in the world. More than a third of the quarter-million field artillerymen in World War II learned their gunnery skills at Bragg's Field Artillery Replacement Training Center.

The sprawling center opened for business in the spring of 1941.

Sixty-five years ago in March, a "faculty" of 2,000 artillery instructors, officers and noncommissioned officers drawn from all over the fast-growing Army welcomed more than 10,000 newly minted soldiers, most of them draftees.

Lt. Col. Edwin P. Parker Jr. was the center's first commanding officer. A Virginian who grew up in Wilmington, he was married to Fayetteville native Mary Hardin and was kin to several old Cumberland County families.

Ground was broken on the center's 500-plus buildings on Dec. 16, 1940, and it took only 98 days to complete them. Since then, the site has been home to the 1st Corps Support Command and to medical units.

Stretching for more than a mile on the east side of Bragg Boulevard — from Randolph Street to Gruber Road — the center was laid out like a giant parade ground formation, with five regimental blocks of 75 buildings each, six buildings wide and 13 deep.

At the east end of the axis of four blocks, the center's headquarters building faced down the axis. From its second-floor balcony, the commanding officer could take in the entire center.

Streets in the area today have names such as Watson, Goldberg, Medic and Service. In 1941, the long north-south streets were simply numbered. The connector streets had alphabetical designations.

With its look-alike rows of barracks — punctuated by mess halls, offices, classroom buildings, clubs, chapels, sheds and warehouses — the center was the ultimate in military town planning.

It was the largest of three such centers where the Army's World War II field artillerymen trained.

Sixty-five years later, you can grasp the size of the center by cruising around the area, although modern buildings have largely replaced the white wooden barracks.

Each of the five regiments concentrated on particular weapons in the artillery inventory. Two trained on the basic 105 mm cannon. Two fired the brand-new 155 mm gun — known in World War II as the "Long Tom" — and the biggest gun of them all, the tractor-drawn 240 mm howitzer.

A fifth regiment (officially the Fourth Training Regiment) trained support troops and had separate battalions for signal communications, cooks, clerks, survey teams, radio operators, auto mechanics and repair for both "gun and general."

In the racially segregated World War II Army, black artillerymen trained at the center but lived in a separate battalion area. Their unit was designated the 16th Battalion.

The best description of what a new soldier could expect at the Field Artillery Replacement Training Center was the advice in a little red-backed book prepared by then-Lt. Temple Fielding, a young officer who after the war pioneered the tour guidebook publishing industry.

This was not high-level strategic information or a field manual about weapons. Instead, the pamphlet was a breezy compendium of word pictures describing what days and nights would be like for a recruit at Fort Bragg.

The information was practical, detailed and written in an informal style. It covered everything from the difference between a battery, a battalion and a regiment, to where to do your laundry, how to check in at a prophylactic station after frequenting a red light district, and how to get a taxi or a bus. The bus cost 20 cents for the trip to Fayetteville.

The guide said: "The place you have come to is the largest of its kind in the Army. The main idea here is to teach you how to soldier. You'll stay for several weeks.

"You'll be taught to drill, to salute, to fire a rifle. ... You'll learn a specialty and learn it well. You're here because it's the best place to train you."

A month after training began, thousands of Field Artillery Replacement Training Center soldiers got their first pass time and most headed for their initial taste of off-duty life in Fayetteville.

The result of this early experience — for the soldiers and the town — can pretty well be deduced from this item in a mimeographed newsletter from the commanding officer of the Fourth Training Regiment.

"Hundreds of selectees were shipped back from town last weekend for being improperly dressed. Be sure you're in proper uniform and that you have a pass with you before you leave the Post.

"Give the MPs in town a break. They're only human. Walk facing the traffic on the highway, and DON'T hitchhike. And for Pete's sake, don't whistle at the gals. There's been too much of that stuff and it's got to stop. More Does and Don'ts later."

Fort Bragg's training role expanded rapidly during the war years with the arrival of several divisions, including the 82nd Airborne, and other field artillery groups.

After World War II, the War Department ordered that the training center handle overflows from other replacement centers, and the center would no longer be primarily for the field artillery, The Fayetteville Observer reported on July 15, 1946.

Young People Get A Chance To Work

Originally published June 1, 2006

In 1940, you had to have a guide to keep up with them.

There was the WPA, and the NYA, as well as the CCC.

They were the "alphabet agencies," government organizations created by President Franklin D. Roosevelt to give folks a way to fight the Great Depression.

When the Army mobilized to expand Fort Bragg into a military city in 1941-42, it called on all three.

The WPA, the Works Progress Administration, was a builder. It paid for crews whose work dotted the American landscape with school bridges, irrigation ditches, hospitals, roads, dams and forest trails.

For the Army at Fort Bragg, the WPA underwrote the project that expanded the highway between the post and the small city of Fayetteville seven miles away. We've known it since 1941 as Bragg Boulevard.

The CCC, the Civilian Conservation Corps, was the Depression-era savior of thousands of men, mostly in their 20s and 30s, who signed up for rugged camp life and work in forests, on beaches and in mountains.

Peacetime Fort Bragg was a center for CCC activities. The Army post became the command center for 34 CCC "camps" established throughout North and South Carolina where more than 34,000 enrollees were housed, fed and directed in their projects.

Young Army officers got hands-on experience in organizational skills, planning, logistics and command as operators of CCC camps.

When the war came, thousands of CCC "boys," as Roosevelt called them, put on uniforms and used their camp experience to become soldiers.

Hundreds of Army officers from CCC offices used their experience to become top-level wartime soldiers.

And then there was the alphabet agency spotlighted in this report. Less known among the agencies but vital to the Army days of the Fort Bragg expansion was the NYA, the National Youth Administration.

The NYA was created to give white-collar work and promise to teenagers whose Depression-era employment prospects were bleak.

In the spring of 1941, the organization was sending young people to Fort Bragg as civilian workers at the desks and behind the counters in the proliferating offices and service buildings serving the tens of thousands of men in uniform.

That's how 18-year-old Louise Gibson (later Louise Griffin) got there in March 1941.

Here is her story, in her own words:

"I believe there are lots of us still around who would enjoy reading about our service to the military before and during the war. And other readers would like to know what we were doing.

"We were teenagers, high school graduates, looking for jobs.

"I began training under the NYA program on March 18, 1941.

"Five girls from the special green bus transporting NYA youth from Fayetteville were selected for the telephone office at the 9th Division. The 9th was the largest outfit in training as Fort Bragg expanded to its wartime size.

"They told us we were the first female telephone operators on a U.S. military post and it was an experiment to find out if girls would be able to relieve the soldiers for other training!

"The boys didn't think we would last. They thought girls couldn't learn Army lingo.

"The NYA trainees started out being switched to other jobs on the post every two weeks. But soon, we same four girls (the fifth one left for employment elsewhere) were on the 9th Division PBX to stay.

"My diary says 'Lt. Bond chose me to be the leader.'

"Later we were transferred to Main Post Signal Building. The other girls training with me at the 9th Division were Margaret West, Elizabeth Adams and Opal Lewis.

"The visual map of my beloved 9th Division had a special place in my heart for many years.

"My diary is incomplete, skipped months at the time. But in late summer of 1941 we had many more girls on army switchboards scattered all over the post and we became civil service employees at $105 a month. There was a picture and small write-up about us in the Fayetteville Observer at the time."

Fateful Sunday

The work week included Sunday, and Griffin used the 30-minute training period between shifts to organize Sunday school classes and a singing session.

And so it was on Sunday, Dec. 7, 1941, that she was at her switchboard when radios began broadcasting from far-off Hawaii. Japanese planes were bombing the Navy base at Pearl Harbor. World War II began for the United States.

She was at the switchboard from that morning until the next afternoon.

"After Carolina Telephone Co. took over part of the telephone communications at Fort Bragg, I transferred to their commercial department at Rocky Mount and 'left home'. I had picked up an odd strain of malaria while working on post, but it was not diagnosed as malaria until mid-1944. In the meantime, I had to take sick leave from the telephone company and finally resigned and recovered at home."

More pioneering work

That was the end of her telephone story. But Louise Gibson Griffin went on to another form of pioneering wartime work for a women.

She tells it:

"There have been many stories written about women replacing men in the factories and elsewhere during the war. And I am also one of them.

"Burlington Mills considered their cost clerk position a male job. It was a general consensus that females were not apt at mathematics.

"I had to be tested, and I didn't record what tests they were, but was told that my results were the second highest in the state.

"Their surprise was evident, but I really don't remember much about it.

"I have a strong faith and am sure God gifted me for the work He chose for me. I was cost clerk at the Lakedale plant (in Massey Hill) only about two years and resigned for further education."

She concludes:

"I don't expect a story about me. It's just an example of how things were back then."

She might not think it was much.

But when a pioneer tells her story, history ought to sit up and take notice.

A High Note During Troubled Times

Originally published August 31, 2006

The month of August 1941 was a time of triumph and tragedy for black soldiers at Fort Bragg.

In that hot summer 65 years ago, the best-known unit of black men in the entire U.S. Army — the 41st Engineer General Service Regiment — celebrated its first birthday.

The "Singing Engineers" got national attention when featured in a Life magazine spread profiling the explosive expansion of Fort Bragg, which in a year grew from a small peacetime artillery post into the largest training area in the country.

A few days later, the goodwill of the birthday was dampened by violent tragedy, which heightened racial tensions that were lying close to the surface that summer.

It happened on a bus loading in downtown Fayetteville. A shouting-and-shoving match between white military policemen and black soldiers erupted into deadly gunplay.

A black private from the 76th Coast Artillery grabbed the pistol of a 20-year-old MP and fatally shot him in the heart. Several other shots were fired, including one that killed the 27-year-old black draftee from Asheboro who had started the shooting.

And then, as was often the case in the racially segregated armed forces of 1941, the post provost marshal overreacted.

He ordered a roundup of black soldiers who were not already in barracks, including some of the 41st Engineers, and confined several hundred to a stockade next to the post guardhouse for several hours.

The particular officer was later relieved of his command.

But tensions were high and rumors flew, including one from a radio station in Raleigh alleging that black soldiers at Fort Bragg were to be moved "nine miles from town."

A few days later, Brig. Gen. Benjamin O. Davis, the only black general officer in the Army, visited the black troops to hear their comments and complaints.

It was a War Department response to the growing record of racial discrimination and incidents on posts across the country.

As the headlines of racial violence faded, men of the 41st packed up for a historic role as the Army prepared for World War II.

On Aug. 13, 1941, the 41st was on the move to a temporary home in a tent camp on the outskirts of the little town of Wadesboro.

The engineers were the vanguard of tens of thousands of troops who would play at war in the Carolina Maneuvers, a vast training exercise that pitted "Blue" and "Red" opposing forces in a test of battle skills across hundreds of miles of Carolina landscape.

The main "battle" would not take place until November, and it would hardly end before the U.S. was in a real war after the Japanese attack on Pearl Harbor on Dec. 7, 1941.

The Singing Engineers were first on the ground in the maneuver area. They were responsible for building and maintaining miles of Army-style infrastructure, especially temporary bridges so military vehicles could get around the battlefields.

The men of the 41st quickly gained more publicity when journalists covering war games got a taste of the music that prompted the regiment's nickname.

A publicity shot by an Army photographer depicted the 41st band in formation with their tent camp in the background. They were nattily dressed in crisp summer khaki, leggings and floppy fatigue caps.

In those days before television, band music was the country's popular entertainment, and the 41st band as well as its smaller dance orchestra left lasting memories by performing in small towns all over the maneuver area.

The Army publicity noted that nine members of the band were "formerly members of leading colored dance orchestras," including bands of Cab Calloway and Duke Ellington.

Music became an early and widely acclaimed specialty of the regiment.

The regimental band was informally organized within weeks after the 41st was constituted Aug. 1, 1940. As early as September 1940, a picture in the regimental history depicts "Sgt. (Frank) Ruffin leading field music at retreat" on Fort Bragg's Main Post Parade Field.

Ranked behind Ruffin's saluting figure is a smartly moving drum and bugle team.

Ruffin was a veteran military musician. He had come to Fort Bragg from the all-black 24th Infantry Regiment at Fort Benning, Ga., where he was a member of the regimental band known throughout the Army.

Formally activated in November 1940, the band was led then by an already highly respected military bandmaster, Warrant Officer John Jordan Brice (1900-1948).

For 15 years after serving overseas in World War I, Brice had directed the band of the ROTC unit on the country's premier black campus, Howard University in Washington.

In those years, he was a frequent guest conductor of the U.S. Army Band.

Brice and the chaplain, Capt. James R.C. Penn, were the only black men who were commissioned officers in the 41st.

The regiment, like all units of black soldiers, was led by whites.

Lt. Col. John E. Wood, a North Carolina native and West Point graduate, was keenly aware of his unit's pioneering role as one of the largest all-black commands in the mobilizing Army of 1941.

Wood, a much-respected engineer veteran, wrote in an anniversary letter to his troops: "You occupy a unique position in the history of our regiment."

The 41st, Wood told his men, was the first Regular Army engineer regiment of colored troops, and it was destined to be "the mother regiment of all future engineer regiments of colored soldiers."

Brice's band was the crown jewel of the outfit and quickly became esteemed throughout the Army. Brice composed music for a regimental march and coached the whole regiment in singing in unison.

Wood proudly noted in his anniversary letter that the 41st was "the first unit to have official permission to sing its regimental song while marching at attention."

The Life article emphasized the singing, quoting words of the regimental march and describing the repertoire as "spirituals, old marching songs, patriotic and Southern tunes."

Brice found new musical fame at Fort Bragg. And he found a wife in Fayetteville.

He married Malissa McNeill, daughter and granddaughter of the town's first undertakers.

"Mallie" McNeill was a teacher with a graduate degree from Boston University. Today she lives in retirement in suburban Detroit, where she was a teacher and a civic and church leader.

In the summer of 1941, she and other 41st wives proudly watched as the Singing Engineers pulled out for their pioneering role in the war effort of 65 years ago.

60 Years Ago, Bragg Went Airborne

Originally published November 15, 2001

Every paratrooper, and many who never jumped from an airplane, know the famous events of airborne history.

Like Normandy or Holland. Like Bastogne or the Rakkassans over Korea.

But who knows what happened 60 years ago this month?

On Nov. 19, 1941, to be precise?

Not to keep you in suspense.

Nov. 19 is the 60th anniversary of the coming together of Fort Bragg and the airborne soldier.

On that day, less than three weeks before the Japanese attack on Pearl Harbor brought the U.S. into World War II, a battalion of paratroopers jumped from 42 transport airplanes onto Pope Field.

It was a demonstration staged during the huge war games of 1941 known as the Carolina Maneuvers.

In the games, two opposing "armies," the Red and the Blue, tested the country's growing military might, including the fledgling parachute forces.

The men who jumped on Nov. 19, 1941, were the first airborne soldiers ever to see the place that would wind up being "the Home of the Airborne."

Here is the event, as related in an unsigned front-page story in The Fayetteville Observer:

"Skies over Pope Field were white with parachutes of Uncle Sam's newest and most spectacular fighting unit this morning as the Red Fourth Army Corps sought to capture the biggest air base of the Blues.

"But the ranks of the parachutists were theoretically decimated by the fire of machine guns with which Col. Neal Creighton had ringed the big airfield.

"And then his blue-dungareed defenders staged a sensational charge through a smoke screen to capture the survivors of the Air Task troops.

"Umpire Col. Rook ruled that the attack was a failure, but that forty percent of the planes which had been based at Pope Field were destroyed in the aerial bombardment which preceded the descent of the parachutists.

"Descent of the parachutists was at once one of the most beautiful and one of the most terrifying events to which the unprecedented war games in this vicinity have given birth.

"It was preceded by an intensive air bombardment.

"Then 42 huge transport planes flying in waves of from three to six suddenly appeared over the field and commenced spilling hardened men of the 502nd Parachute Battalion commanded by Major George Howell.

"Hundreds of snow white chutes blossomed instantaneously as gun and rifle fire cracked from the defenders. It seemed that they fell interminably, but as a matter of fact they were all out and down in the matter of a few minutes, 360 men, 36 officers, and 10 umpires.

"As they fell, the attacking planes tossed out bundles of ammunition and guns which were supported by parachutes of different colors.

"The men on the ground rallied around these as smoke pots belched billows of blackness in simulation of smoke bombs.

"Later, the planes returned with three or four more Red troops on the presumption that the field was captured.

"Both the parachutists and the air-borne troops were Air Task troops under the command of Col. William C. Lee."

Successful mission

Umpires were more kindly toward the second and third jumps of Lee's command during the war games.

On Nov. 26, 1941, a unit of 119 men of the 502nd, this time acting for the "Blue" army, jumped near Camden, S.C., and successfully blocked roads.

The commanding officer set an airborne tradition by issuing a terse report: "Our mission was accomplished."

On Nov. 27, another batch of 502nd paratroopers "captured" an airfield at Maxton.

On that same day, a flight of heavy bombers theoretically destroyed Creighton's defenses at Pope Field that had been rated so highly on Nov. 19.

The jump at Pope Field got national attention only a week before Pearl Harbor, when Life magazine in early December used a full-page image of the "blossoming" parachutes as its popular "Picture of the Week."

The caption read: "In a sudden blizzard of parachutes, 400 Red soldiers drop in on a Blue Carolina airfield."

By then, Lee, "the Father of the Airborne," was back at Fort Benning, Ga., then home of his fast-growing force.

Despite the low opinion of the umpires on Nov. 19, the airborne had enthusiastic champions, especially Gen. George C. Marshall, the Army chief of staff, and Lt. Gen. Henry "Hap" Arnold, head of the Army Air Corps, who personally witnessed the jump.

By the summer of 1942, two divisions of the Army were "airborne," the 82nd and the 101st. The 502nd battalion went on to become a highly decorated unit of the 101st Airborne Division in World War II, jumping in Normandy, fighting in Holland, battling at Bastogne.

Known as the "Falcons," the unit is now a brigade of the division at Fort Campbell, Ky., with a history that includes service in Vietnam and the Gulf War.

The moment 60 years ago when airborne and Fort Bragg first met gets little mention in books of airborne history. But for the paratroopers, and for the place where airborne now lives, Nov. 19, 1941, has to be a defining moment in that history.

As in so much of paratrooper history, "Bill" Lee was prophetic about the coming together of jumpers and the place. Two days after the history-making jump, the Observer reported:

"Before leaving, Col. Lee told Col. Neal Creighton, commander of the air base, that Pope Field and Fort Bragg were ideal spots for parachute troops to work."

Indeed.

Japan Fails To Crash City's Bash

Originally published December 6, 2001

It was planned as the biggest party Fayetteville had ever seen.

The date set was Wednesday, Dec. 10, 1941.

Sixty years ago, the city was gathering the stuff for a bash welcoming at least 20,000 guests.

The guests were to be the soldiers who would be spending the holiday season at Fort Bragg.

Welcome Neighbor Day, under the chairmanship of A.G. Murchison, would feature food, free movies, music and a street dance on Hay Street.

The menu, as promised in the colorful language of reporter Bob Gray of The Fayetteville Observer, would include "15,000 pounds of hog meat, sufficient to make a sausage three miles long, 12,000 pounds of chicken, enough to scratch up all the gardens and flowers in Fayetteville."

Churches would be "wide open," and anybody in an Army uniform could walk right into the movie theaters, no ticket needed.

The motto, Murchison said, would be: "A soldier's money is no good today."

Columnist Allene Moffitt of the Observer said there had been a street dance way back in 1925 when the American Legion held a state convention in Fayetteville.

"But let's forget that one," she continued, "and think about the one coming Dec. 10. You can bet your frog skin it will be a hot time in the old town."

Tom Sutton, chairman of the dance, reported there would be "3,000 pretty girls from Fayetteville, Raeford, Red Springs, Lumberton and all around the section."

In addition, the program called for music from bands, glee clubs and orchestras comprised of soldiers from the post that by December 1941 was the largest military training place in the country.

Fort Bragg had more than 60,000 men crowding a sprawling city of barracks that had not even existed 12 months earlier.

Many of the troops had just returned to the post after participating in the Army's largest peacetime training exercise, the Carolina Maneuvers. Many of them were anticipating holiday furloughs to spend Christmas with families across the country.

And so, on Sunday, Dec. 7, 1941, people in Fayetteville and the troops at Fort Bragg were taking it easy, enjoying an unseasonably warm late autumn day, looking forward to a week of fun and fellowship.

As civilians began gathering for church services, the first bus loads of soldiers on weekend passes arrived in downtown Fayetteville, and many headed for the movie theaters that opened at 9:30 a.m. to accommodate the weekend crowds.

At the post information office on Fort Bragg, a young soldier-newsman, Frank Jeter, was checking the wire machines for weekend news.

Sometime between 1 and 2 o'clock, the wire machine clattered with news that changed everybody's plans.

Japanese planes were bombing the U.S. naval base at Pearl Harbor on Oahu in the Hawaiian Islands. The United States was in World War II.

Quickly the word was spreading around the post, and official actions were already being taken. Shortly after, military police were posting notices in downtown Fayetteville for off-duty soldiers to "return to your barrack at once."

Motion pictures were interrupted to bring the news and the notices.

Within an hour, hundreds of soldiers were milling around the bus station trying to get back to Fort Bragg.

Variations on a theme

A knot of 60 men in uniform gathered around the Market House two blocks away, and according to the Observer, "began singing variations of 'California, Here We Come.' "

Variations included such themes as "Yokohama, Here I Come" and "Fujiyama, Here I Come."

By early afternoon, the new post commanding officer, Col. John T. Kennedy, had mobilized garrison forces to guard "strategic points" on the post.

He met with the commanding officers of major units, the 9th Infantry Division, the Field Artillery Replacement Training Center and the 34th Coast Artillery (Anti-Aircraft) Brigade, and made plans for blackout drills. He reported that "all critical areas are well-guarded."

By then, too, the word was spreading among civilians. As they sat on porches in the unseasonably warm weather digesting their Sunday dinners, radios blared the news.

At some houses on Haymount, radios were brought out on porches, and off-duty soldiers who had been heading for the bus station downtown were hailed and invited to listen to the latest bulletins.

By evening, everyone knew that the country was at war and that Fayetteville and Fort Bragg were destined to play a big role in the conflict.

By Monday afternoon, the first units from Fort Bragg called to actual wartime duty were boarding trains.

They were the soldiers of the three regiments of the 34th Coast Artillery (Anti-Aircraft) Brigade, outfits formed in February from troops called in a year ago in the first wave of peacetime mobilization.

Two of the regiments were composed of black enlisted men and white officers.

They were the 76th Coast Artillery (Anti-Aircraft), which was put on trains for Philadelphia with orders to set up its batteries of rapid-firing anti-aircraft guns around naval installations there, and the 77th, dispatched to Wilmington, Del., for similar duty.

The white 67th was posted to Mitchell Field, the big Army air base on Long Island.

Other small contingents of Fort Bragg soldiers were quietly trucked to several National Guard armories in North Carolina as temporary guards for military gear stored there.

In the midst of the bustle, did they forget the party?

No, even as the anti-aircraft units shipped out, Kennedy notified the Welcome Neighbor committee to go on with the festivities.

The Observer reported on Dec. 10, 1941, that "thousands of Fort Bragg soldiers and Fayetteville citizens jammed Hay Street for the mammoth Welcome Neighbor program." A photograph depicted a shoulder-to-shoulder crowd squeezed into the three blocks west of the Market House.

The festive day would be remembered.

In the week before Christmas in 1941, the disaster at Pearl Harbor came home to Cumberland County.

In their residence at a small neighborhood of Army families in Spring Lake, Master Sgt. and Mrs. John Hochstrosses received a telegram.

It was from the Navy Department.

It confirmed that their son, Navy Seaman Johnny Hochstrosses, 19, a former student at Fayetteville High School, had been killed in action at Pearl Harbor on Dec. 7, 1941, one of the 2,400 sailors, soldiers and Marines who died 60 years ago tomorrow on "a date which will live in infamy."

USS Dobbin Joins Fight At Pearl Harbor

Originally published December 7, 2006

The historical marker for James C. Dobbin on Raeford Road in Fayetteville tells of a 19th-century native son who was secretary of the Navy.

Another historical plaque halfway around the world tells of another Dobbin, the USS Dobbin, a destroyer tender of the U.S. Navy and listed among the hero vessels at Pearl Harbor when Japanese planes struck on that fateful Sunday morning, Dec. 7, 1941.

The Pearl Harbor Dobbin was, of course, named for the Fayetteville Dobbin.

Sixty-five years ago today, the USS Dobbin was tending her "nest" of five destroyers at moorings a few hundred yards off the northeast corner of Pearl Harbor's Ford Island.

Just around the corner of the island — which forms a rough bull's-eye in the Pearl Harbor roadstead — the entire south shore was lined with battleships of the U.S. Pacific Fleet. The nearest of the great battlewagons to the Dobbin was the USS Arizona.

On that sunny morning, the Dobbin's crew was busy as usual meeting the demands of the sleek fighting ships of her nest. The bulky tender, more than 480 feet long, was a veritable floating supply and repair "mother ship."

There was even a Navy band on board to provide live music for the crews of the small ships. On Sunday mornings the band turned out for the hoisting of the national ensign by playing the "Star-Spangled Banner."

On Dec. 7, 1941, the music was interrupted abruptly by the noise of bombs raining down on the battleships just across the low-lying north end of Ford Island.

Veterans of that day recalled that the concussion of the huge explosion that shattered the Arizona "knocked several men to the deck." The Naval Historical Center's summary of vessels at Pearl Harbor tells Dobbin's story in brief:

"Dive bombers singled out this nest, and fragments from near misses killed three men and wounded several others on board the tender.

"Concentrated antiaircraft fire from Dobbin and the destroyers broke up a second attack. Throughout the attack, Dobbin's boats plied the waters off the harbor, rescuing survivors from burning and sinking ships."

Cmdr. H.E. Paddock, captain of the Dobbin, filed this after-action report:

"At 0910 three enemy planes, identified by yellow disks painted on their wings, attacked. They came in low approaching from the starboard quarter.

"The heavy fire from this ship and from the destroyers of Division One alongside, caused the plane to swerve and cross just astern.

"Three bombs were dropped, resulting in near misses on the starboard quarter, astern and port quarter. They appeared to be 300-pound bombs.

"Fragments from these bombs struck the stern of the ship, causing the personnel and material damage described below.

"All personnel casualties were members of the Number 4 three-inch antiaircraft gun located on the after end of the boat deck. No other attack was experienced by this nest of ships.

"The personnel of Number 4 AA gun were badly hit. But in spite of this, Coxswain H.A. Simpson, U.S. Navy, in charge, reorganized his crew, got them in hand, and continued to fight against planes which approached within gun range."

The men of the Dobbin whose names are among the more than 2,000 killed in action on that day are Coxswain Fed H. Carter, who died instantly when bomb fragments struck his antiaircraft gun position; Torpedoman's Mate 3rd Class J.W. Baker; and Fireman 1st Class Roy Arthur Gross.

Another Dobbin crewman, Gunners Mate 1st Class Andrew Michael Marze, was killed in action by a bomb blast on board the battleship Pennsylvania "while assisting in effort to repeal air attack."

Marze, Baker and Gross are buried in the vast National Memorial Cemetery of the Pacific in Honolulu overlooking the place where they died.

While the Dobbin's own battle history doesn't mention it, other ship commanders in the destroyer nest credited Dobbin antiaircraft fire with shooting down a Japanese dive bomber.

For its fight that day the ship received the Pearl Harbor battle star. The combat performance of the men of the Dobbin's 14 gun crews was exemplary.

But just as awesome was the matter-of-fact way that even as Japanese planes buzzed overhead, the 400 other crewmen went about their "mother ship" duties, the work of serving the vessels under its watch — hooking them up to electricity, arming them, helping fire up boilers and assembling machinery.

One of the five destroyers in the Dobbin's nest got under way within 10 minutes of the air attack that struck the Dobbin's stern.

In his after-action report, Paddock described the performance with undisguised pride:

"All boats of the Dobbin were sent into the landings immediately and have been in continuous service where ever needs ever since.

"The ships alongside started clearing at 0920, the last one getting under way at 1450. Repair personnel from this ship were turned to continuously during this period and during the attack to assist them in getting machinery reassembled. Dobbin issued replacement ammunition to destroyers alongside so that they left filled up.

"The commanding officer is pleased to report that the performance of all hands was excellent and the conduct most commendable, characterized by a strong will to fight and to turn to enthusiastically in any and every way possible to assist in servicing the destroyers and in protecting their own ship."

Destroyer commanders uniformly praised the crew of their tender. One reported "a single example."

"A Dobbin welder was over the side welding up air ports when a bomb hit 25 yards from his post. Not for a moment did he stop welding."

The USS Dobbin would soon leave Pearl Harbor for a long World War II history of serving destroyers in Australia, New Guinea and Subic Bay on Luzon in the Philippines.

She sailed from there to San Diego in the autumn of 1945, arriving four years to the day after her momentous role at Pearl Harbor.

Ten months later, in September 1946, she was decommissioned and left the fleet, ending a career 22 years and one month after she first flew her commission pennant on July 23, 1924.

War Mobilization Crowds Out Celebrations

Originally published December 21, 2006

Pearl Harbor was a surprise.

But 65 years ago, Fort Bragg and Fayetteville went to a war footing only hours after Japanese planes bombed U.S. military installations on Oahu in the Hawaiian Islands on Dec. 7, 1941.

At Fort Bragg, thousands of troops were looking forward to a holiday trip home. For many, leaves were canceled or postponed.

For a couple of weeks, the War Department was uncertain about just what to do with nearly a million men in the draftee Army, many of whom were expecting to get time off for the holidays.

By New Year's Day, the plans had been sorted out and thousands did finally get on the road home. For many it would be their last holiday home before World War II was over.

But not everybody went home.

Barely 24 hours after the news of Pearl Harbor, on the afternoon of Dec. 8, 1941, the first units from Fort Bragg called to wartime duty were boarding trains.

They were soldiers of the three regiments of the 34th Coast Artillery (Anti-Aircraft) Brigade. The units were formed in February from troops called in a year ago in the first wave of peacetime mobilization.

Two of the regiments were composed of black enlisted men and white officers.

The 76th Coast Artillery (AA) Regiment was going to Philadelphia with orders to set up its batteries of rapid-firing anti-aircraft guns around naval installations there. The 77th was on its way to Wilmington, Del., for similar duty.

The white 67th was posted to Mitchell Field, the big Army air base on Long Island.

Few of the soldiers of the anti-aircraft units ever saw Fort Bragg again in World War II.

The African-American battalions wound up in the Pacific theater, and many of the soldiers wound up as stevedores at remote island supply depots.

The white troops of the 67th served in the European theater in North Africa and Italy.

Wartime brought changes to Pope Field. In the week after Pearl Harbor, a new commanding officer, Maj. Sam Price, took over from Capt. Ward "Crash" Robinson, a World War I veteran.

Robinson had been called out of retirement from his home in Spring Lake to hold the fort after the departure in late November of Maj. Neal Creighton, the energetic commanding officer of the field's major tenant, a squadron of observation balloons.

Within days of Pearl Harbor, the observation balloon was dropped from the Army's arsenal. Pope Field suddenly found itself home base of a squadron of the Army's largest aircraft, the B-18 Bolo bomber, forerunner of the most famous bomber of World War II, the B-17 Flying Fortress.

In response to Pearl Harbor, the Air Corps was dispersing planes to previously unused fields up and down the East Coast.

Like the swift dispatch of anti-aircraft units, the move was partly in response to worries about sabotage.

The big planes of the 29th Bombardment Squadron also were tasked with anti-submarine patrols off the Carolina coast.

In the latter years of World War II, they would use these early experiences as a major anti-submarine force in the Caribbean and later in the Pacific.

In the second week after Pearl Harbor, Fayetteville's civilians were busy mobilizing for war.

First, the schools held their first wartime air-raid drill.

At the junior high school, several hundred children took only 90 seconds to move from their classrooms to the basement.

At the new white high school on Robeson Street, 400 students took a minute and a half to make the move.

Meanwhile, Cumberland County set up a Defense Council and called for volunteers to serve in an array of civil-defense tasks.

Committees were named for fire protection, public safety, emergency medical services, public utilities and "air raid precautions."

More than 300 people signed up for the committees.

The Fayetteville Fire Department took on a dozen new volunteer firemen.

The local chapter of the American Red Cross started a fund drive to raise its quota of $12,000 for war relief.

A few days later, a full-scale air-raid warning drill was judged to be much more effective than a drill held before Pearl Harbor.

In that early exercise, when a Hay Street merchant left his store lights on, several off-duty soldiers broke his window and put out the light themselves.

While Fort Bragg, Pope, and the civilian community coped at home, the "war news" told of momentous events in faraway places.

Japanese forces seized Wake Island, invaded the Philippines, swept through Malaysia and captured Hong Kong.

On the day after Christmas, the sad cost of war came home to Cumberland County.

The Fayetteville Observer reported that 19-year-old Navy Seaman Johnny Hochstrosses, son of Master Sgt. and Mrs. John Hochstrosses of Spring Lake, was killed in action at Pearl Harbor on Dec. 7, 1941.

1942 Finds Post Preparing For Battle

Originally published February 28, 2002

The first months of 1942 found Fort Bragg and Fayetteville joining the rest of the country following events of a World War II in which the United States was now a fighting part.

In these first weeks after the Japanese attack on Pearl Harbor on Dec. 7, 1941, some troops from the big Army post, mostly black anti-aircraft units, left for war duty, guarding military installations in Philadelphia and New York.

But in the winter months 60 years ago, the tens of thousands of soldiers at Fort Bragg were still in training.

Just like the civilians of Fayetteville, for these soldiers the war was still only in the headlines as Japanese forces overran the Philippines and German armies won victories in North Africa.

In January, a major piece of Bragg's training, the Field Artillery Replacement Training Center, celebrated its first birthday.

From a cadre of 27 officers in early 1941, the center was a city of thousands of soldiers and hundreds of barracks which sprawled for more than two miles on the east side of Bragg

Boulevard.

Soon after its birthday, the center's original commanding officer, then Lt. Col. Edwin P. Parker Jr., by now a one-star brigadier general, got word that he was receiving his reward for successfully birthing the center.

He was promoted to two-star major general and given command of the newly formed 78th Infantry Division.

Entertainment

For most of the year before Pearl Harbor, the thousands of fledgling soldiers who poured into Fayetteville on weekend passes looking for recreation could hope for little more than crowded beer joints, roadhouses and cafes.

For those whose appetites went further, there were hastily established houses of prostitution in former residences in town or deserted farmhouses in rural areas.

But in February 1942, the Army finally opened an array of off-post service clubs that offered a more wholesome atmosphere in Fayetteville, by now a city of more than 20,000 trying to cope with the wartime role that had been thrust upon it.

The big new USO center on Ray Avenue offered the standard snack bars and reading rooms and also a large hall for dances and athletic events.

Its opening ceremony, which drew visiting War Department and United Service Organization (USO) officials, featured music by the Field Artillery Replacement Center's orchestra.

The USO centers in downtown could handle only a small part of the 60,000-plus soldiers at Fort Bragg, and the main providers of recreation were the on-post centers, also run by the USO. In February, these places got even more popular when the administration at Woman's College in Greensboro agreed that students from the state-run campus could attend weekend dances at Fort Bragg. Enthusiastic crowds of off-duty troops were waiting when buses brought the first contingent of 100 to a dance in February.

The Woman's College students were great, but they didn't provide the frenzy that surrounded a famous visitor who arrived in January.

Mickey Rooney, the boyish superstar of the movies, came to entertain the troops. He was mobbed by reporters, who really wanted to see his new bride of one month, another young superstar, North Carolina's own Ava Gardner.

The newsmen were disappointed. Ava stayed with her folks over near Smithfield, and let Mickey get all the attention.

The Fayetteville Observer reported that "surrounded by a cloud of reporters and public relations men, he rode a mule on Smoke Bomb Hill."

Roadhouse fights

If there were famous visitors and dancing and innocent fun, there was also violence at the roadhouses and beer joints.

The Observer in a single February week reported a series of such altercations, including deadly ones.

An African-American soldier of the 96th Engineer Battalion fought with another soldier and "killed him with a pancake flipper."

A soldier suffered a slashed face in a knife fight at the notorious Wagon Wheel roadhouse on Clinton Road.

Then, in a roadhouse melee at the Town Tavern on Murchison Road, a soldier of the

41st Engineers, the famous African-American "singing engineers," was killed in a gunfight that "wrecked the entire place."

Two other soldiers and a woman were arrested in connection with his death.

The next day, another African-American soldier was shot and seriously wounded during a brawl at Truckers Inn on Lumberton Road.

For civilians, these first months of the war meant such events as an air-raid drill and blackout, a war bond drive in the schools, and Red Cross fund drives to support war activities.

Help with blackout

An air-raid drill in January 1942 wasn't a total success. One downtown jewelry store forgot to turn off its window display lights during the blackout. A crowd of off-duty soldiers and civilians broke the store window and did it themselves.

The schools announced that $22 in contributions for war bonds and stamps was collected from a total of 22 teachers and 265 kids.

The Red Cross reported raising $14,000 for its work.

Fort Bragg, as the biggest Army post in the country, had for a year been a magnet for writers from big-city newspapers and magazines, who came to report on conditions of the troops and observe military training.

With the United States actually in the war, the post also attracted a notable playwright who came to gather material for a play about America at war.

Maxwell Anderson's fame was built on his World War I drama, the long-running "What Price Glory?"

Anderson arrived in March 1942.

He was given VIP treatment. A 22-year-old Army public information soldier, Marion Hargrove, was assigned as his on-post escort.

Anderson would write a short sketch, "From Reveille to Breakfast," that became a favorite of Army camp shows.

And he wrote his serious drama, "The Eve of Saint Mark," about a young soldier killed early in the war. Its poignant message came in the midst of the country's second year of war in 1943. It ran on Broadway for 291 performances.

But Hargrove would soon be the most famous of all World War II home-front authors.

Hargrove had been writing a series of columns about the life of a hapless new soldier caught up in the sprawl and bustle of Fort Bragg.

Anderson took a sample of the columns to his New York publisher.

The result later in the year was "See Here, Private Hargrove," which was the best-selling book of 1942.

The book offered the folks back home an ironically appealing glimpse of what their loved ones in uniform were experiencing 60 years ago as the country prepared millions of its men and women for the sterner part of war that was to come.

Airborne Leaps Into War History

Originally published June 26, 1997

The huge shape of the Queen Mary towered over the Manhattan skyline as the line of gawking men moved up the gangplank in the darkness of a June night in 1942.

Only a day earlier, the largest vessel any of them had ever seen was a fishing skiff.

Two days before they had been sleeping in their barracks in the far corner of the Army's biggest World War II installation, Fort Bragg in North Carolina.

Now, in a movement shrouded in secrecy, they were joining the first wave of American fighting men headed overseas in that springtime 55 years ago.

At the time, they were the 2nd Battalion of the 503rd Parachute Infantry Regiment.

A few weeks later in their camp 25 miles south of the great university town of Oxford in England, the name of the outfit was changed to 509th Parachute Infantry Battalion (Separate).

They would make history five months later as the first U.S. soldiers to go to war by parachuting into combat.

At Fort Bragg, only a few top officers knew the future history-makers were gone.

The place was too busy to wonder where a few hundred men might be.

In that first spring of World War II, as headlines told of the surrender of Bataan in the Pacific and of German offensives in Russia, Fort Bragg was bustling as the Army's largest training ground.

The peaceful 1930s post of 3,000 to 4,000 artillerymen had grown to a sprawling military city of as many as 70,000 soldiers.

The largest unit was the 9th Infantry Division of nearly 15,000 soldiers who lived in a new divisional cantonment that 55 years later has long been known as the "old division area."

The 9th had been activated at Fort Bragg in the late summer of 1940 and spent its first months in a temporary tent city.

Tucked in a corner of the 9th area was the new home of the fledgling Airborne Command, two small regiments of parachutists headed by Col. William C. "Bill" Lee, a native of the nearby town of Dunn and already known as the "father of the airborne."

Another large cantonment nearly equaling that of the 9th Infantry area was that of the Field Artillery Replacement Training Center. It "graduated" a battalion-sized class of gunners nearly every week from its cantonment east of the main north-south highway through the post.

The Airborne Command came to Fort Bragg officially on March 25, 1942, when Lee brought his troops from the 2-year-old parachute center at Fort Benning, Ga., to begin molding what he expected to be a veritable airborne army.

On May 9, 1942, less than a month before the 503rd men boarded the Queen Mary, Gen. George C. Marshall, the Army chief of staff, and Secretary of War Henry L. Stimson led an entourage of U.S. and British generals and admirals to Fort Bragg for a firsthand look at the airborne soldiers and the 9th Infantry Division.

By then, high-level decisions had been made that would send the parachutists and the 9th to war in November in the Allied invasion of North Africa.

If the 9th Infantry Division and the airborne were the most celebrated Fort Bragg units in the first wave of Americans ticketed to go overseas in World War II, they were not the first to leave.

The first soldiers from Fort Bragg to head for the battle fronts pulled out only four days after the Japanese attack on Pearl Harbor.

On Dec. 11, 1941, the anti-aircraft artillerymen of the 76th and 77th Coast Artillery (Anti-aircraft) (Colored) Regiments boarded trains, the former for Philadelphia, the latter for New England.

After standing guard over military installations for several weeks, these two units of black soldiers and white officers received overseas orders that made them among the first American servicemen to leave the country for overseas.

The 77th left New York on April 2, 1942, and arrived a month later in the Tonga Islands of the far South Pacific.

Both anti-aircraft regiments, like those that were to follow in the first months of 1942, were old-timers from the artillery post in the Carolina Sandhills.

Another Fort Bragg artillery unit got to the South Pacific before the 77th, however.

The 97th Field Artillery Battalion, a 105mm howitzer outfit organized in January 1941, left the West Coast on March 15, 1942 and arrived in New Caledonia in the South Pacific on April 5, 1942.

A month later, on May 4, 1942, another Bragg artillery unit arrived in the island of New Hebrides.

It was the 4th Field Artillery Battalion, a big-gun unit with 155 mm weapons. It had been a noted outfit at Fort Bragg in the peacetime Army of the 1930s, when it was armed with the 75 mm pack howitzer of World War I vintage. Five days after the 4th reached the South Pacific, on May 9, 1942, another unit of black soldiers from Fort Bragg arrived in Australia.

The 96th Engineer Battalion (Colored) sailed from New York for its trip to the Pacific, leaving on March 4, 1942.

The 96th was formed out of the most famous of all black units, the 41st Engineer General Service Regiment (Colored), which had organized at Fort Bragg in August 1940.

Major elements of the 41st, known as "the singing engineers," soon followed the 96th Battalion, headed in the opposite direction.

The 41st boarded trains for Charleston, S.C., on May 22, 1942. Three weeks later, the regiment landed in Liberia.

There it went to work building airfields along the route through which U.S. war material flowed to Egypt, where British troops were holding off German offensives.

Like their comrades of the anti-aircraft and engineer units in the Pacific, the soldiers of the 41st would spend more time overseas than any men in Army uniform in World War, not returning to the United States until late 1945 or even 1946.

Women Were Pioneers In 1942

Originally published June 17, 2004

Fort Bragg during World War II was a man's place. By the tens of thousands, they came to learn how to be soldiers.

But it was also a woman's place.

Hundreds of young women found careers on the sprawling post that, by the summer of 1942, was home to nearly 100,000 troops, the largest military population in the United States.

In December 1942, almost a year after the Japanese attack on Pearl Harbor, the post welcomed its first contingent of uniformed females in the new Woman's Auxiliary Army Corps, first known as WAACs, then as WACs.

But two stories from The Fayetteville Observer from the summer of 1942 reveal that women in uniform were already on post and that women were playing a big role in the day-to-day life of wartime Fort Bragg.

Here is the first story:

"Fort Bragg is one Army post that won't have to wait for the ladies of the Women's Auxiliary Army Corps to see girls in uniform.

"They already have a whole company of them.

"They are the Signal Corps lassies who are largely responsible for the operation of the vast network of telephone communication lines serving the post.

"The girls are not in the Army but are part of it.

"A year ago, Maj. Martin Moody, post signal officer, decided that girls would make fine switchboard operators and release Signal Corps soldiers for other duties.

"There were no funds with which to hire the girls, so Maj. Moody obtained his first girls through the National Youth Administration (a New Deal agency that provided jobs for the unemployed young).

"After a few months of on-the-job training, the girls were able to pass Civil Service examinations and enter regular government service.

"Fort Bragg became the first Army post to use girl operators in an Army-owned telephone office."

After several months, the women "began to think about acquiring uniforms in order that they might present an appearance as military as their work."

"Betty Larkins of Clinton and Julia Baggett of Whiteville took upon their slim shoulders the job of getting together with the more than half-a-hundred others, into uniform.

"Now these girls work in their military-style uniforms, a light blue skirt, blue coat on the military cut, and crisp white shirt."

This report of pioneering women on Fort Bragg reminds us that by the end of World War II, several hundred young North Carolina women took their first career step by manning telephone switchboards or working in other offices.

And then there were the hostesses of the club and recreational branch, which by war's end was presiding over a dozen services clubs and an array of recreational facilities.

In the summer of 1942, the women of the services branch had their own "Army look."

This is another story from that summer:

"This largest Army post now has its coeds. Apparently appearing out of nowhere, and quickly swarming over the ramparts of this vast camp, have come the 'Braggadears,' a band of fair warriors bent upon providing entertainment and pleasure for homesick soldiers during their leisure hours.

"Braggadears are composed of girls who are employed on Fort Bragg's main post.

"Organized recently, they are patterned along Army lines with their leader boasting the title of 'master sergeant' and the other officers holding some sort of sergeant rank.

"Committee members are corporals, and unlike the male Army, new recruits are all PFCs. If one of these latter commits some infraction of the rules, however, she can be 'busted' down to a buck private.

"Miss Sarah Seawell of Greensboro is master sergeant. Miss Eulalah Lowder of

Albemarle is vice president, or first sergeant.

"Miss Pat Abernathy, junior hostess at the Post Service Club, acts as general advisor of the girls, and it is through her effort that the activities of the club are promoted.

"Another pioneer of the organization is Mrs. Helen Emmert, who assists in the management of the club's affairs.

"The club meets every evening in the spacious Main Post Service Club, and after dinner and a business meeting, they attend the weekly broadcast, 'Fort Bragg on the March.' After that, they hold a dance in the service club, which is attended by soldiers galore.

"Sometimes there are hayrides, or a watermelon party or a swimfest.

"On Friday evening, they attend the weekly dance at the service club.

"Besides the regular meeting, they promote other social events for the soldiers from time to time.

"Later there will be one-act plays and other forms of entertainment."

Entertaining visitors

"Another fun function of the Braggadears is the entertainment of visitors to the camp who are staying at the Main Post guest house.

"The object is to make these guests feel at ease so they will carry away with them a warm feeling for Fort Bragg and its soldiers.

"All in all, the Braggadears have gotten off to a splendid start and bid fair to be an integral part of the social life of soldiers on this huge post, Fort Bragg men report."

In this summer 60 years later when we are so focused on the veterans of World War II, I wonder does anyone have word out there of the later history of such pioneer veterans as Betty Larkins and Julia Baggett of the telephone operators or Sarah Seawell, Eulalah Lowder, Pat Abernathy and Helen Emmert of the post services organization?

Submarine Warfare Deadly Off N.C. Coast

Originally published July 13, 2000

This week on enchanting Ocracoke Island on the North Carolina Outer Banks, the summer season is in high gear.

The big ferries from Cedar Island disgorge hordes of happy tourists who savor the village walk along sandy Howard Street, under overarching live oak trees. Cyclists and surfers head for the ocean beach, where the National Park service campground resembles a circus with brightly colored tents dotting the sand dunes.

Fifty-eight years ago this week in these same sparkling July waters, desperate men were fighting and dying in an early sea battle of World War II.

On July 15, 1942, the battle joined between German submarine U-576 and the U.S. Navy guardians of a plodding small fleet of merchant ships known as convoy KS-520. The scene was barely 20 miles from Ocracoke's small harbor.

As was the case so many times in the preceding six months off the East Coast of the United States, the submarine struck first and with success.

In quick succession, torpedoes exploded on the U.S. bulk-carrying freighter Chilore, the Panamanian freighter J.A. Mowinkal, and the Nicaraguan merchantman Bluefield.

Limping toward the coast, the Chilore blundered into a minefield laid by the Navy as a haven for merchant ships plying the submarine-invested Atlantic waters. Two crewmen of the unlucky vessel drowned when a lifeboat capsized while being lowered.

The abandoned ship would drift for several days and finally capsize far off to the north at the mouth of Chesapeake Bay.

An unarmed tugboat chartered by the Navy was busy maneuvering the other stricken vessels into the minefield haven.

Up to now, the day seemed to have been like so many before it since the German submarine war was launched in January 1942, only a few months after the United States entered World War II following the Japanese attack on Pearl Harbor on Dec. 7, 1941.

U-boat packs on prowl

In the winter, spring and early summer of 1942, the coastal waters of the Atlantic off the U.S. East Coast became a happy hunting ground for German U-boat packs.

Scores of sailors drowned or were killed in torpedo explosions as dozens of helpless merchant ships went down under the relentless attacks.

No area was more deadly than the waters off Hatteras and Ocracoke on the North Carolina Outer Banks.

In the shallow sea lanes where Cape Hatteras had already long been know as the Graveyard of the Atlantic, sinking vessels could be seen from the shore as residents of isolated Outer Banks villages such as Ocracoke and Hatteras were eyewitnesses to some of the earliest carnage of U.S. World War II history.

Beginning on Sunday, Jan. 18, 1942, when the tanker Allen Jackson was torpedoed off Hatteras with the loss of 22 of its crew, more than a dozen merchantmen had gone down within a few miles of the Ocracoke or Hatteras beaches that 58 years later sparkle in a peacetime July sun.

That many more were victims in the ocean off Wrightsville Beach or Morehead City to the south.

The U-boats often acted with deadly boldness.

On that same Sunday afternoon in January, the U-123 had surfaced to shell the tanker Malay off Oregon Inlet north of Hatteras. Four crewmen were killed by shellfire even as the crippled ship was being towed toward Norfolk.

Other U-boats often surfaced to check on the condition of merchant sailors in lifeboats from the ships they had just torpedoed.

By summer 1942, the toll from the submarine blitz had provoked reaction at the highest level of the government.

Unarmed merchant ships on the coastal lanes were being gathered in makeshift convoys, protected by a growing fleet of destroyers and smaller armed vessels, and by land-based airplanes from naval stations and Air Corps fields near the coast.

Merchant ships armed

Among the jerry-built fleet of guardian vessels were so-called “q” boats, merchant ships armed with hidden guns designed to decoy surfaced submarines into firing range.

The defense system paid off first along the North Carolina coast on May 9, 1942, when the plucky Coast Guard cutter Icarus dropped killing depth charges that sunk the U-352 off Cape Lookout near Morehead City.

So it was on July 15, 1942, as the victims of U-576 milled in the ocean off Ocracoke.

As the U-boat surfaced to savor its victory, the seemingly peaceful merchant vessel Unicoi unleashed its hidden guns.

And overhead, a Navy Kingfisher bomber swooped in, the first time in the submarine war that the air defense system paid off.

U-576 was fatally crippled and then sunk.

It turned out that the loss of U-576 was a sort of straw that broke the camel's back in the submarine war off the U.S coast.

Adding up his losses and noting the growing defense system of the U.S. Navy, German Adm. Karl Doenitz, the submarine boss, called off his six-month operation, ordering the six remaining submarines still prowling the area to deploy to other hunting grounds.

On the same day Doenitz issued his order, the final chapter of the action off Ocracoke came to an unhappy conclusion.

For three days, the chartered tugboat Keshena of the Navy Rescue Tug Service had waged a dogged struggle to salvage the torpedoed Panamanian tanker A.J. Mowinkal, which was wallowing helplessly in the ostensibly friendly haven of the minefield.

Suddenly, the little vessel was wracked by an explosion.

A mine.

In a few minutes, Keshena went to the bottom. Two of its crew of 16 perished in the explosion.

Fourteen others of the unique crew of the Keshena, one of whom was a female member of the tugboat owner's family, were rescued from the summer ocean by Coast Guard launches from the Ocracoke Station.

The submarine war along the Outer Banks would continue off and on for more than another year before the fury of the U-boats was finally tamed.

While it doesn't rank with the great World War II battles in the Pacific or the mighty invasions of Europe, the struggle along today's playground beaches of the North Carolina Outer Banks was no less a time of courage and perseverance worthy of an honored chapter in the history of that war.

Engineers Become Esteemed

Originally published August 3, 1995

Roger Thornhill, now retired in Lumberton, was an 18-year-old high school graduate from Pittsburgh when he arrived at Fort Bragg in the summer of 1940.

He was in the first batch of several hundred young men who would open a new chapter in the history of black Americans in the military.

Mobilizing against the threat of World War II, the U.S. Army activated units of black enlisted men commanded by white officers.

Fort Bragg was home to three of these pioneering outfits — the 41st Engineer General Service Regiment, and the 76th and 77th Coast Artillery (Anti-Aircraft) Regiments.

They were officially activated Aug. 1, 1940. By late August, the hundreds of young black men were busy learning to be soldiers.

Fort Bragg that summer was a scene of busy confusion.

The Army had chosen the small artillery post for major expansion, but thousands of barracks had yet to be built. Ten thousand brand-new soldiers were living in tent camps.

Separate camp

Black soldiers were in tent camps on the northwest corner of the post, near present-day Spring Lake.

There was a shortage of everything, even uniforms.

Thornhill remembers that because of the shortages and confusion, he waited for several weeks before the Army was ready to swear him in as a soldier.

Nonetheless, because he could write and type, he was tapped as a company clerk even before he took the oath.

Within a few more weeks, he was holding informal literacy classes for recruits who were products of the miserable schools of the South.

"It was my second job. I was to teach them how to sign their pay forms," he recalls.

Later, Thornhill's informal classes grew to a formal program using faculty members from Fayetteville State Teachers College (now Fayetteville State University).

Moving forward

Despite the shortages and confusion, recruit training went forward in the hot Carolina summer and fall.

The 41st Engineers marked the end of the recruit period with a Regimental Field Day in late October. The anti-aircraft battalions finished with a demonstration day and review by the post commander.

In November, everybody pitched in to get ready for their first Carolina winter. Tents were "winterized." Winter clothes and blankets were issued, and mattresses replaced the hay-stuffed ticks of summer. As part of their training, the engineers erected temporary wooden mess halls amid the tents.

In the 76th area: "Salvaged two-by-fours and saplings cut on the reservation were used to make frames and rafters for the tents and to eliminate poles. Pine needles were gathered to spread on company streets to hold down the dust."

In training, 12-mile overnight hikes were kept up even as night temperatures dropped to the 20s. Equipment began to arrive — guns, searchlights, signal equipment for the artillerymen; bridging material, construction machinery, trucks for the engineers.

Growing outfits

In February 1941, as the troops moved to the new barracks, both anti-aircraft outfits grew to authorized two-battalion strength and became part of a provisional brigade that included the white 76th Regiment.

From the beginning, there was continual turnover of personnel as trained soldiers left to form the core or "cadre" of other units.

The 41st Engineers quickly lived up to the prediction of its commanding officer, Lt. Col. John E. Wood, that it would be the "mother regiment" of all other all-black engineer units.

By August 1941, more than 800 men had gone out from the regiment to form the cadre of other units. Among a dozen such outfits was the 96th Engineer Battalion (Separate) (Colored), which remained at Fort Bragg until it went to the Pacific in the summer of 1942.

At the same time, nearly 400 men left the 76th Coast Artillery to form the cadre of two other anti-aircraft regiments, as well as quartermaster and artillery replacement centers elsewhere in the country.

Expecting the best

Wood was a tough, experienced Regular who was determined that his troops would not be relegated to the menial dig-and-haul jobs that had been the lot of most black soldiers during peacetime.

Despite shortages and heavy turnover, he instituted a rigorous training program that assumed his men would rise to the challenge.

He was especially proud of their bridge-building skills. He took the entire regiment to the brand-new Marine Corps base at New River in Onslow County where his troops provided pontoon-bridging for a joint Marine-Army training exercise.

The men of the regiment made good on his goals.

National prominence

By June 1941, when Life magazine profiled Fort Bragg as the Army's largest post, the men of the 41st got a half-page spread with a text describing them as "expert at building bridges, storming pillboxes, building roads and airfields."

The regimental history boasted that in rifle firing, more than 87 percent of the men qualified as marksmen, "a record not equaled by any regiment at Fort Bragg."

In the autumn of 1941, the 41st was the largest all-black outfit participating in the Carolina Maneuvers, the most extensive peacetime war games of the newly mobilized Army.

Music became an early and widely acclaimed specialty of the regiment. The 41st has gone down in history as the "Singing Engineers."

The regimental band was informally organized within weeks. As early as September, a picture in the regimental history depicts "Sgt. (Frank) Ruffin leading field music at retreat" on Fort Bragg's Main Post Parade Field.

Ranked behind Ruffin's saluting figure is a smartly moving drum and bugle team.

Capable leader

Activated in November, the band was led by an already highly respected military bandmaster, Warrant Officer John Jordan Brice (1900-1948).

For 15 years after serving overseas in World War I, Brice had directed the band of the ROTC unit on the country's premier black campus, Howard University in Washington. In those years, he was a frequent guest conductor of the U.S. Army Band.

Brice and the chaplain, Capt. James R.C. Penn, were the only black men who were commissioned officers in the 41st.

Brice's outfit quickly became esteemed throughout the Army. He wrote music for a regimental march and coached the whole regiment in singing in unison.

Col. Wood proudly noted in his first anniversary letter that the 41st was "the first unit to have official permission to sing its regimental song while marching at attention."

Life magazine's article emphasized the singing, quoting words of the regimental march and describing the repertoire as "spirituals, old marching songs, patriotic and Southern tunes."

Brice not only found new musical fame at Fort Bragg. He found his wife in Fayetteville.

He married Malissa McNeill, daughter and granddaughter of the town's first undertakers. "Mallie" McNeill was a teacher with a graduate degree from Boston University. Today she lives in retirement as a longtime resident of suburban Detroit, where she was a teacher and a civic and church leader.

Anniversary review

The first birthday of the 41st Engineers on Aug. 1, 1941, opened with a review. The departing commanding general of the 9th Division, Maj. Gen. Jake Devers, delayed his departure long enough to present streamers to each company for training excellence.

The Observer reported: "In a full review yesterday, the 'Singing Engineers' put on a show that was almost completely overlooked in the rush of the 9th Division's 'Organization Day' program."

The regimental history recounts a full day of recreational and military activities for the engineers and their families. It concluded: "The 41st Engineers Swing Band ended the day with music under the stars and a hundred gaily colored lanterns."

Tensions explode

A few days later, the goodwill of the birthday was dampened by a violent tragedy that heightened racial tensions always lying close to the surface that summer.

It happened on a bus being loaded in downtown Fayetteville. A shoving match between white military policemen and black soldiers turned deadly.

A private from the 76th Coast Artillery grabbed the pistol of a 20-year-old MP and shot him fatally through the heart. Several other shots were fired, including one that instantly killed the 27-year-old black draftee from Asheboro who had started the gunfire.

The post provost marshal overreacted by ordering a roundup of all black soldiers not already in barracks, and confined several hundred in a stockade next to the post guardhouse for several hours.

He was later relieved of his command. But tension was high and rumors flew, including one from a radio station in Raleigh alleging that black soldiers at Fort Bragg were to be moved "nine miles from town."

A few days later, Gen. Benjamin O. Davis, the only black general officer in the Army, visited the black troops to hear their comments and complaints. It was part of a War Department effort to counter the growing record of racial discrimination and incidents on posts across the country.

New facilities

That autumn, the tensions of summer were eased by the construction of new recreational facilities for the troops and housing for the families of black noncoms.

The USO center on Murchison Road became a favorite mingling place for black soldiers and for the black community of Fayetteville.

Washington Square was a $1.5 million array of apartments also on Murchison Road.

While some families occupied the quarters in October, official dedication day was set for Dec. 9, 1941.

Forty-eight hours before that event, Japanese planes attacked Pearl Harbor and the United States was in World War II.

The Army's pioneering black outfits quickly went on a war footing.

Among the earliest units from Fort Bragg to leave on a wartime mission were the 76th and 77th anti-aircraft regiments. On Dec. 9, troops and guns were loaded in trains and sent to duty in Wilmington, Del., and Philadelphia.

17TH'S GUNS SILENCED IN TUNISIAN DEBACLE

Originally published November 27, 2003

On a chilly January day in 1921, when the stubby "Schneider" howitzers rolled into Camp Bragg, they were the latest thing in big field artillery.

Twenty-two years later, when the same guns took positions on a rocky Tunisian plain near a notch in the hills called Faid Pass, they were described in a new history of World War II as "eighteen World War I-vintage 155 mm howitzers."

Moreover, the guns and men of the 2nd Battalion of the 17th Field Artillery Regiment were, in the phrase of the history, "somehow forgotten in the confusion."

The "confusion" was the major World War II baptism of battle for the U.S. Army, when in the early morning of Feb. 14, 1943, German armored columns boiled out of Faid Pass in central Tunisia.

In a day, the Germans broke over the American defenders, capturing several thousand prisoners and sending the Americans reeling back 50 miles west to Kasserine Pass, the place that gave its name to the weeklong battle that unfolded after the disaster just west of Faid Pass.

According to the account in "An Army at Dawn" (Henry Holt, 2002) by Rick Atkinson, a staff officer reported:

"The attack on the battalion erased it, getting every gun and most of the men."

In the larger historical annals of the U.S. Army, the very word "Kasserine" has come to mean a humiliating defeat, an example of how not to use green troops in battle, of a tragic comedy of errors by commanders.

For the 17th Field Artillery, the ordeal at Faid Pass was a World War II baptism opening a long odyssey of combat in North Africa, in Sicily, Italy, France and on to final victory in Germany in 1945.

But for many men of the 17th, the war meant long years in German prisoner-of-war pens.

POW's tale

The best account of the fate of the 2nd Battalion on Valentine's Day 1943 is the story by James F. Bickers, one of those POWs, written for his veterans' organization.

Bickers was a 24-year-old second lieutenant who joined the 17th Field Artillery at Fort Bragg in August 1941.

The 17th, with two other other regiments of the 13th Field Artillery Brigade, went overseas first to England and then to North Africa in December 1942.

In a bittersweet irony tinged with the humor of a survivor, Bickers described what happened two months later:

"At dawn on the 14th, we received march orders and pulled out of our gun positions.

"The guns that we pulled out were box-trail 1917 Schneider 155 mm howitzers, real state-of-the-art! The name Schneider came up again soon, for that was the name of the German Commandant at Oflag 64, Oberst Schneider, a jolly old elf.

"As we were pulling out of position, we received orders to go back. We went back. Again we were ordered out. This time we formed a Battalion column and headed down a road to some place in Tunisia. We stopped, then suddenly the column started up again; this time across country.

"There were many interesting things going on around us: bombs bursting in the air, and stuff like that.

"We chugged along in the desert until we came under machine-gun fire.

"We thought that it was from the air, stopped, and took cover.

"To our dismay, we realized that the fire was from tanks coming from both sides.

"At this historic moment, Captain McCord pulled up in his command car, paused long enough to say 'adios,' and cut out. I haven't seen him to this day.

"As the new leader of Battery F, I managed to get two howitzers back-to-back, and attempted direct fire against the tanks.

"That seemed to irritate the Germans; instead of using machine guns, they shelled us.

"Their Mark IV tanks had 77 mm cannons, and the lads in the Afrika Korps knew how to use them.

"Soon we were looking down the barrels of those 77s, and a kid with long blond hair hanging over the turret of his tank actually said, 'For you, the war is over.'

"The situation was hopeless and the survivors were captured."

The ordeal of the 2nd Battalion and its attached company of the 1st Field Artillery Observation Battalion, also a longtime prewar Fort Bragg outfit, directly touched homes around the big post in the Carolina Sandhills, where many of the soldiers were peacetime residents.

Listed as missing

A month after the battle, The Fayetteville Observer reported that among the men missing at Faid Pass were Pfc. Belmont Carroll of Massey Hill, Cpl. James D. Harrell of St. Pauls, Tech. Pvt. Van D. Martin of Fayetteville, Pvt. Carl Riddle of near St. Pauls, Pvt. Weldon P. Shoe of Fayetteville, Pvt. Ray T. Stephenson of Wade, and Pvt. Shellie B. Woodall of Pembroke.

Belmont Carroll wound up spending 26 months as a prisoner of war before being liberated by U.S. troops in Germany. He came home in May 1945. He had enlisted at Fort Bragg in 1940 and had been in Tunisia only 10 days before being captured at Faid Pass.

The observation detachment was captured almost in its entirety; 38 enlisted men and five officers went off to POW pens.

When the Army mobilized in 1940, the 17th Field Artillery was the senior unit in the post complement at Fort Bragg.

It and two other understrength field artillery regiments constituted the 13th Field Artillery Brigade, which since the 1920s was the post garrison command as well as a tentative combat unit.

The 17th arrived at Camp Bragg with much fanfare in the early winter of 1921, wearing battle colors from the earliest fighting of the American Expeditionary Force in France in World War I.

At the time, the sprawling post in the pine woods was barely two years old, finished only a few months after the armistice of Nov. 11, 1918, ending that earlier war in Europe.

The World War II observation battalion at Faid Pass was descended from 1st Observation Flash Battery, organized at Bragg in 1922 by Lt. L.E.W. Lepper.

The squat, stub-nosed Schneider howitzers were horse-drawn when they arrived at Camp Bragg.

In the 1930s, their trailers were modified to be hitched to a standard Army truck, the so-called "high-speed carriage," and that is the way they went to battle in World War II.

French field piece

Photographs from the 1930s depict a seven-man gun crew, wearing World War I "tin hat" helmets, preparing to fire the two-wheeled gun from a position in a broomstraw field at Fort Bragg.

The rugged field piece, like nearly all U.S. artillery from World War I, was acquired from the French army.

By the time the U.S. entered World War II, it was rapidly being replaced by the long-barreled multiwheeled gun known as the "Long Tom."

But on that rugged plain beyond Faid Pass on Valentine's Day 60 years ago, the "vintage" howitzer, long familiar on the sandy hills of peacetime Fort Bragg, had its unique, if unhappy, moment in the history of the big guns of the Army.

Airborne From Bragg On Attack

Originally published July 8, 1993

For airborne soldiers who trained for war at Fort Bragg, July 1943 was one of the great defining moments of World War II.

The time for training was over; the time for fighting began.

In the hours around midnight of July 9-10, 1943, several thousand young men of the 82nd Airborne Division began jumping into the dark from lumbering C-47 transport planes carrying them to the Allied invasion of the Mediterranean island of Sicily. It was a major step on the road to victory over the Axis.

Back in the United States, at Fayetteville and Fort Bragg, civilians and soldiers read Saturday afternoon headlines telling of divisions of American and British soldiers storming ashore at such places as Gela and Licata.

There was no hint in the early news accounts that the troopers of the 82nd Airborne were involved.

Yet people by then knew that the division, which had shipped out of Fort Bragg in great secrecy less than 90 days before, was destined for early action.

The second weekend in July 1943 was hot in Fayetteville, with the temperature typically close to 100 degrees.

Thousands of soldiers from Fort Bragg were on weekend pass in the town. Movie theaters were packed. They offered such films as "White Savages" starring Maria Montez and John Hall, "It Ain't Hay," with Abbott and Costello, and "Carson City Cyclone," with Don Red Barry.

As the weekend continued and the sidewalks of Fayetteville sweltered, far away on a rocky height above the Gulf of Gela in Sicily a force of several hundred young paratroopers of the 82nd Airborne Division was shooting it out with enemy forces in a bitterly contested firefight.

It was a struggle that would be known as the "Battle of Biazza Ridge."

Led by the regimental commander of the 505th Parachute Infantry Regiment, Col. "Jumpin' Jim" Gavin, the paratroopers held the ridge against repeated attacks and were still there when infantrymen of the 1st Infantry Division pushed up from the Sicily invasion beachhead. Many of the fighters were from a battalion of the 504th PIR that had jumped as part of the invasion force under Col. Gavin's command. A few hours earlier, another gathering of paratroopers, led by Lt. Col Arthur Gorham of the 1st Battalion of the 505th

PIR, discovered they were more than 65 miles from their drop zone.

Nonetheless, they found something to do. Using bazookas and hand grenades, they managed to disable four German tanks on their way to attack the Allied beachhead.

These were the highlights of the first days of action for the 82nd Airborne, as widely scattered paratroopers managed to gather in fighting groups to play their part in the Sicily invasion, to make history for themselves and for the airborne idea.

On July 17, the War Department officially acknowledged that the 82nd was part of the Sicily invasion force.

On the other hand, more than seven months would pass before there would be any announcement of what happened on July 11, when two more battalions of paratroopers, men of the 504th PIR, flew toward Sicily in the C-47s.

Jittery gunners on U.S. Navy ships fired on the planes. Hundreds were killed and wounded. A third of the force was crippled.

If folks back home didn't hear of that disaster at the front, their attention was brought back home on July 24 by news of a freak train wreck in Fayetteville that left three dead and six injured.

It happened in the early dawn at the busy Milan marshaling yard just north of downtown Fayetteville. The rear section of the crack Tamiani Champion passenger train, gathering steam for its run toward Florida, crashed into the rear of its first section, still waiting to be switched and picked up.

The observation car of the train was crowded with servicemen and civilian travelers. The dead included an Army Air Corps sergeant from Charleston, a sailor from Massachusetts returning to his station in Florida, and a 19-year-old woman who was returning to her job in Washington after visiting her soldier-husband in South Carolina.

Rescue workers and railroad crews worked for several hours to untangle the wreckage.

The Sicily invasion headlines also vied in local interest with news of Cape Fear region men who were in the battle zones.

Air Force 1st Lt. Robert Butler of St. Pauls was already a veteran of more than a year of the air war in the Pacific. He was awarded the Distinguished Flying Cross for 200 missions from August 1942 through March 1943. Another veteran of the air war, 1st Lt. James K. David of Raeford, had come home "to visit his wife and baby daughter" after being overseas for 14 months. He had just been awarded another cluster for his Air Medal for flights during the fighting in North Africa.

The Cumberland County community was abuzz over a case in the county recorder court. A popular farmer, M.D. "Doc" Bennett of Lumberton Road just south of Fayetteville, was charged with selling illegal whiskey to soldiers.

State and local liquor officials thought they had him cold, because the prosecuting witnesses were ABC agents who had been allowed to pose as soldiers by Fort Bragg officials. A taxi driver who had delivered soldiers to the rural area had already pleaded guilty.

Nonetheless, a recorder court jury agreed with the contention of Bennett's lawyer that "masquerading" liquor law agents as soldiers amounted to unfair enforcement tactics. Bennett was acquitted.

As July 1943 came to its end, the battered units of the 82nd Airborne Division took part in a final offensive that cleared western Sicily. It was "little more than a road march," according to an unofficial history.

A picture of the division's commanding officer, Maj. Gen. Matthew B. Ridgway, still wearing his peacetime garrison cap headgear, appeared in The Fayetteville Observer on July

27. The caption noted that "the old Bragg division is making a fine record on the island."

The war had become real for the airborne soldiers.

A day later, and shrouded in wartime secrecy, Gen. Ridgway would be calling on the Allied headquarters of Gen. Dwight Eisenhower, getting word of the next assignment for the division.

*Major units of the 82nd Airborne Division taking part in the July 9-10, 1943, invasion of Sicily were the 505th Parachute Infantry Regiment, the 3rd Battalion of the 504th PIR, the 456th Parachute Field Artillery Battalion, and Company B of the 307th Airborne Engineer Battalion. Actions and places in Sicily have long been marked on Fort Bragg by the names of Gela and Biazza streets and Sicily Drive.

Nadzab Ties 41st To Airborne Milestone

Originally published March 8, 2007

Few marriages last longer than that of the 41st Airlift Squadron and Pope Air Force Base.

When the unit moved out last month for a new home in Arkansas, it was the end of a 36-year stay for a unit that for all those years ferried Fort Bragg soldiers on missions around the world.

But the link between airborne and the 41st actually goes back so much further, and includes a moment of unique historical significance for paratroopers.

It happened Sept. 5, 1943, over a sprawling landscape of jungle and savanna on the far-off island of New Guinea in the South Pacific.

On that day, lumbering C-47 aircraft of the 41st flew low to disgorge their sticks of jumpers from the 503rd Parachute Infantry Regiment.

It was the first daylight combat parachute jump of World War II, and it went off with spectacular success.

Jumping in midmorning on a muggy day, 1,800 paratroopers hit a cloverleaf drop zone in a flyover that took two minutes. A fleet of 87 C-47 aircraft from the 41st and a half-dozen other transport squadrons flew the historic mission.

The paratroopers' goal was to link up with Australian ground troops to seize a Japanese airfield and cut off a Japanese infantry force.

By 3 p.m., the paratroopers of the 503rd marched the mile and a half to the jungle-fringed airstrip and took control of it, suffering only three casualties.

The place was called Nadzab in the Markham Valley near the village of Lae on the north coast of New Guinea.

"Nadzab" became the official name of the history-making combat jump and earned commendations for pilots, paratroopers and entire units.

The whole operation was the carefully rehearsed brainchild of Gen. Douglas MacArthur, who considered the 503rd his personal paratrooper force.

If anything, the project was an exercise in overkill.

The C-47 transports that ferried the few hundred paratroopers were layered between fleets of heavily armed fighting planes.

This fighter escort alone numbered 100 aircraft.

Then there were the light bombers armed with 50-caliber machine guns and fragmentation bombs — six squadrons of them; more than 60 aircraft — to whip up the ground in advance of the jumping paratroopers.

As the transports unloaded, six fast A-20 fighter-bombers scurried close to the ground to lay a giant smokescreen over the steaming tropical terrain.

Above the entire formation, like a cherry on top of a deadly layer cake, flew MacArthur's personal command fleet, a formation of three B-17 bombers bristling with machine guns.

In one of them, the commander of Southwest Pacific Forces looked down on the whole show.

And above the Flying Fortresses, a buzzing flight of six P-47 fighter planes kept constant guard lest any Japanese planes try to jump the command fleet.

The operation got maximum publicity.

Photographers in the air fleet took newsreel-ready film looking down on parachuting troopers as they floated in tight formation toward the drop zones as well as images of the formation of the transport fleet as it approached the drop zones.

You can see segments of the official film photography in the Pacific Theater exhibit at the Airborne & Special Operations Museum in downtown Fayetteville.

First Lt. Monte Kleban, MacArthur's assistant public relations officer, jumped in with the 503rd PIR troopers, and he described how the airborne soldiers and Australians "hugged and embraced like lost brothers."

Tolson link

Two days later, an Associated Press photo depicted MacArthur patting the shoulder of a smiling young airborne officer, "Lt. Col. J.J. Tolson of New Bern, N.C., who in his jump gear was just before the take-off."

Twenty-nine-year-old Jack Tolson was a battalion commander who would go on to be a postwar pioneer of Army helicopter aviation and end his career as commanding general of Fort Bragg and the 18th Airborne Corps. He died in 1992. The youth center at Bragg is named in his honor.

In the perspective of history, the feat of the 41st and its load of paratroopers on Sept. 5 1943, is more symbolic than it was tactically decisive.

Even the first account of the landing in 1943 pointed out that the airfield at Nadzab was really not much of a threat. It had "not been used for months and was overgrown with Kunai grass four to six feet high." In the official history of American Army units published after World War II, the 503rd PIR's role was described as "landed unopposed ... relieved on Sept. 17."

Nonetheless, the jump at Nadzab was a key milestone in the history of airborne forces.

Initial airborne missions in North Africa and Sicily were fiercely contested and marred by terrible "friendly fire" incidents and scattered drops. There was talk of scrapping the whole airborne endeavor.

News of Nadzab came at the critical moment, reinforcing the strong support for airborne of such World War II leaders as Gen. George C. Marshall, the chief of staff of the Army.

Within a few days of the good news from the Pacific, airborne forces in the Mediterranean theater began to show the stuff that would forever redeem the airborne concept.

The Nadzab lessons of careful planning and realistic training would become more firmly embedded as key principles of airborne doctrine. And for the 41st Airlift Squadron, it was the pioneering first chapter in a link with airborne that remains firm today.

Today, the Nadzab-Lae area is a popular tourist scene for Australians, who come to enjoy the spectacular north coast beachfront and explore the towering mountains of the Owen Stanley Range. Resort hotels dot the coast.

The wartime landing strip near Nadzab village is now a modern airport with a 5,800-foot runway and easy flights from Australia and other places in New Guinea.

Italy Success Erased Doubts About Airborne

Originally published September 23, 1993

It was one of the defining moments of World War II for airborne soldiers of the U.S. Army.

In the final two weeks of September 1943, the 82nd Airborne Division came of age as a fighting unit, blowing away the doubts about the concept of having soldiers jump from airplanes and nailing down the World War II airborne role in the war against Germany.

It began with a harried, quickly mounted jump of several thousand parachute soldiers into the beachhead of Allied forces invading the coast of Italy at Salerno.

It ended 16 days later when a column of truck-riding paratroopers roared into the Italian city of Naples, led by a jeep with a two-star general riding shotgun with his personal Springfield rifle, ready to take on any snipers that fired at his troops.

It actually began back in May 1943, when an invasion of Italy was on the Allied planning boards.

In earliest planning for the invasion of Italy, airborne forces were to jump in behind the landing forces.

But events changed the plans. As Allied forces, including the 82nd Airborne Division, ended the campaign for the island of Sicily, Italy began to totter.

By early September, the Allied high command had ordered the 82nd to prepare for a daring drop on Rome, as part of a scheme to take advantage of the fall of Mussolini's government.

When that was scrubbed even with some airplanes warming up, the invasion fleet had already sailed for Salerno, and the airborne part of the attack had been whittled down to practically nothing.

Then, only hours after the Allied seaborne forces had gone ashore along the Bay of Naples in the vicinity of Salerno, German counterattacks were threatening to break the scattered landing forces into small pockets.

Gen. Mark Clark, commander of the invasion forces, called on the airborne to do its thing.

Gen. Matthew Ridgway, commanding general of the 82nd, threw together a mission, bits and pieces of previous plans, to drop reinforcing paratroopers into the beachhead.

So quickly was the mission mounted that some pilots of C-47 transport planes were told of the route by briefing officers using flashlights and maps spread against the wings of the planes.

It was the night of Sept. 13-14, 1943.

To greet the columns of C-47s, 31-year-old Lt. Col. Bill Yarborough, who had been sent out of the 82nd to become the airborne adviser to Gen. Clark, devised a gigantic ground marker for the drop zone.

Already on the beach, Yarborough had commandeered engineering forces to build a T-marker of lighted gasoline cans, with each leg of the T a half-mile long!

To guard against any "friendly fire" disasters like the one that had cost scores of lives in the Sicily invasion, Yarborough had been dispatched to every battery of artillery then in the Salerno beachhead with orders from Clark that they cease firing during the time the C-47s would be coming in. The commanding admiral of the Navy landing force had also given such instructions to ship commanders.

It worked. The 504th Parachute Infantry Regiment came down without a single friendly fire incident. More than 1,300 paratroopers were quickly gathered together by Lt. Col. Reuben Tucker.

Within a few hours, the paratroopers were in the middle of the climactic fight of the Salerno beachhead, helping British troops turn back the fiercest of the German attacks aimed at splitting the Allied forces.

On the night of Sept. 14-15, the 505th Parachute Infantry Regiment, commanded by Col. James Gavin, jumped into the beachhead, losing one plane to friendly fire and somewhat more scattered than the 504th.

The final jump of the airborne in the invasion of Italy was one of the most bizarre of all paratrooper missions in World War II.

More than 600 paratroopers of the 509th Parachute Infantry Batallion took off on the same night to jump into the mile-high Italian village of Avellino in rugged mountains more than 20 miles behind the northern shoulder of the Salerno beachhead.

The troopers were scattered over 30 miles of mountain terrain. The 60 or so paratroopers who managed to rendezvous in the village found themselves in battle with German tank crews.

Ultimately, 520 of the 640 men in the jump made their way back to Allied lines, some after stints in German prison camps.

The Avellino jump had been part of an earlier airborne plan for the Italian invasion, and probably would have been scrapped if the commanders and planners had not been so busy stitching together and directing the missions that were called up at the last minute.

The 509th was an independent unit that had been in England. It was so shattered by the Avellino mission that thereafter it went out of the jumping business and became the elite headquarters guard for Gen. Clark's 5th Army.

Meanwhile, back in the beachhead, Allied forces began to break out and drive for Naples.

On Sept. 16-17, Tucker's 504th Parachute Infantry Regiment crashed into an Italian hill town named Altavilla and a ridgeline named Hill 424 and fought a fierce two-day battle with German forces.

The fight cost 30 dead and more than 130 wounded among the paratroopers. It won Tucker a Distinguished Service Cross and a reputation as the airborne's most aggressive regimental commander.

For the next two weeks, paratroopers pushed along with other Allied forces in a ground drive for Naples. Gavin's 505th was part of the grouping for the final push.

So it was on the morning of Oct. 1, 1943, that a column of trucks loaded with men of the 82nd Airborne rolled into Naples as the vanguard of the forces that would liberate the largest city in Europe yet to fall to Allied forces.

And out front in his jeep was the 82nd's commanding general, 48-year-old Matthew Ridgway, packing his trademark Springfield rifle and ready to do battle with any sniper who dared fire on the liberators.

Christmas 1943: Frightful, Delightful

Originally published December 9, 1993

As Christmas 1943 approached, the people of Fort Bragg and the surrounding Cape Fear region prepared to observe the third holiday season of World War II.

War training didn't take a big holiday.

In the first week of the month, the Airborne Command at Camp Mackall planned a big nighttime maneuver, testing lessons that paratroopers and their Air Force transporters had learned in the invasions of Sicily and Italy during the previous summer.

Bad weather caused cancellation of part of the operation, but not before highway crashes resulted from civilian cars smashing into the trucks of blacked-out military convoys.

When training was curtailed for the few days before Christmas, the Fort Bragg command published the menu for its traditional Christmas meal for the troops.

It read like a Norman Rockwell painting of a holiday feast:

"Creamed celery soup, croutons, roast turkey, sage dressing, giblet gravy, cranberry-orange relish, snowflake potatoes, baked squash, creamed corn, head lettuce, Russian dressing, hot rolls, butter, mince pie, chocolate nut cake, assorted fruits, candy, nuts, and coffee."

The extent of Fort Bragg's wartime recreational services was revealed in the announcements of holiday events for soldiers who would remain on post for the holidays. Parties were planned at "Service Club Number 1" and "Service Club Number 2." At "Service Club Number 4," the 79th Field Artillery Group would have its own affair.

And at "Service Club Number 5," the club for black soldiers in Spring Lake, dance partners would include "girls from Fayetteville."

With the chaotic experience of two previous Christmas seasons in mind, post and civilian officials had gone all out to find accommodations for the visitors who would flood the area for the holidays, families of soldiers who remained on post.

Fort Bragg praised the efforts of the civilian community in lining up rooms in private homes. For their part, civilian officials sent out one more call for people with extra beds to list them, "so visitors won't have to spend Christmas night sitting up in local USO's and other public places."

Just as the holiday schedule picked up, however, a railroad disaster pushed aside other matters.

On Dec. 17, 1943, two passenger trains loaded with holiday travelers collided near the little Robeson County siding of Buie.

For four days, military and civilian rescue teams worked at the wreckage.

Seventy-two passengers were killed, another 123 were injured. The list of dead included 51 military people and 21 civilians.

Among the dead were soldiers, sailors, a Navy woman enlistee, a chaplain of the Harvard campus military unit and a family headed for Florida to visit a serviceman father.

The awful scenes of railroad carnage were softened when one of the area's rare December snowfalls fell that week. Photographers rushed to record scenes of Fort Bragg's handsome main post chapel in a winter coat.

For those who stayed at Fort Bragg, there was a record holiday mail call. In five days just before Christmas, the Fayetteville post office handled a million pieces of mail, three train carloads in and three out each day.

The civilian post office was a major holiday employer, with 125 workers. At Fort

Bragg, military postal offices used 100 clerks, many of them from the ranks of the Women's Army Corps — the units of women soldiers that were handling more and more of the office work on the wartime post.

At Pope Air Field, the command thanked the women with "AAF WAC Day," and gave a party with music from the Pope Field Dance Orchestra. The women had been at Pope since July 16, 1943, and already were involved in "personnel, statistics, legal, and administrative functions at HQ, and medical, field clearance, QM, and operations."

Fort Bragg noted a Christmas greeting from a personality from the past. The warden of the Cook County Hospital in Chicago sent a card. He was none other than retired Brig. Gen. Manus McCloskey, who had been the commanding officer of Fort Bragg from 1931 until 1938.

Not all such messages were happy.

The M.M. Britt family on Branson Street received a War Department message saying their son, Sgt. Marvin Britt, an Air Force gunner, was missing after a Dec. 1 bombing raid over Germany.

As Dec. 25 approached, the racial tensions that often lurked just under the surface of wartime life in a small Southern town exploded.

It happened on a Saturday night when a black military-police officer tried to control a group of white soldiers near the bus station and the notorious night spot known as the Town Pump on Donaldson Street in Fayetteville.

Tear gas and live ammunition were brought into play, a white soldier was wounded and several women complained of being "mistreated" in the melee.

The Fayetteville Police Department stayed out of the fray.

The most newsworthy thing to The Fayetteville Observer was that a black MP had shot a white soldier.

As Christmas passed, many plans for a happy New Year were probably dashed Dec. 27 when police and ABC officers broke into the flooring under an air-conditioning unit on the top floor of the Prince Charles Hotel and found 51 cases of tax-paid whiskey.

A bellhop was arrested and pleaded guilty. He was fined $500 and was given a suspended sentence. He admitted that he was in cahoots with "another party" but declined to reveal his confederates.

The pint bottles sold for $5 each to thirsty revelers on the days when the ABC stores were shut down. The accused said he already had sold 100 bottles.

The huge cache of booze had been brought into the hotel on a freight elevator. The hotel management denied any knowledge of the affair and praised the lawmen who made the catch.

Anzio Landing Was A Time Of Valor For U.S. Paratroopers

Originally published January 20, 1994

Cpl. Paul B. Huff of the 509th Parachute Infantry Battalion grew up in a Tennessee town listening to tales of the World War I exploits of a neighbor, Sgt. Alvin C. York.

As a corporal in 1918, York won the country's highest award for military valor, the Medal of Honor, for his single-handed assault on an enemy machine gun nest that had pinned down his company of the 82nd Infantry Division.

Now, 25 years later, 23-year-old Paul Huff was going into battle for the sixth time in the 30 months since he enlisted in the paratroopers and trained at Fort Bragg. He was remembering again the legacy of the hero from Pell Mell, Tenn.

It was Jan. 22, 1944, 50 years ago this week, and an Allied invasion force was landing on the beaches around the Italian coastal town of Anzio.

Huff's paratrooper battalion as well as the 504th Parachute Infantry Regiment were in the landing force, along with American and British infantry and armored divisions. The mission was to outflank German forces and pave the way for Allied capture of Rome.

Instead, the Germans counterattacked, pinning the invasion forces inside a small pocket of coastal terrain. A mission that was expected to last a few days turned into a two-month slugfest, one of the bloodiest of World War II.

For paratroopers, it would not be jumping from the sky, but rather weeks of vicious combat in a muddy, cold battleground, often behind enemy lines, always within sight of enemy big guns. It would be a time of death and terrible wounds.

And it would be a time for valor and the making of enduring traditions for the airborne forces.

Paul Huff would be a major contributor to the catalog of valor.

Two weeks after his unit landed, Huff was leading a six-man reconnaissance patrol in no man's land when fire from a German machine gun pinned them down.

In an uncanny re-enactment of Alvin York's exploit a quarter century earlier, Huff single-handedly assaulted the enemy gunners. Later that afternoon, Feb. 8, 1944, he was the major actor when a larger patrol engaged 125 Germans in a firefight that left 27 enemy soldiers dead and 21 wounded. Three paratroopers were killed.

For his feat on that day, a few months later Paul Huff was receiving the Medal of Honor. He was the first paratrooper of World War II to win the country's highest military award.

Like Alvin York, Huff became a famous figure back in the United States as he toured the country with other Medal of Honor winners. Unlike York, however, he didn't return long to his Tennessee hometown. After World War II, he went back in the Army, served a tour in Vietnam, and finally retired as a command sergeant major in 1972.

Many heroes

Huff's individual valor was a highlight of the performance in the Anzio beachhead that brought awards to the 509th Parachute Infantry Battalion.

There were others.

Six weeks after Huff's exploit, Company C of the battalion assaulted two houses that were supposed to be lightly held enemy outposts. Instead, the attackers found themselves up against a veritable fortress. "Though nearly destroyed, they finally did it," they captured the houses and then repulsed an enemy counterattack, according to the airborne historian.

For that fight, the company was awarded the Presidential Unit Citation.

Company C's action was the final chapter of the 509th's 73 days of continual battle at Anzio. When 32-year-old Lt. Col. William P. Yarborough led his unit out of the fight, the original force of over 700 men was down to 125 survivors.

As the 509th was forging its unforgettable legacy, so was the 504th Parachute Infantry Regiment.

The unit, commanded by 34-year-old Lt. Col. Reuben Tucker, already had made history during the invasion of Italy in September 1943, by seizing high ground and holding off enemy counterattacks. Company K had won the Presidential Unit Citation.

At Anzio, the 504th became one of the most famous individual Army units of World War II when, thanks to the diary of a German foe, it won a nickname that was spread across headlines: "Those Devils in Baggy Pants."

Used throughout the perimeter of the beachhead, the paratroopers were especially busy at night patrolling and conducting daredevil raids on the enemy.

"Devils in baggy pants"

In one such raid, a captured German diary revealed how the enemy regarded them.

It read: "American parachutists — devils in baggy pants — are less than 100 meters from my outpost line. I can't sleep at night: they pop up from nowhere and we never know when or how they will strike next. Seems like the black-hearted devils are everywhere ..."

An alert unit public information officer seized on the diary to write a hurried history of the 504th's actions so far in World War II. "Devils in Baggy Pants" was among the first battle accounts published during the war.

For the soldiers of the 3rd Battalion of the 504th, the Anzio fighting brought the Presidential Unit Citation for their defense against German counterattacks in early February 1944.

By the end of the fight, however, many companies of the regiment were down to 20 or 30 men. When the unit left the beachhead two months after it landed, it had suffered 120 battle deaths, had 410 wounded in action and had given up 60 men counted as missing in action.

The 504th, which had been part of the 82nd Airborne Division, was so battered by its Anzio experience that it was not rebuilt in time for the division's next mission, the jump into Normandy on D-Day, June 6, 1944.

Greatest tribute

Anzio was ironic in that so many paratrooper legends and legacies were made, and so much paratrooper blood was spilled, in a campaign in which they did not do the thing for which they were trained, but rather got to the fight in an amphibious landing and generally fought as front-line infantry.

In a real sense, however, that they were used in such a way is the greatest tribute of all to the Anzio fighters.

The Allied commanders who had seen the individual fighting qualities of paratroopers in North Africa, in Sicily and in Italy were always eager to use them in the toughest assignments, encouraged by the belief that they would pull off tasks that others could not.

The Anzio campaign wrote a chapter in airborne history that proved the belief.

In A Hot Jungle In 1944, A Private Found Honor

Originally published December 29, 1994

Even the heat of a Robeson County summer had not prepared Pvt. Deormy Ray for the steamy climate of the jungle on the Pacific island of Bougainville.

Yet, in the spring of 1944, the young black soldier from North Carolina was not complaining.

For in a World War II Army in which black men were rarely called on for combat roles, assigned instead as road builders, stevedores, truck drivers and laborers, Ray was in the thick of front-line service.

As a trooper in the 93rd Cavalry Reconnaissance Troop, attached to the 25th Infantry Regiment, Ray was among the few thousand black soldiers in the Army called on to meet the enemy face-to-face.

And when that actually happened, Ray, a former plowboy in fields near the little Robeson County town of St. Pauls, won distinction for bravery — and gave his life.

Ray's unit, part of the all-black 93rd Division, and attached units of the 25th, had arrived in the Pacific in the first months of 1944.

In the rigidly segregated World War II Army, black soldiers who were trained for combat, other than artillerymen and combat engineers, were allowed only in the units of the 93rd Division and its sister, the 92nd, which was sent to the European Theater of Operations.

And in neither theater were the divisions ever actually employed in their entirety.

The 93rd, for instance, was, in the words of its division history, "parceled out to various islands throughout the Pacific for the duration of the war." Its men were used as stevedores, laborers, port guards and supply workers.

Among all the division's units, only the 25th Infantry Regiment and its attachments were ever actually used as fighting forces.

Even the men of these units first seemed destined never to see action. The regiment, along with the headquarters elements of the 93rd Division, were first put to work as stevedores and laborers in the docks and warehouses on the island of Guadalcanal.

But then, in late March 1944, Army authorities agreed to send the regiment to the nearby island of Bougainville to participate in mopping-up missions against the Japanese garrison there.

Reconnaissance patrols

Ray's small specialized unit had a unique job.

Moving around and behind Japanese strong points, it went on reconnaissance patrols to gather information to be used on maps of the big island.

It moved out on such a patrol on May 16, 1944.

Moving around enemy pillboxes, the patrol was ambushed by a Japanese squad and, fighting all the way, it fell back to a ridgeline where artillery fire was called in on the enemy.

After a night in a perimeter defense position, Ray's platoon moved out on the morning of May 17 to continue its map reconnaissance.

In midmorning, it was ambushed again. A firefight began that would last through much of the day.

During the hours of confused back-and-forth fighting, the platoon's commander, Lt. Charles Collins, was repeatedly wounded, partially blinded. He was led out of the fight by his sergeant, Staff Sgt. Rothschild Webb. It was three days later before they emerged from behind enemy lines.

For that, Webb would be one of the few black soldiers to be awarded the Silver Star Medal for heroism in battle.

Deormy Ray's heroism occurred in the same hours.

The citation for the Bronze Star Medal awarded to him posthumously read:

"While on patrol in the Solomon Islands, his unit was ambushed in enemy territory. Without regard for personal safety, he advanced without hesitation to give aid to a fallen comrade. Found that he was too late, and trying to withdraw, he was killed in action by a burst of machine gun fire."

Three days after the firefight in the Bougainville swamps, the 93rd Reconnaissance Troop ended its combat service on Bougainville.

Ray was one of three 93rd Reconnaissance Troop men killed in action on May 17, 1944, and among a dozen 25th soldiers to receive the Bronze Star during the unit's two-month combat service on Bougainville.

It is an irony of history that a deed such as that of Pvt. Ray, if performed by a white soldier in World War II, often resulted in an award of a higher-ranking medal than the Bronze Star.

The Bronze Star ranks fourth in the hierarchy of military medals that soldiers can receive for heroism in ground combat. Higher awards are the Medal of Honor, Distinguished Service Cross and Silver Star.

Nonetheless, there was recognition back home for the young North Carolinian who died in the jungles of Bougainville in the southwest Pacific.

Fifty years ago, in the final hours of December 1944, Gov. J. Melville Broughton made a final official visit to wartime Fort Bragg before ending his four-year term as North Carolina's chief executive.

And in a ceremony arranged by the post commander, Brig. Gen. John T. Kennedy, with Fayetteville city officials in attendance, Broughton presented Ray's medal to his father, Joseph E. Ray of Route 2, St. Pauls.

An adjutant read the citation, which was headed: "For heroic achievement in military operations in the Solomon Islands," and which described the service that had cost Ray his life.

Billy Shaw: A Memorial Day Memoir

Originally published May 29, 2003

W.M. "Billy" Shaw Jr. was 22 years old, a graduate of Davidson College and a private first class in the 5th Infantry Division when he wrote a fateful letter to his mother on a June day in 1944.

Shaw had been a happy-go-lucky kid who lived on Clarendon Street in Fayetteville, where he grew up in a loving family headed by the town's longtime postmaster.

Now, he was waiting for orders to go to France and join the battle against Nazi Germany in World War II.

Billy and his mother, Helen Shaw, shared a special bond. He was the oldest of her four children.

They wrote chatty letters to each other as he went off to basic training and then sailed for Ireland with the 5th Infantry.

But in June of 1944, in the middle of one such letter, Billy Shaw struck a poignantly intimate note. He wrote:

"Even though it does seem like your children are growing up, I am sure Helen (his sister) and I will consent to your continued bossing. I've had such luck with it for 22 years that I'm content always to be guided in part by your generous bossing. Funny thing, I can't ever remember you being far off the right track! Must be the woman's intuition, or something equally as mystic!"

He closed the letter: "Your letters are a constant strengthening of the tie across the water and I can't say how it would be without getting letters from you. Devotedly, Billy."

Less than three months later, on Sept. 12, 1944, Billy Shaw was mortally wounded when a German artillery shell exploded over the foxhole he was sharing with his best Army buddy near the Moselle River in eastern France.

In what turned out to be his last letter to his mother, dated a week before his death, he wrote that his letter-writing had suffered because "everyone feels that now the end is in sight and any delay causes a bit of resistance."

He concluded: "Sometime I am in the mood to write and have no trouble. But now it is no go as you can see. I am all here and in good shape. Will write again soon. Much love. Billy."

Billy Shaw never got to tell his family about his life in the battle zone.

His commanding officer, Maj. William E. Simpson of the 2nd Battalion, 10th Infantry Regiment, wrote about his record, prefacing his account with a tribute:

"Not only was your son a fine soldier, but he was the most outstanding man I've ever had during my five years of Army life.

"His clean living, devotion to duty, and friendly personality had won the highest commendation from my staff as well as all his buddies who served with him."

Shaw, who at the time was the acting operations sergeant of the battalion, had been recommended for a battlefield promotion to second lieutenant.

Wrote Simpson:

"In many instances, I have used him to do an officer's job because of the trust I had in him as well as the knowledge that the job will be handled properly."

On what turned out to be the last day of his life, Shaw had been sent forward to guide tank-destroyer vehicles to beef up the perimeter of a hard-won crossing of the Moselle River.

Simpson continued: "He reported back after doing a swell job and early in the morning he was hit by a tree burst of artillery at the battalion command post. We immediately aided him and rushed him to the aid station, where he died."

Simpson concluded: "Your son was given a military burial by an Army chaplain in a U.S. cemetery somewhere in France."

Billy Shaw's story has been generally told. His family remains prominent in Fayetteville.

His younger brother, Harry Shaw, was a city councilman and is the longtime chairman of the trustees of Fayetteville Technical Community College.

A sister, Gillie (Mrs. Riddick) Revelle, is a lifelong Fayetteville resident.

Mrs. Helen Shaw was still alive in 1986, when a visiting group of French cyclists paid tribute to him at an emotional ceremony at his memorial stone in the family plot in Cross Creek Cemetery. The group brought sand from the Normandy invasion beach and earth from Billy's wartime grave to sprinkle on the plot.

And Billy Shaw figures prominently in Fayetteville's official tie with Saint Avold, its sister city in France.

Groups from Fayetteville, including Shaw's siblings, have visited the U.S. military cemetery at Saint Avold where his stark white cross is among 10,489 there.

Long before the visits, however, Billy Shaw's cross had loving attention.

In October 1945, an officer of the U.S. Logistical Command wrote to Mrs. Shaw about a visit to the stone: "I know you will be interested in knowing that I arranged with a French family in the community to keep an eye on it for any personal touches that might make it more attractive, although the entire cemetery is being well cared for.

"This old fellow and his wife were delighted at the idea of having something to do for someone they knew, who lost his life in the liberation of their community."

It is a great privilege for me to have been allowed to read the lovingly preserved file of letters from Billy Shaw to his mother, and to make from it this memoir of him.

I write this in Memorial Day tribute to Billy Shaw and to the hundreds of thousands of young men and women like him, of every country and of every time, who in the phrase of the poet, poured out the red wine of their youth in the beastliness of war.

82nd D-Day Fighting Legendary

Originally published June 9, 1994

Paratroopers of the 82nd Airborne Division made history on D-Day, June 6, 1944, as the spearhead of the Allied invasion of Europe.

But the accomplishments of the men of the division on that day were only the opening chapter of a monthlong saga of battle for their unit.

Before they were pulled out to return to the camps they had left in England, paratroopers and glidermen of the division would fight as ground-slogging infantrymen with forces driving toward the big French port at Cherbourg.

On D-Day, men of the division had captured the French village of Ste. Mere-Eglise, but had missed the capture of bridges over the Merderet River to the west.

On June 7, the day after D-day, airborne forces continued to fight off strong German counterattacks as infantry units from Utah Beach and glider reinforcements poured into the beachhead.

By the end of the day, the perimeter around Ste. Mere Eglise was firmly established and troops were poised to drive across the causeway bridge over the Merderet River.

On June 8, Gen. Matthew Ridgway, the division commander, joined his parachute infantry forces with those of the 4th Infantry Division in an attack northwest from Ste. Mere Eglise, while others attacked again to secure the Merderet River causeway to the west.

Once more, the attack on the Merderet failed.

Finally, on June 9, in a bloody fight that produced heroic deeds among all ranks, the bridge at la Fiere — which had been a D-Day objective of the division — was finally captured.

Although the paratroopers might have been pulled out then, Allied commanders thought too highly of their fighting abilities not to use them in the drive for Cherbourg.

They were given a variety of attack missions.

From June 11 to June 13, the 508th Parachute Infantry Regiment drove south to the village of Beuzeville-la-Bastille, linking up with paratroopers of the 101st Airborne Division.

On June 14, other units of the 82nd joined a drive westward to cut the Cotentin Peninsula. The battered airborne units were heavily reinforced by tanks and artillery from the 7th Corps.

By the evening of June 16, the 505th and 508th Parachute Infantry Regiments had captured the important road center of St. Sauveur-le Vicomte.

Two days later, columns from the 9th Infantry Division drove through the 82nd positions and reached the west coast of the Cotentin.

Another job for the division was on the night of June 18 and the early hours of the next day, when men of the 325th Glider Infantry Regiment pushed across a bridge at the village of Baupte south of where the 508th regiment had attacked a week earlier.

A final task was accomplished on July 3, when the 82nd paratroopers were attached to the U.S. 8th Corps and made part of a three-division attack south out of the Cotentin toward the town of la Hays du Puits.

By morning of the 4th of July, the All-American Division was on the outskirts of the town, where, in accordance with the attack plan, it stopped and went over to the defensive.

Five days later, the orders arrived. The 82nd's Normandy service was over. It was relieved. Its survivors boarded landing craft at the Utah Beach and headed back to the Midlands of England, to Quorn and Leicester, Nottingham and Ashwell.

The division had paid dearly for its accomplishments.

Of the nearly 12,000 soldiers of the 82nd who had gone to France 33 days earlier, 5,245 — nearly 46 percent — were listed as casualties. About 1,300 had been killed, and 2,373 had suffered wounds or injuries serious enough to be evacuated from the battle zone. Many companies that jumped on June 6 with more than 100 men were down to a dozen.

Despite its losses the division's performance had, in the words of its historian, become "the stuff of instant legend."

Among the accolades received when it returned to England was a letter from Gen. Troy Middleton of the VII Corps, who wrote:

"The 82nd Division has fulfilled in an admirable manner every requirement asked of it. Furthermore, during the current campaign the division has acquitted itself in a manner such as to mark it as one of the outstanding units of the United States Army. The personnel of this division should feel proud of its achievement. Those of us who have watched her work have been proud to associate with you."

Remembering D-Day

Originally published June 6, 2002

By a quirk of the calendar, June 6 falls on this week's date for this column.

And so how can I not write about another June 6?

That would be the day in 1944, D-Day, when Allied armies poured ashore on the beaches of Normandy in the World War II invasion of Europe.

About the time you will be opening your morning paper, 58 years ago many young men were engaged in a desperate, unforgettable endeavor.

Fifty years later, then-President Bill Clinton immortalized them in an anniversary tribute, speaking from near the Normandy beaches, saying: "We are the children of your sacrifice."

Among the first to put his feet on Normandy that day, 20-year-old Kenneth J. "Rock" Merritt of Fayetteville came in with the 82nd Airborne Division, a rifleman in the 508th Parachute Infantry Regiment.

Merritt fought across Europe with the 82nd. He came home to eventually become command sergeant major of the 18th Airborne Corps and Fort Bragg, and retire as one of the legends of the airborne.

You can bet that at least for a moment today, he'll be thinking of that moment 58 years ago.

So will others.

Harry "Doc" Davis cuts my hair and keeps me straight on the history of old-time Massey Hill.

On D-Day, 20-year-old Doc Davis rode to the Normandy battle on a tiny anti-aircraft vessel named simply LCF-12.

On board LCF-12 with 66 other young crewmen, Davis spent D-Day seeing the giant panorama of the invasion of France a few hundred yards from where Rangers scaled the heights of Pointe du Hoc, where the landing boats of the 1st and 29th Divisions crashed ashore on the landing zone known as Omaha Beach, where German artillery blasted nearby vessels into oblivion, where there was confusion and death, heroism, and stubborn endurance.

Doc shows snapshots taken from the deck of LCF-12 as she plunged just off the Normandy beaches during World War II's "longest day."

The official name of this tiny fleet was 1st Gunfire Support Group. It was divided into three parts, each with four vessels. Four were rocket platforms. Four had two 47 mm naval cannons that fired directly at beach targets. Doc's LCF-12 and three others bristled with anti-aircraft guns, available to guard either the other vessels or nearby landing craft from whatever German aircraft managed to get over the invasion armada.

In the first hours after H-Hour of June 6, LCF-12 paced along Omaha Beach, and Davis and his shipmates saw the chaos that nearly led the Allied high command to call it off.

"I only saw one man moving," Davis recalls. "No one left the beach until noon."

In the choppy waters near shore, German artillery battered the landing fleet.

"We saw an LST (landing ship, tank) just blow up," Davis remembers. Later the crew of LCF-12 recovered the bodies of four dead Americans floating amid the terrible debris off the invasion shore.

After 19 days off Normandy, the 1st Gunfire Support Group was ordered back to England. Leaving four hulks beached or sunk and the others limping across the channel, LCF-12's crew maneuvered by stopping and starting its engines.

For Davis, D-Day was his first taste of war and also his last. But you can bet he is thinking about it this morning.

Combat veteran

My old golfing buddy who died two years ago, then-Lt. Phil Haigh, was also in the naval force.

He was a 20-something executive officer on a bulky LST that wallowed through the heaving seas and disgorged its cargo of fighting men and vehicles into the bloody surf at Omaha Beach.

In later years, Haigh was reticent about describing the D-Day spectacle.

In fact, he never saw a minute of the big day.

He heard it.

His duty required him to be below decks, below the water line actually, herding men and cargo toward the landing ramp.

Reticent maybe. But few World War II survivors saw as much war as Phil Haigh.

Normandy was his fourth invasion.

After that, the Navy figured he had done his duty, and he came home to be married a year after Normandy.

A D-Day anniversary without mention of Al Alvarez would be unthinkable.

He retired as a lieutenant colonel of paratroopers and had a distinguished second career as a public servant. He and his wife, Flo, are among Fayetteville's most famous retiree couples.

On June 6, 1944, "Smiling Al" was a 20-year-old artillery forward observer who came ashore with the 7th Field Artillery Battalion of the 1st Division on a bullet-swept stretch of Normandy beach known ironically in the invasion plan as "Easy Red."

Half the howitzers of his battalion never made it to the shore. But Alvarez and his comrades pushed inland, among the first to get off the beach, pressing up a gully or "draw" known as Colleville-sur-Mer.

His unit would receive a Presidential Unit Citation, an award given to an entire unit for heroism in extremely difficult and hazardous conditions. The degree of heroism required is the same as that which would warrant an individual receiving the Distinguished Service Cross, the second highest U.S. military decoration.

Two dozen of its men would receive medals for their tenacity and heroism on that day.

Alvarez keeps the flame of World War II memory alive by writing gripping narrative histories of his part in it. D-Day was just the opening chapter of his odyssey. He fought with the 1st Division, "the Big Red One," across Europe.

These were some who came home, who got to make a life after the sounds of war had ceased.

Others, in the phrase of the war poet, gave up the red wine of their youth, dying in the struggle.

I think often of Robert G. "Jonesey" Jones, a second lieutenant in the U.S. Air Force, who winged his barrel-shaped P-47 over the invasion beaches, flying in low to deliver the close air support that was the specialty of the 512th Fighter Squadron of the 406th Fighter Group.

Of that day, Jonesey Jones wrote home to his sisters in Fayetteville:

"I never saw so many ships on the sea at one time."

Jones didn't come back.

Exactly two months after the invasion, on Aug. 6, 1944, Jones was killed in action when German flak exploded his P-47.

This brash young pilot, winner of the Air Medal with three clusters in a battle career that last only four months, is honored under a cross at the Normandy American Cemetery at the village of St. Laurent-sur-Mer on the headland overlooking the beach near where Alvarez fought on D-Day. Jones' cross is at Plot C, Row 11, Grave 23.

Too many to mention

I can't mention them all, but there were others, such as:

• 1st. Lt. Augustus Hamilton Jr., 23 years old, a P-47 pilot who had been captain of Fayetteville High School's state championship golf team in 1940, missing in action after July 12, 1944.

• Pfc. Casper D. Waybright, 29 years old, of Fayetteville, who was killed in action on Aug. 9, 1944, in France with the 47th Infantry Regiment of the 9th Division. He was a four-year veteran, and Normandy was his second invasion. He had been overseas since the North African invasion in November 1942.

• James B. Bolton Jr., 19 years old, and Tech. Sgt. Ernest "Butch" Grissom, also 19, both of Fayetteville and classmates at Fayetteville High School. They were killed in action July 12, 1944, after surviving the June 6 invasion.

• Pvt. Dwight Miller Jr., "a popular member of the Fayetteville High student body," landed alive in France on D-Day, but he was missing in action after Aug. 8, 1944.

And so it is, that for those who survived to come home, and for those who did not, the day 58 years ago today lives on in either living or honored memory.

737th Tanks Were The Point Of Patton's Spear

Originally published May 31, 2001

The survivors of the 737th Tank Battalion call themselves "Patton's Spearheaders."

The 750 men of the outfit fought across Europe in World War II with Lt. Gen. George S. Patton's 3rd Army.

When more than 100 of these veterans meet in Fayetteville on June 7-9 for their 54th annual reunion, they might also aptly call themselves the 3rd Army Rescue Squad.

Twice in their long combat history, 737th soldiers rode their M4 Sherman tanks to break through German lines encircling other American soldiers.

In early August 1944, only a few weeks after landing across the D-Day invasion beaches in France, the 737th was engaged in rescue operations.

The 737th tankers fought alongside the infantrymen of the 35th Infantry Division to relieve the battered 2nd Battalion, 120th Infantry Regiment, 30th Infantry Division, surrounded for four days on a rocky eminence named Hill 317 near the French town of Mortain.

The bloody fight cost the 35th Division and its attached units more than 700 men killed and wounded in action.

It also won the 737th Tank Battalion a Presidential Unit Citation from Harry Truman and the unit Croix de Guerre from the government of France.

Four months later, just before Christmas 1944, the 737th rode again to the rescue, this time fighting with the tankers of the 4th Armored Division as they punched apart German lines circling the Belgian village of Bastogne.

Caught inside the German ring were paratroopers of the 101st Airborne Division and assorted artillery units, fighting to contain the last-gasp German counteroffensive of World War II, the Battle of the Bulge.

The reunion next weekend brings together what is rated as one of the largest World War II veterans organizations representing a unit that never included more than 750 soldiers at any one time.

They recall a history that included 299 days in combat from the first days after arrival in France on July 13, 1944, until the war ended in May 1945.

The 737th paid a price.

The survivors who meet in Fayetteville will remember 64 of their comrades killed in action. Twenty-one were reported missing.

The battalion armament consisted of 76 medium tanks. In the course of the war, the battalion counted 74 tanks lost to German fire.

There were also many battle honors. In addition to the unit awards received for the fight at Mortain in 1944, the men of the 373rd received two Distinguished Service Crosses, 22 Silver Stars, 188 Bronze Stars, more than 400 Purple Heart medals and two Croix de Guerre from France. Three enlisted men received battlefield commissions.

Fayetteville is an especially appropriate place for the reunion of the 737th veterans.

On a chilly day in January 1943, 20 young men from Fayetteville whose draft numbers had come up were sworn into the Army at the reception center at Fort Bragg. A few days later, they were on their way to Fort Lewis, Wash., where they trained as tankers in the newly formed 737th.

Nearly all who survived the war managed to stay in the unit throughout its World War II battle history.

Five are still in Fayetteville.

They are Jennings Johnson, Thomas Haywood, Thomas Williams, Willie Beard and Howard Bloom.

Jennings Johnson, who was a 22-year-old Fayetteville native when he was sworn in, is the local host for the gathering next week. Like many of his comrades, he is a walking encyclopedia of the history of his battalion, a unit that Patton called on repeatedly when the 3rd Army stormed across river after river in its drive from just back of the D-Day beaches to the far reaches of a defeated Nazi Germany.

The battalion's impressive historical sketch on its Web site puts it this way:

"General Patton sent a letter in late 1945 to the officers and men of the 5th Infantry Division, to whom we were attached.

"He wrote: 'To my mind, history does not record incidents of greater valor than your assault crossings of the Sauer and the Rhine. You crossed so many rivers I am persuaded many of you have web feet.'

"The 737th was the first tank battalion of the 3rd Army to cross the Moselle and Meurthe rivers, the first armored unit in 12th Corps to touch German soil, and the first armored unit of the 3rd Army to cross the Rhine River."

Among the men of the 737th who didn't get to share in all the history was young Cpl. Crawford L. Hedgpeth, who grew up in a house at the corner of Hay Street and Bragg Boulevard on a site where a veterans memorial park was dedicated last week.

In October 1944, his parents got a letter from his twin brother, Robert, saying that Crawford had been killed in action.

The department telegram came two days later.

The Crawford twins were together in the battalion only a few days before Crawford's death. Robert joined the unit after briefly attending an Air Corps training school.

Memories of Crawford Hedgpeth, and of many others, will be on the minds of the veterans of the 737th Tank Battalion next week as they meet at the Holiday Inn on Cedar Creek Road.

Operation Dragoon Overlooked

Originally published September 1, 1994

Everybody knows about Operation Overlord, the Allied Invasion of Normandy on June 6, 1944.

But you have to be a World War II history specialist to know about Operation Dragoon, which occurred 10 weeks later, on Aug. 15, 1944, when Allied forces came by landing boat, parachute and glider onto the Mediterranean coast of France.

This year the 50th anniversary of June 6 drew tens of thousands of veterans, as well as kings, queens and presidents to the Normandy shores.

It was a media event rivaling a half-dozen Super Bowls combined.

By contrast, Dragoon's 50th birthday was a big day only on the French Riveria, where thousands of enthusiastic French men and women attended ceremonies and four U.S. veterans parachuted in.

And by contrast to the tidal wave of media attention to Overlord, Dragoon's anniversary coverage was a ripple. Even this newspaper's Military section played it on the fifth page.

Despite its modest attention in the larger scheme of World War II, in the history of airborne soldiers, Dragoon was of major significance.

For one thing, the size of the airborne operation rivaled that of Normandy. Nearly 10,000 paratroopers and glidermen jumped and rode to the ground from the skies in the 4 a.m. opening phase of Dragoon.

For another, the organization of the attack was unique. The "1st Airborne Task Force" was made up of "15 disparate U.S. units and one British," of which "half were not airborne trained, half had not seen combat, and most had never operated together before," according to an account of the force. In code jargon, it was "Rugby Force."

The command was assembled from all over the Mediterranean theater and from the United States.

For instance, a planeload of staff officers from the Airborne Training Center at Camp Mackall next to Fort Bragg was plucked from offices in mid-July 1944 and sent to help at the command's headquarters.

Hardened core

At the core of Rugby Force, however, was a battle-hardened outfit, the 509th Parachute Infantry Battalion, which claimed lineage as the premier combat unit of the Army's airborne forces.

Elements of the unit had landed in North Africa in November 1942 and had fought in southern Italy and in the Anzio beachhead, winning a Presidential Unit Citation in that latter bitter struggle.

Among newcomer units were the 517th Parachute Infantry Regiment, the 550th Glider Battalion, and the 1st Battalion of the 551st Parachute Infantry Regiment, as well as associated airborne artillery, signal, ordnance, medical and engineer units.

In the early hours of Aug. 15, 1944, this "disparate force" pulled off a spectacularly successful mission.

By dawn, when tens of thousands of U.S. and French infantrymen of the 7th Army waded ashore along the sunny south coast of France, the airborne forces were ranging far beyond the beaches, breaking up German attempts to organize for a counterattack and seizing key ground.

By the end of the second day of Dragoon, the airborne task force "held seven towns, had taken 1,350 prisoners, had decimated enemy movements and defense far beyond the 7th Army's wildest dreams."

Among the feats of the paratroopers was the "inadvertant" capture by the 509th of the Riviera resort town of St. Tropez.

The town, famous today for its topless female sunbathers, was thought at the time to be too heavily defended to be seized from the air.

That accomplishment won the unit front-page coverage in The New York Times, under a headline: "American Parachutists Take St. Tropez by Lucky Error."

While the invasion of the Cote D'Azur of southern France has often been referred to as the "champagne campaign" because of the ease with which the 7th Army subsequently proceeded to drive German forces up the Loire River valley, the opening hours of the invasion were a major test of airborne fighting.

The War Department's opinion was highly favorable. It was, a report said, "the most successful airborne operation in the war."

The major units that pulled off this success went on to hard fighting and many honors in the European theater.

The 509th Battalion, 517th Regiment, and 551st Battalion participated in the Battle of the Bulge in December 1944, when airborne forces were thrown across the path of the enemy's Ardennes Offensive.

Presidential citation

The 509th won another Presidential Unit Citation for that fighting. The 517th won its first.

Many of the young officers and men of the Dragoon invasion went on to become major postwar figures, in uniform and out.

Among young officers who didn't remain in uniform after the war was 26-year-old Lt. Terry Sanford of the 517th, who went on to be governor of North Carolina in 1960.

Among professional soldiers who stayed in service were two young commanders of battalions in the 517th, Melvin Zais and Richard Seitz.

After the war, Zais wore four stars as commander of Allied Forces Southeastern Europe in Turkey. He retired at Fort Bragg in 1976 and died in 1981 at age 64.

Seitz, now 76, commanded Fort Bragg and the 18th Airborne Corps from 1973 until his retirement from the Army in 1975. He now lives in Kansas.

Perhaps the young officer who most influenced future airborne history was Lt. Col. William P. Yarborough, who was 32 when he led the 509th spearhead on the Riviera.

Yarborough had earlier designed the first airborne insignia and much of the original airborne equipment. He was in charge of Special Forces at Fort Bragg in 1961 when President Kennedy authorized the green beret as the unit's distinctive headgear and gave those soldiers a key role in fighting communist insurgency.

Yarborough, who is now 82 years old and lives in retirement in Southern Pines, is active in plans for the Airborne and Special Operations Museum to be built at Fort Bragg.

Although Special Forces were a postwar development, the pioneers of Green Berets' fighting style were actually part of Dragoon.

Veteran members of the 1st Special Service Force, a joint U.S. and Canadian version of the British Commandos, were among the first Dragoon soldiers in action, seizing islands off the harbor of Toulon.

The 1st SSF is recognized now as the World War II parent of all other Special Forces.

By the time of Dragoon, the organizer of this pioneer outfit, Brig. Gen. Robert T. Frederick, had gone on to other things. He was a major general, and he was none other than the commanding general of Rugby Force.

Did 82nd Lose The Moment?

Originally published September 15, 1994

The afternoon of Sunday, Sept. 17, 1944, was one of those crisp blue days that can bathe the green plains of Holland with a special glow.

For the 7,250 men of the 82nd Airborne Division flying over that landscape in a stream of 480 C-47 transport planes, there was little time to admire the beauty of the Dutch landscape.

At 1 p.m. on that day 50 years ago, the paratroopers and glidermen of the 82nd began the opening phase of an audacious operation aimed at driving the Germans out of Holland and perhaps ending World War II before 1944 was over.

Operation Market Garden involved the largest airborne force ever assembled, as well as a powerful ground force of armor and infantry.

The plan was for the paratroopers to seize a series of bridges over the many waterways that course across the Dutch plains, paving the way for the ground force to drive across the Rhine River at the city of Arnhem and on into the heart of Germany itself.

British Gen. Bernard Law Montgomery, who conceived the plan, set a timetable of three days for Market Garden.

The great daylight airborne jump, the Market part of the operation, involved the 82nd and its U.S. sister, the 101st Airborne, as well as the 1st British Airborne Division and the independent Polish Airborne Brigade, carried in 1,545 planes and 478 gliders.

The U.S. divisions jumped to seize bridges short of Arnhem, the 101st just across the front lines near Eindhoven and the 82nd farther out, at the Meuse River and near the Waal River at the city of Nijmegen.

The British and Poles jumped at Arnhem, 64 miles beyond the front lines, with the mission of seizing the great highway and railroad bridges there.

The Garden part of the operation involved the British 30th Corps, spearheaded by the Guards Armored Division and supporting infantry divisions. It jumped off for the 30-mile drive even as the paratroopers filled the afternoon sky.

For many paratroopers of the 82nd, Market Garden was the fourth combat jump of World War II. They and the men of the 101st had last been in action in France in June and July, first in the great June 6 D-Day invasion and then in the fighting in Normandy before the Allied breakout that had brought U.S. and British forces to the borders of Holland and Germany.

Since then, there had been a change in the lineup of the 82nd. The 507th Parachute Infantry Regiment, which had participated in D-Day, had been shifted to a new division, the 17th Airborne, to make room for the return of the 504th Parachute Infantry Regiment, which had been left behind in Italy the previous winter.

The other D-Day units were still there, the 505th PIR, veteran of four jumps, and the 508th PIR, as well as divisional artillery, engineer and logistical units.

One of the features of the afternoon jump was that men of the 307th Airborne Engineer Battalion jumped as a complete unit for the first time, coming down a few minutes after the 505th, second in line in the divisional stream.

The skills of combat engineers would be needed early on.

The paratroopers quickly seized a key bridge over the Meuse at the town of Grave. But two of four other target bridges south of Nijmegen were destroyed before they could be captured.

As the British armor struggled up the single road that was the axis of the operation's drive, and as British paratroopers were hemmed in short of the Arnhem bridges, 82nd paratroopers fought unsuccessfully for two days to grab the railroad and highway bridges over the Waal at Nijmegen.

By Sept. 20, British armored forces had entered the 82nd's area, but so had fresh German forces.

A German counterattack temporarily overran the divisional drop zones on Groesbeek Ridge southwest of Nijmegen, and men of the 505th and 508th were busy fighting back.

That afternoon came one of the most daunting missions performed by the 82nd in World War II.

Riding in 26 canvas boats provided by British engineers, 400 men of the 3rd Battalion of the 504th PIR paddled across the Waal under a storm of German artillery fire to seize the other ends of the Nijmegen bridges.

The Waal River crossing was proposed by Brig. Gen. James Gavin, the 82nd commander, who had trained in hostile river crossings with the division back at Fort Bragg in 1942.

However, there was quite a difference in the Waal compared to the sluggish streams of Fort Bragg. The Dutch river was a half-mile wide, with a current of 10 miles an hour.

The mission cost over 200 casualties in Maj. Julian Cook's battalion. And it, in a sense, came too late.

By the evening of Sept. 20, British paratroopers in and near Arnhem were surrounded by a tightening ring of German forces. British armored columns crossed the newly captured Nigmegen bridges, but then stopped for the night just beyond the Waal, 10 miles short of Arnhem.

For four more days, British forces would batter at the German perimeter around Arnhem, but they could not break through.

On Sept. 26, the few hundred surviving men of the 1st British Airborne were given permission to escape across the Rhine and try to gain the lines to the south. Market Garden was over.

In postwar analysis of why Market Garden fell short of its objective, historians quoted the British general who commanded the airborne assault, Gen. Frederick Browning. In a remark at the pre-attack planning session, he said that the plan "might be going a bridge too far."

For U.S. paratroopers, they also quote the notion of "the unforgiving moment," which holds that in a military operation, there arise momentary, sometimes unexpected opportunities that, if not taken, lead to increasing difficulties.

They apply that to the failure of the 82nd's commanders to go all out to seize the Nijmegen bridges in the first hours of their assault, at a time when German defenses were largely not formed. Instead, they opted to secure their drop zones, which were just inside the Dutch-German border.

Two days later, when a major effort was made at Nijmegen, the German force had grown from a handful to more than 500 defenders. Then came the daring river crossing, but at a time when Market Garden was already slipping far behind its planned schedule.

But that is postwar quarterbacking.

For the men of the 82nd, Market Garden went down as a success, if a costly one.

And it wasn't over yet. Paratroopers would remain in the battle line on the Dutch plains until Nov. 11, 1944, fighting as infantry alongside British units.

Montgomery, like U.S. commanders, didn't pass up the opportunity to hold on to airborne soldiers whenever he had them.

1944 Became The Year Of The Beer

Originally published October 20, 1994

In the fall of 1944, the Allies were rolling back the dictators on the battlefronts of World War II.

Back in Fayetteville, there was the great beer war.

It began when the Office of Price Administration rolled back the prices for the amber stuff that was the favorite off-duty beverage of the tens of thousands of soldiers at Fort Bragg and adjoining Camp Mackall.

The order hit like a thunderclap when it came down in September 1944 from the price-control office in Raleigh.

OPA had surveyed the beer scene in 54 eastern counties and discovered that nearly half of the beer sold in the region was sold in Cumberland County, which had but 6 percent of the region's population.

Despite the huge market for brew in this reigning beer capital of the state, OPA found that prices generally were higher than anywhere else.

It ordered a 3-cent slash in the 20-cent price of a bottle, and a 1-cent reduction in the 10-cent price for a glass of draft.

That would bring Cumberland's 111 beer-license holders (roughly one license-holder for every 600 civilians in the county) in line with those in other counties.

Beer dealers didn't take it lying down. They leaped to the barricades. They even counterattacked.

Splashing big advertisements in The Fayetteville Observer, and calling "mass meetings," they vowed to "strike." They would sell no beer before its time, or at least at those prices.

Seventy-eight license-holders, from corner beer joints to the big beer halls such as the Town Pump and the Brass Rail, as well as fancy restaurants such as the Prince Charles Hotel, signed an advertisement claiming that higher rents, higher wages and higher taxes necessitated the higher prices.

For a tumultuous week, the battle raged. But the dealers found few allies.

Brig. Gen. John Kennedy, the commanding officer at Fort Bragg, called on soldiers to refuse to pay more than the new prices. The state beer dealers' association condemned the strike and warned of sanctions.

After a week, the dealers capitulated. At another mass meeting, they vowed to follow the order, meanwhile appealing through regular OPA channels.

In a sense, the great beer war of 1944 was inevitable.

Like the battle against dictatorship abroad, it had its beginnings several years earlier when the sleepy town of Fayetteville awoke to become, in the phrase of the 1944 OPA, "North Carolina's first war-boom town."

By the hot summer of 1941, more than 60,000 men in uniform were swarming through the town on weekends, trying to enjoy some of the civilian-side pleasures they had so lately given up, such as beer, a good hot dog or a ham and cheese sandwich.

This weekend army created a beer bonanza.

In the autumn of 1940, Fayetteville beer dealers marketed 2,500 cases a week. By July 1941, weekly sales were 25,000 cases. The return on a case had gone from one dollar to $2.20.

But all was not hunky-dory. Some of the beer aficionados had complaints for their hometown newspaper.

The Chicago Tribune sent reporter John Thompson, who later would jump with paratroopers in the invasion of Sicily, to survey the pricing scene in Fayetteville.

The headline on his report said: "Welcome Mat Out for Draftee If He Can Pay."

Chicago men who were accustomed to 10-cent beer were paying twice that much out of their $21-a-month Army pay, he reported.

Thompson painted a picture of "Fayetteville merchants as a group of Shylocks," according to a rebuttal article in the Observer.

But then the Fayetteville newspaper carried out its own investigative reporting survey, and it echoed much of what Thompson had said, at least about beer.

The Observer found that "beer originally sold at 12 cents is now 15 or 20."

Sellers told the reporter that wholesalers had raised the prices, "but this isn't so, two beer wholesalers told us."

Others claimed that because their competitors raised prices they were forced to follow suit. "An obviously ridiculous statement," wrote the Observer reporter, "since lower prices should by all lights sell more beer."

The most devastating information about prices came from an insider source:

"The waitresses will tell you that the price of beer was upped simply because cafe owners knew they could get it. Sales have not dropped off."

Waitresses said that despite the price hikes, their income was unchanged, the usual dollar a day in wages and tips.

The "most tenable" argument of the sellers, said the Observer, was their claim that weekend beer-drinkers were taking up table space that would ordinarily be occupied by diners.

While backing the complaints about beer, the Observer found that Chicago men were paying the same 10 cents as they did back home for the four lowest-priced sandwiches — cheese, ham, egg and hamburger.

But then, in what it said was "the most uncalled for" price hike of all, the newspaper reported that the 5-cent hot dog was now the 10-cent hot dog.

Wiener sellers blamed higher wholesale prices.

That, concluded the Observer report, was "balderdash."

Heroism Of Black Units Recalled

Originally published December 18, 2003

It was a perfect picture postcard of a winter wonderland.

Two weeks before Christmas in 1944, the Ardennes region of Belgium was a snow-covered vista of fir forests and rocky hills.

For the men of the 578th Field Artillery Battalion, it was a new experience.

The 578th, you see, had spent the past two holiday seasons in the comparatively balmy clime of Fort Bragg, where it was activated in 1942 as a field artillery regiment.

The gunners of the battalion were more familiar with the green forests that covered the rocky ridges and coves of the Ardennes, not unlike the sandy pine woods of the American South.

And this was a uniquely "Southern" outfit, one of nine artillery battalions in the Ardennes composed of black artillerymen, commanded mostly by white officers.

Many of the gunners of the racially segregated 578th were natives of the Carolinas, and they had spent two summers training at Fort Bragg.

The cold was bitter on the ridge known as the Schnee Eifel, and the snow was everywhere.

But the troops were enjoying it.

For nearly two months, they had been in the same position, near the village of Burg Reuland, and there had been little call for the big 8-inch guns.

A welcome calm

After a summer and autumn of campaigning across France, following landing in the Normandy beachhead in July, the men of the 578th welcomed the winter calm, even if it was a frigid one.

By the second week in December 1944, the battalion's gun lines were behind the outposts and foxholes of the 106th Infantry Division, the greenest big unit in the European Theater of Operations, having arrived from the United States less than three weeks earlier.

Four of the seven artillery battalions supporting the newly arrived 106th were black outfits.

Every four days, the big guns would blast away with precisely 250 rounds at the German lines to the east, following orders that the Ardennes part of the western front was to be a "quiet" one.

In the Ardennes, fought-out or brand-new U.S. divisions could rest or learn the ropes for further battles.

The long-range artillery, like the 578th's tractor-drawn howitzers, was fired mostly for effect, required as the soldiers were to husband scarce ammunition.

Billeted in houses, log cabins and winterized tents, the men of the 578th thought it was "as tranquil as garrison conditions."

They made the most of the time by constructing a recreation center, a beer hall, a bowling alley and even a badminton court.

There were United Service Organizations shows and movies, short trips to nearby villages (one was a road center named Bastogne), and for a few lucky ones, furloughs to Paris.

By Dec. 16, 1944, men of the battalion were "rehearsing a choral play" for Christmas Day.

And then, 59 years ago this week, it came.

Out of the misty forests, German infantry and tank columns swarmed over the Schnee Eifel in what forever after would be known as "the Battle of the Bulge."

Before it ended three weeks later, this last-gasp offensive of Hitler's Germany bent but did not break the British and American lines, inflicting 80,000 U.S. casualties, including 19,000 killed in action.

When it began in the early hours of Dec. 16, the men of the 578th found themselves fighting as infantry.

Dug in with the startled soldiers of the 424th Infantry Regiment of the 106th Infantry Division, they fought with small arms against German troops swirling through their positions.

774 rounds fired

On that first day, as Germans lurked only yards away, the guns of the battalion managed to get off 774 rounds.

When German infantry closed in on the divisional observation post at the crossroads of Heckhuscheid, gunners of the battalion, led by Lt. Col. Thomas F. Buckley, were in the defensive perimeter, with Buckley personally gunning down a half dozen of the enemy.

During the next day and a half, as the 106th collapsed (two of its three infantry regiments would surrender almost to a man), the 578th found itself linked to the all-white 559th Field Artillery, a 155 mm outfit, in a scratch gun line known as "Groupment Buckley."

On Dec. 19, the unit's guns were ordered to pull back on the Houffalize Road, where, after-action reports said, the unit fought with "steadiness and determination" as the U.S. lines stiffened around the flanks of the German offensive.

From the Houffalize position, the guns of the 578th fired in support of the men of the 101st Airborne Division, the unit made famous for its defense of the surrounded town of Bastogne.

Three days later the battalion moved again, this time to positions behind the 3rd Corps of Lt. Gen. George S. Patton's 3rd Army as it struck north against the flank of the German offensive.

On Christmas Day, 1944, their play rehearsals long forgotten, the 578th's gunners fed shells into the pieces aimed at the crumbling German perimeter around Bastogne.

Often moving every half-day as the U.S. counteroffensive picked up steam in the two weeks after their peaceful winter interlude was so violently interrupted, the gunners of the 578th managed to fire 3,445 rounds of their big-blast artillery ammunition.

A few days after the New Year in 1945, the battalion was assigned to the veteran 333rd Field Artillery Group, a higher headquarters for several black gun outfits. Buckley took command of the group.

For the next four months, the 478th moved forward with the victorious American forces as they first broke the crust of German resistance and then overran Germany.

The significant story of the 578th and other black artillery outfits in the Battle of the Bulge is often obscured by the more famous tales of the beleaguered paratroopers in Bastogne and of the stout defenses of outnumbered infantrymen and engineers who held the "northern shoulder" of the bulge in U.S. lines.

Maj. Gen. Maxwell Taylor, commanding general of the 101st Airborne, didn't forget one such unit, however.

He recommended a Presidential Unit Citation for the 969th Field Artillery Regiment, whose gunners stayed in the Bastogne perimeter, firing guns abandoned by another unit.

It was the first award of a distinguished unit citation to an black combat unit in World War II.

Airborne Courage Saved Day

Originally published December 11, 1997

The first truckloads of paratroopers rolled to a stop in the narrow streets of a Belgian village clinging to the sides of high bluffs overlooking the meeting of the Salm and Ambelve rivers.

The time was midday of a cold, gray Dec. 20, 1944, and the most famous of World War II fights, the Battle of the Bulge, was raging across a 55-mile front of the Allied lines.

Four days earlier, several hundred thousand German troops, spearheaded by tanks, had roared out of the night to smash through thinly-held U.S. lines in the rugged hills, forests, streams and villages of the Ardennes.

The village of Trois Ponts was in the path of the German spearhead, a tank-tipped force driving to cross the Meuse River and reach the great port of Antwerp.

The paratroopers who hastily unloaded into the cobbled streets were men of 2nd Battalion of the 505th Parachute Infantry Regiment of the 82nd Airborne Division. They were ending a two-day convoy march from camps in northern France.

Their orders were straightforward: Hold the line of the narrow, rocky Salm River against the oncoming Germans.

Other regiments of the 82nd were rolling to other blocking points along the so-called "northern shoulder" of the great bulge created in the Allied line by Hitler's winter legions.

All the paratroopers would fight critical engagements with the Panzer soldiers.

But when the struggle for the Bulge was over, the fight at Trois Ponts would rate as the "high water mark" of the enemy's Ardennes offensive, according to a new history of the Army's World War II service in Europe, "Citizen Soldiers" by Stephen Ambrose.

Unexpected help

As the paratroopers arrived, there was a piece of unexpected help. A small party of engineers under Lt. Robert P. Yates had already dropped two of the town's three bridges and wired the other for demolition.

Lt. Col. Ben Vandervoort, commander of Trois Ponts paratroopers, set up his forces to take advantage of the high bluffs through which the Salm wandered, lined by a railroad and an auto road.

He sent Company E, under Capt. William Meddaugh, across the river to the bluff on the other side, facing east. With a single 57 mm anti-tank gun and six bazookas, the paratroopers dug in to meet the men and tanks of the 1st Panzer Division who were inching their way from the east.

The battalion's other two companies took up firing positions in village buildings on the lower western bluffs. A four-gun battery of the airborne's tiny 75 mm pack howitzer was in artillery support.

Shortly after 3 a.m. Sept. 21, 1944, the fight got under way.

Two German armored scout vehicles appeared out of the woods to Meddaugh's front. The paratroopers let them come down the road, then bushwhacked one with a bazooka shot. The other was stopped by an anti-tank round. The German crews were driven off.

The dug-in paratroopers promptly moved their small minefield out beyond the wrecks, using them as a come-on for the heavier stuff they knew would be arriving.

Dawn came, and then the big attack.

At 8:30, five German tanks and companies of infantry pressed out of the woods.

Bazooka and 57 mm rounds stopped all the tanks. The infantry wavered. But then others came on. The big German tanks maneuvered in the roads that narrowed as they became twisting streets.

From the heights on the west side, Vandervoort could clearly see Panzer vehicles, tanks and infantry massing for an even larger attack against Meddaugh's thin line.

"Disaster seemed eminent," he wrote later. "But not one man of Company E left his position."

Vandervoort sent Company F across the remaining bridge to counterattack in the jammed streets. But the weight of the German attack still grew.

Vandervoort himself then jumped into his jeep, dodged wrecked vehicles on the bridge, scrambled up the 30-foot bluff, and got to Meddaugh's command post just as tanks maneuvered only yards away.

It was 3:30 in the afternoon.

"Pull out, and do it now!" he yelled.

"You're surrounded"

A German officer called: "Halt, Americans, you are surrounded!"

The paratroopers listened to Vandervoort.

Using what they called "walking fire in reverse," they fell back.

As Vandervoort jeeped back, he saw the hardworking 75 mm anti-tank cannon jackknifed in a ditch. It was abandoned.

Dodging fire from the streets, Meddaugh's men dashed for the eastern bluff.

Company E survivors got to the bluff even as German tanks finally cleared a path through a street leading to the remaining bridge and were gunning for its eastern terminus.

The paratroopers didn't hesitate. They literally jumped from the three-story high sheer bluff, several suffering broken ankles. They ran 100 yards to the narrow river, waded across, sprinted over the road and railroad tracks and hauled up the opposite bluff.

Vandervoort later said:

"They were a tired, ragged, rugged looking bunch. But what I saw was beautiful. About 100 troopers, with weapons and ammunition." They were still ready to fight.

Thirty-five of Company's E' s little band had been killed or wounded.

By the time survivors reached the line of the other companies, the last bridge was blown, and artillery fire from divisional batteries several miles back was dropping on the German columns piling up on the roads and fields east of the Salm.

In a last daylight attempt to drive Vandervoort's defenders off the positions on the west side, a giant Tiger tank, bearing the skull-and-crossbones insignia of the SS Panzers, took position amid the jammed streets of "east Trois Pont" and fired its big gun point-blank, smashing buildings with a single shot.

Infantrymen huddled behind its flank, waiting to dash for the ruin of the bridge.

Bazooka rounds merely bounced off the huge machine.

But then a paratrooper mortar got off a phosphorus round that exploded on the tank turret, its spectacular shower of white hot fragments raining on the infantrymen. The Tiger backed out of the street.

The Panzer infantrymen shortly tried one more time to get across the Salm.

But the attack was half-hearted and ended in the looming darkness of evening.

By 6:30, the fight was essentially over for the day.

Most of the German tanks had turned away, probing for somewhere else to get across the Salm.

After the war, the commander of the German spearhead, Lt. Colonel Jochem Peiper, told a U.S. interviewer that if his troops could have gotten through Trois Ponts that day, they would have been at the Meuse by nightfall.

Second-guessing

Years later, Vandervoort and his 82nd Airborne commander, Maj. Gen. Jim Gavin, wondered why Peiper didn't go all out the first night, using infantry to seize bridges and hit the still-arriving paratroopers.

Even they agreed it probably would have succeeded.

In which case the stand of the 2nd Battalion at Trois Ponts would have been just another of the fights of the early days of the Battle of the Bulge when U.S. forces lost ground but bought time.

Trois Pont never succumbed.

Special Operations Had Noble Beginnings

Originally published January 26, 1995

It was a black day in U.S. military history, that May morning in 1942 when battered U.S. and Filipino defenders surrendered to the Japanese who had them bottled up on the Bataan Peninsula of Luzon Island.

It was a bright day 32 months later, on Jan. 30, 1945, when U.S. and Filipino forces liberated several hundred Bataan survivors in a daring daylight raid on a Japanese prisoner-of-war camp at the village of Cabanatuan.

People and military units with ties to Fort Bragg participated in these famous World War II events.

The Bataan surrender was followed by the infamous "Death March" in which more than 10,000 POWs were herded for 70 miles to prison compounds. Many died from heat, starvation and the brutality of their escorts.

Among those who survived that ordeal was Brig. Gen. Edward P. King, who 23 years earlier — as a young officer in World War I — had surveyed a site for a new artillery post in the North Carolina Sandhills to be named for Confederate commander Braxton Bragg.

There was Col. Edmund J. Lilly (1895-1978), a Fayetteville native who as a young officer had designed the "Screaming Eagle" shoulder patch that became the emblem of the 101st Airborne Division.

And there was 22-year-old Jesus Rabano, an enlisted man in the ranks of the crack Filipino Scouts. He came to the United States after the war, retired as a master sergeant and became a noted civilian employee at Fort Bragg.

None of these three were in the Cabanatuan compound that memorable day 50 years ago when men of the U.S. Rangers, the Alamo Scouts and Filipino guerrillas burst out of the underbrush to quickly overwhelm Japanese guards and free 511 Bataan survivors.

Today, the men of the units who pulled off the raid are honored in memorial stones on the Memorial Plaza recently dedicated adjacent to the new headquarters of the U.S. Army Special Operations Command at Fort Bragg.

They were men from a company of the 6th Ranger Battalion and teams from the

Alamo Scouts.

Those soldiers of 50 years ago are considered ancestors of today's special operations soldier.

The Cabanatuan raid remains as the most spectacular among history's small list of successful POW liberation operations.

The raiding force, with Alamo Scouts out front, marched 24 miles behind Japanese lines, set up an observation post in a hut across the road from the prison compound, and then stormed it, killing or wounding 200 Japanese defenders in a 30-minute firefight in which not a single POW was hurt.

The force of 107 Rangers, a dozen Alamo Scouts and Filipino guerrilla fighters then successfully escorted the liberated prisoners to American lines.

The 6th Rangers and Alamo Scouts were unique outfits of the 6th U.S. Army, the main ground force in the Southwest Pacific.

The Rangers, patterned on units already serving in the European Theater of Operations, had been constituted in New Guinea in January 1944 by converting a long-idle artillery outfit, the 98th Field Artillery Battalion, into a three-company force.

The unit was trained for Ranger-type missions, specifically to seize small islands in the roadstead of the U.S. landings on Leyte Island in the Philippines in October 1944.

They carried out that mission to perfection. One of the successes was that of Company B, led by young Capt. Arthur D. "Bull" Simons. Men of the unit scaled cliffs to take out a Japanese garrison defending a lighthouse.

Twenty-five years later in 1970, Lt. Col. Simons would lead a notable special operations mission of the Vietnam War, a 400-mile air-drop raid on the Son Tay POW camp just outside the North Vietnamese capital of Hanoi. U.S. POWs had been moved away, but the raiding force wiped out a 25-man garrison and got away without casualties.

Before executing the Cabanatuan raid, the 6th Rangers spearheaded U.S. ground forces landing on Luzon Island. They continued in small-unit missions and cooperated with Filipino guerrilla forces.

The Alamo Scouts were formed in November 1943 in New Guinea. Even more than Rangers, they were a direct forerunner of today's Green Beret-type soldiers.

Their job was defined as "strategic intelligence" and "covert operations." Volunteers, they were organized into teams of one officer and six or seven enlisted men. More than 20 teams were in place at one time or another, carrying out a variety of missions for 6th Army headquarters.

Among the successes was a liberation mission that freed 32 natives from a Japanese prison pen in New Guinea, wiping out the garrison of 25 without losing a man.

In the Philippines, the Scouts often operated as far as 250 miles ahead of the main U.S. forces, cooperating with the widespread Filipino guerrilla operations.

Leading the 6th Ranger rescuers to Cabanatuan was the highlight of the Alamo Scout record, however.

Within weeks after the surrender of Japan in September 1945, the Rangers and the Scouts were disbanded. Their role as spearheads of an invasion of the main islands of Japan wasn't to be needed.

Varsity Assault A Lesson In Courage

Originally published April 13, 2000

For military spectacles, it could hardly be topped.

A half-million troops pouring over the Rhine River on a dozen bridges thrown across by combat engineers.

Three thousand artillery pieces banging away in support.

And in the air overhead, a great stream of 1,500 transport planes and 1,300 gliders loaded with 16,000 British and U.S. paratroopers and glidermen.

It was the morning of March 24, 1945. Operation Varsity was under way.

The day was crisp and clear, and in less than a half-hour the men in the sky were on the ground, more than 8,000 dropping by parachute, nearly as many coming in by glider.

Within 72 hours, the force commanded by British Field Marshal Bernard Montgomery was across the Rhine in overwhelming force, leaping the last natural barrier between Allied forces and Berlin, where Adolf Hitler crouched in his underground bunker as his Nazi empire crumbled in what would be the last weeks of World War II in Europe.

Last of the war

For U.S. airborne forces, the jump across the Rhine at Wesel 55 years ago this spring would turn out to be the last of the war. It ended a string of combat jumps that started in North Africa 29 months earlier, followed by Italy, Normandy, Southern France and Holland.

The 8,000 U.S. airborne soldiers wore the shoulder patch of the 17th Airborne Division.

Military pioneer

Their commander was 45-year-old Maj. Gen. William "Bud" Miley, who along with his friend, Maj. Gen. William C. Lee, had been a pioneer organizer and trainer of early paratrooper units.

The assault forces were men of the 507th and 513th Parachute Infantry Regiments and the 194th Glider Infantry Regiment.

They were a mix of veterans of past battles and of others nearly new to war.

Their division was less than two years old, activated at Camp Mackall in the North Carolina Sandhills in the spring of 1943.

In early 1944, they sailed to England, where Allied forces were assembling for D-Day, the June 6 assault on Europe.

By a fluke of the war, the new 507th PIR found itself temporarily assigned to the 82nd Airborne Division and given a D-Day assignment in the Normandy invasion.

Six months later, on Christmas Day 1944, the reunited division went into action in the snowbound Ardennes Forest in Belgium, helping stem the German attack known as the "Battle of the Bulge."

Now three months later, in the spectacular Varsity assault, the 17th carried the colors of the U.S. 18th Airborne Corps into a giant military operation that ironically turned out to be less significant than expected.

By the time the British-led operation began, U.S. forces farther south along the Rhine had grabbed a bridgehead across the river at the city of Remagen.

A day before Varsity, Gen. George S. Patton launched a full-scale drive into Germany.

Rather than being the main punch of final Allied victory of Hitler's Germany, Varsity became the northern pincher of a double-pronged Allied attack that ended the war.

No lack of courage

If the 17th Airborne Division's moment in World War II history turned out not to be as strategically significant as it might, there was as usual no lack of courage and gallantry among the paratroopers and glidermen who participated.

The exploits of three men of the assault would win them the Medal of Honor, all posthumously.

Pfc. Stuart Stryker of the 513th PIR was a platoon runner.

After his unit was pinned down, he single-handedly rushed a German headquarters building, waving a carbine to rally the platoon.

He died leading them in a charge that broke the enemy resistance.

Pvt. George L. Peters was a radio operator in the 507th PIR who landed practically on top of a German machine gun nest.

While his buddies struggled to untangle from their gear, he charged the enemy with rifle and grenades.

Mortally wounded, he crawled far enough to heave a final grenade into the German position.

Tech. Sgt. Clinton M. Hedrick of the 194th Glider Infantry Regiment survived the first two days of combat, but was killed in another single-handed charge during an assault on a fortified castle near the town of Lembeck, firing his semiautomatic weapon from his hip.

1,400 died

They were among the 1,400 men of the 17th Airborne Division killed instantly or fatally wounded in World War II.

The names of Stryker and Hedrick are memorialized at Fort Bragg by a golf course and a stadium, respectively.

"Bud" Miley's division lasted barely four months after the war ended in Europe.

It sailed home in September 1945 and was deactivated at the debarkation port.

The great arena along the Rhine River where operation Varsity unfolded 55 years ago periodically is the scene for a commemorative ceremony uniting veterans of the Allied airborne forces and of the German parachute-soldier division that formed a major portion of the defense overwhelmed by the giant assault in 1945.

1948 A Year Of Troubles For Bragg

Originally published August 6, 1998

Fifty years ago this summer, Fort Bragg was the pits.

In 1948, the sprawling Army post in the Carolina Sandhills, where as many as 75,000 World War II soldiers trained at a time, had a postwar uniformed population of fewer than 15,000.

But even with that small complement, there was a severe shortage of housing, especially for the families of married soldiers. And Cumberland County's old and crowded public schools were inadequate for military children as well as civilian.

Fort Bragg's troubles in the summer of 1948 were endemic in the Army three years after the end of World War II.

Even greatly reduced from the millions of men who served in the war, the Army was still 100,000 short of its manpower requirements. Military service held no attraction in a country absorbed with peacetime pursuits.

Fort Bragg was still seeking a postwar mission.

In 1947, a parade of top Army brass visited the post, and despite the rapid drawdown of the World War II Army, they voiced hope that Fort Bragg and Pope Field would have a major postwar future.

Gen. of the Army Dwight D. Eisenhower made a visit as Army chief of staff in late March of 1947. Gen. Omar Bradley, who was to follow Eisenhower as Army chief of staff, came in early December.

Both of the famous World War II commanders joined Maj. Gen. James Gavin on the reviewing stand to view the 82nd Airborne Division's march-by and parachuting demonstrations. The word was that the division would be expanded to 16,000 officers and men, a third larger than its wartime size.

But despite these brave promises, the new year brought sober reality.

Fort Bragg in 1948 was headquarters for the 5th Army Corps. The corps commander, Maj. Gen. Leroy Irwin, was also commander of the post. Irwin was World War II commander of the 5th Division and the 12th Corps under Gen. George S. Patton.

The largest outfit on post as 1948 began was the famed 82nd Airborne Division. It had come home in late 1946 from its illustrious battle career in the European Theater of Operations. It was still commanded by Gavin, its famous World War II commander.

But while there were hundreds of veterans of the Normandy invasion and the Battle of the Bulge still wearing the All American shoulder patch as nomcoms or officers, most of the soldiers in the division's understrength units were postwar recruits.

The most experienced unit on the post was the 555th Parachute Infantry Battalion, the Army's only World War II paratrooper outfit for black soldiers.

Relegated to fighting forest fires during the war years, many of the "Triple Nickel" men stayed in after the war in hopes of seeing the combat service that was denied them.

1st integrated unit

In December 1947, Gavin had ordered the 555th absorbed into the 505th Parachute Infantry Regiment of the 82nd Airborne, the first racially integrated unit in the Army's history.

But Gavin was gone after February 1948, stepping down on the same day Eisenhower announced he was leaving the Army to become president of Columbia University in New York City.

Gavin was replaced in command of the 82nd by Maj. Gen. Clovis E. Byers, whose war career was in the Pacific, first as commander of the 32nd Division in New Guinea and then longtime chief of staff of the 8th Army before and after it became the Army of Occupation in postwar Japan.

In that same winter, Army hopes for a second permanent airborne division were dashed with the deactivation of the 13th Airborne Division, which had returned to its wartime training ground at Camp Mackall next to Fort Bragg.

Mackall was closed, and its tarpaper-covered barracks were quickly demolished.

"Slummy housing"

The problem of housing for peacetime military families was dramatized early in the year when Col. R.C. Mallonee, the post executive officer, spoke to the Fayetteville Kiwanis Club.

The newspaper account of his talk said:

"He punctuated his remarks by passing around photographs showing the slummy housing which was the only living room many Army men with families could find.

"Among the photos were those of a remodeled garage without plumbing which sheltered two families, a six-room cottage housing two families, and having as plumbing but one back porch water tub tap, a crude latrine, and cold water shower, and an auto court inhabited by 12 families forced to share one open air water faucet and one or two outdoor privies."

Mallonee said 1,255 Army families lived on post, many of them in converted office or barrack buildings. But 2,000 families were living in the off-post conditions graphically depicted in his photographs.

The veteran officer, a survivor of the wartime Death March of American and Filipino soldiers after the fall of Bataan in the Philippines, said the Army wanted to regard Fort Bragg as a "permanent part of the Fayetteville community," but that the civilian community should join in improving housing conditions.

Mallonee's plea to civilians was coupled with his effort to get Army help.

In August 1948, there was a ray of hope.

The Army announced it would allocate $473,000 for construction of family housing at Fort Bragg, and Mallonee was named deputy commanding officer of the post to oversee the building program.

The appropriation would not be forthcoming until many months into 1949, however.

Tragedy at Eglin

Meanwhile, the news of impending relief for housing was overshadowed when disaster struck 82nd Airborne paratroopers participating in a rare joint service training exercise at Eglin Field in Florida.

A shower of live bombs from a flight of B-29 bombers fell among the black paratroopers of the 505th Parachute Infantry Regiment and of the attached 758th Tank Battalion.

Four were killed and dozens injured in what the Triple Nickel unit history called one of the saddest days in the history of the battalion.

The disaster at Operation Combine III put a damper on all postwar joint training.

Soon after, Irwin stepped down as 5th Corps and post commander.

His replacement was Lt. Gen. John R. Hodge, a veteran of the war in Europe and most recently commanding U.S. Forces in Korea.

Putting the final bleak touches on a year of troubles was the news from Fort Bragg in the final months of 1948.

Three soldiers were caught trying to steal the Army payroll, and a half-dozen other soldiers confessed they were planning to shoot an officer in another attempted robbery.

Fifty years ago, Fort Bragg was looking for better times.

CHAPTER FIVE
Korea & Vietnam

Area Got Geared Up For Korea

Originally published October 12, 1995

In 1950, October was a month of triumph for U.N. forces in Korea.

Forty-five years ago, spearheads of U.S., South Korean and Allied columns raced across North Korea to the Yalu River, the border with Manchuria.

It was a stunning recovery from the previous June, when North Korean armies poured across the 38th parallel, the boundary between the two Koreas, in a surprise invasion. They smashed hastily assembled United Nations defenders, pinning them into a small corner of the Korean peninsula around the port of Pusan.

The tide had turned on Sept. 15. That's when Gen. Douglas MacArthur threw a "left-hook" amphibious invasion against the port of Inchon, and the U.S. 8th Army and Republic of Korea forces went on the offensive north out of the Pusan Perimeter. North Korean forces reeled back toward the Yalu.

It looked like the Korean War might be ending in triumph as quickly as it had begun so disastrously.

The headlines from the battle fronts were especially good news back in the Cape Fear region and at Fort Bragg and Pope Air Force Base.

There, thousands of men and women were wearing uniforms and training as part of President Harry Truman's mobilization order. It increased the armed services to 3 million soldiers, sailors, airmen and Marines.

Steady stream

Fort Bragg and Pope had gone into virtual mothballs when World War II ended five years earlier. They were humming again with the arrival of a steady stream of draftees, recruits and reservists.

The stream included hundreds from the Cape Fear region.

The first 101 draftees to be sworn in held up their hands Sept. 25, 1950. They were sworn in by Brig. Gen. Robert Cannon of the 82nd Airborne Division.

Fifty-nine white men and 42 black men were in the group. They were from Robeson, Lenoir and Harnett counties.

The draftees were destined for the infantry, immediately bundled off to Fort Jackson, S.C., for basic training with the 8th Infantry Division.

The Army and Air Force recruiting station in Fayetteville handled Selective Service and voluntary sign-ups for 31 eastern North Carolina counties. By the end of October, it reported giving pre-induction exams to 3,701 draftees, passing 2,260 of them. It said 599 were "already in uniform."

The same station had signed up more than 1,000 men as voluntary recruits in the five months since the outbreak of the Korean War.

In the last week of October, 85 Harnett County men and 82 from Sampson County reported for pre-induction exams.

Mirroring the bad record of World War II 10 years earlier, only 49 percent of the potential draftees passed the mental and physical exams. All but one of those rejected were turned down for failing the mental test.

The military-minded Sampson County town of Roseboro, which had 1,300 people in 1950, continued its tradition when 11 young town men showed up together to enlist in the Air Force. The recruiting office said it was the largest group ever accepted for enlistment from a town of that size. Six Sampson County men joined up with the Roseboro ll.

"Life is easier"

When the Korean War erupted, Fort Bragg was home to 12,000 men of the 82nd Airborne Division and to 6,000 soldiers, mostly of the 5th Corps artillery and headquarters. That was a far cry from its World War II peak, when more than 60,000 troops would be on the post at any one time.

By October 1950, dozens of Reserve and National Guard units, mostly artillery and service forces, had joined the post population.

An early arrival was the 540th Field Artillery Battalion from the High Point-Thomasville-Monroe area.

It was a typical call-up outfit. It had a smattering of officers and noncommissioned officers who wore campaign ribbons from World War II. But it was largely made up of 18- and 19-year-old recruits and draftees.

One old-timer contrasted the two wartimes: "It's a whale of a lot better than the old Army. The life is easier, chow is better and the food is tops."

As the newcomers poured in, the 82nd Airborne Division expanded its training. For the first time, Fort Bragg had its own basic parachute-training center, and by late October several hundred airborne volunteers were going through the 14-week course.

Pope Air Force Base had a new role: It was a test site for new troop-carrying aircraft under development in the huge buildup of U.S. forces that would flow from the mobilization triggered by the Korean War.

In mid-October, the service's Development Board lined up a display of new aircraft on the Pope runway.

They were the Chase XC-120, a twin-engine "assault transport"; the Chase YC-122, a "powered version of an attack glider"; the Northrup C-125, a "tri-motor assault transport"; and the Chase G-20, an "all-metal glider."

Those were experimental. The primary workhorse airplane for carrying troops and their equipment was the C-119, a twin-tailed behemoth known as the "Flying Boxcar."

Two soldiers treated

So far, none of the old or new soldiers or airmen at Bragg or Pope had seen war.

But the Fort Bragg hospital had.

In late September, the Pentagon ordered the hospital to prepare 500 beds for Korean War casualties.

Thirteen wards were set up. Recruiting of nurses began hastily.

By the second week of October, a front-page story in The Fayetteville Observer depicted two young North Carolinians in Bragg hospital beds. In the accompanying story, they talked about "the fear, the hunger, the rain and cold," and also about ferocious Korean mosquitoes and ants.

As President Truman called for an "ironclad" peace as soon as the war was over, interest back home turned to football and homecoming. The Army football team dominated the collegiate season, and Old Bluff Presbyterian Church in rural Cumberland County observed its 192nd birthday.

Rakkasans' Efforts Go Unnoticed

Originally published October 19, 1995

When 2,600 paratroopers of the 187th Airborne Regimental Combat Team floated down to attack two North Korean villages 45 years ago this month, they pulled off what has been called the best executed combat jump in airborne history.

But like some other airborne operations, the accomplishment of the paratroopers nicknamed the "Rakkasans" on Oct. 20-21, 1950, was destined to be swept away by the rush of other events.

When they jumped, it looked as if the Korean War might be over as swiftly as it had started five months earlier when North Korean forces invaded South Korea.

In October, the tables were turned. Allied forces of the United Nations command were sweeping into North Korea, and Gen. Douglas A. MacArthur was ordering them to go all the way to the Yalu River, the border with Manchuria.

MacArthur personally ordered the jump of the 187th, the only airborne unit in the Far East.

The 187th was the postwar descendent of the 187th Glider Regiment, which was part of the 11th Airborne Division under MacArthur's command in the Philippines during World War II. That division had been formed and trained at Camp Mackall next to Fort Bragg.

Process speeds up

The original plan set the Korean jump for Oct. 21, 1950.

But as was the case in several World War II jump plans during the 1944 German retreat across France and Belgium, events on the ground outran the plans.

Allied ground forces moved so fast that by Oct. 19 they were almost to the jump area. The mission was moved up to Oct. 20.

The objectives were the towns of Sukchon and Sunchon, 30 miles north of the North Korean capital of Pyongyang.

The attack plan called for two battalions of Rakkasans — about 1,300 men — to jump at Sukchon. Their mission was to block two roads and the railroad leading north out of Pyongyang.

Another 1,300-man battalion and artillery would drop on Sunchon and take up similar blocking positions across two other roads from Pyongyang.

The paratroopers would hold for one day before infantry and tank columns pressed north for a link-up.

Between the pinchers, MacArthur expected to nab leading officials of the North Korean government fleeing from Pyongyang. He hoped to free hundreds of Allied POWs believed to be moving north with the fleeing enemy.

The air fleet carrying the airborne soldiers consisted of 71 new C-112 "Flying Boxcar" transports and 40 World War II C-47 "Skyraiders."

The attack plan went off without a hitch, except that early morning rain at Kimpo airfield forced a wait until early afternoon. The air armada rose in tight V formations and turned briefly over the Yellow Sea before wheeling north toward jump zones.

MacArthur joins

As the transports streamed by, they were joined by MacArthur's own private command aircraft flown over from Tokyo.

Traveling with the general and his staff to view the attack was a gaggle of "favored Tokyo reporters."

The mission was an uncanny repeat of MacArthur's role seven years earlier, when he flew with the 503rd Parachute Battalion in the first World War II airborne mission in the Pacific, the seizure of a Japanese airfield near the village of Nadzeb in New Guinea.

In that September 1943 afternoon jump of only two and a half minutes, the paratroopers went down with little opposition and captured the airstrip. Three parachutists died in the jump, but none on the ground.

The Nadzeb air armada consisted of 82 transports layered between 100 escorting fighters, accompanied by 60 light bombers armed with 50-caliber machine guns and fragmentation bombs for whipping up the ground in advance of the jumping paratroopers.

As the Nadzeb transports unloaded, six fast A-20 fighter-bombers scurried close to the ground to blanket a giant smoke screen over the steaming tropical terrain.

Above the entire formation flew MacArthur's personal command fleet, a formation of three B-17 bombers bristling with machine guns. Over him buzzed a flight of six P-47 fighter planes keeping constant guard lest any Japanese planes try to attack the command fleet.

The 1950 Sukchon-Sunchon jumps took about 20 minutes, and everything went splendidly. Historians describe the 187th's Korean War feat as "the best combat jump the Army ever staged" and "a magnificent exhibition."

A mission first

Only one paratrooper was killed in the jump. For the first time in a mission, artillery pieces, Jeeps and three-quarter-ton trucks were air-dropped from the open rear bays of the Flying Boxcars.

As at Nadzeb seven years earlier, however, pickings on the ground turned out to be slim.

At Sukchon, there was no enemy. A South Korean division racing north from Pyongyang had liberated the town.

At Sunchon, paratroopers found the bodies of 75 American POWs massacred by their captors and 21 wounded prisoners who were still alive.

As he had done for the 503rd at Nadzeb, where he claimed 30,000 Japanese troops were cut off, MacArthur put a glowing public relations face on the Rakkasans' mission.

At a news conference on the afternoon of the jump, he assured reporters that the mission had cut off half the remaining North Korean army, a big step toward the "complete annihilation of the enemy."

Actually, the only North Koreans in the 187th's trap were men of a regiment fighting a rear guard action.

But they put up a fierce fight when the paratroopers and units of the Commonwealth Division caught the regiment south of Sukchon on Oct. 21.

Honor for battle

In a chaotic struggle that raged all night on Oct. 21-22, men of the 3rd Battalion of the 187th virtually wiped out the North Koreans, killing 800 and capturing 680. The paratroopers "held their ground magnificently," said a Korean War historian. For that fight, the 3rd Battalion was awarded the Presidential Unit Citation.

Even as the paratroopers buttoned up their mission and Allied columns streaked within 55 miles of the Yalu, events were about to overtake the seeming triumph of their arms.

On Oct. 25, "volunteer" divisions of the Chinese Communist Army were pulling into place south of the Yalu in front of the widely separated and unsuspecting Allied columns.

On Oct. 26, a regiment of the same South Korean division that had liberated Sukchon before the airborne assault was ambushed and virtually destroyed near the Yalu.

It was the first wind of a hurricane as 300,000 Chinese troops eventually attacked in snow and fog to blunt and then turn back the Allied advance.

The struggle would go on for another two years. The Rakkasans would fight more ferocious fights and win more honors as the only U.S. airborne outfit in the Korean War.

Psy Ops Got Start At Bragg

Originally published January 30, 2003

In the Korean War, it was the latest thing in the latest sort of warfare, "psychological warfare."

They were leaflets, "paper bullets," dropped by the thousands urging North Korean soldiers to surrender to onrushing United Nations forces. Or they were radio broadcasts over a dozen stations, making the same appeal.

Fifty years later, in the buildup to an invasion of Iraq, the fledging propaganda effort of those days has become a major weapon in the U.S. arsenal.

Psychological operations, with its home base at Fort Bragg, has expanded to include a dazzling array of Space-age techniques and equipment aimed at influencing the other guy to give up before the fighting starts.

We are hearing of huge leaflet-dropping and electronic campaigns, first in Afghanistan and now in Iraq.

And in the latter case, the target is not just the common soldier. The U.S. propaganda campaign is aimed at the highest levels of the Iraq military, hoping to trigger rebellion even among the generals who surround President Saddam Hussein.

The campaign also involves efforts to get to the civilian population, employing all the techniques of propaganda the military has adapted from the civilian world of advertising and market manipulation.

Fifty years ago, the commander of United Nations forces in Korea, Gen. Mark Clark, credited "the bloodless battle" of psychological warfare with causing thousands of North Korean soldiers to lay down their arms.

A report from Clark's headquarters said that a study of the prisoner-of-war camps concluded that 65 percent of the POWs "were influenced directly by the intense U.N. propaganda campaign."

Eighty percent who had "voluntarily surrendered" were influenced by the leaflets, according to Clark's glowing estimate.

The officer who directed the Army's first large-scale psychological warfare campaign took a more measured view of the success of the effort, and of how it works.

Col. J. Wendell Greene put it this way:

"No one surrenders simply because of a leaflet. We simply put the surrender idea in his head and keep pounding at his morale whenever we can.

"When he is hungry or sick or worried about his family, disgusted with his Red Bosses, harassed by our riflemen, pounded unmercifully by our planes and artillery, or half a hundred other things, then maybe he will give up. At any rate, he won't be bad to beat."

Greene was more enthusiastic when describing how the U.N. propaganda arm countered a clumsy North Korean effort to accuse the United States of using "germ warfare" on North Korean troops.

The effort failed because it was so false and diffuse, with no evidence.

In what may be called an early echo of a debate about Iraq's biological-warfare capability, Greene said "the Reds haven't been able to identify a single bug with USA written on it!"

The Korean "bloodless war" was history's latest chapter in the use of propaganda to influence the course of warfare.

In World War II, the Germans used the broadcasts of "Lord Haw-Haw" to plant the seeds of defeatism with British soldiers and civilians.

Other broadcasters with names such as "Axis Sally" and "Tokyo Rose" were soldiers in the propaganda war.

Both sides used "surrender leaflets."

The history of this modern use of the ancient tools of war propaganda had its beginnings, of course, at Fort Bragg.

The story begins on a hot day in June 1952, when the volunteers for the Army's first "Special Forces group" assembled in a few buildings in the array of World War II barracks on a site known first as "Artillery Area" but by 1952 generally referred to as "Smoke Bomb Hill."

Under a directive from the Pentagon, the Army's chief of psychological operations was organizing the Army's first in-house outfit shaped for what was coming to be called "unconventional warfare."

Fort Bragg was home to the Psychological Warfare Center, the training post of the fledgling command for unconventional warfare warriors.

The center itself was only five months old, formally dating from April 10, 1952.

Col. C.H. Karlstad commanded the center. The executive officer, Lt. Col. Aaron Bank, moved over to command of the newly activated 10th Special Forces Group. Lt. Col. Carlton Fisher took Bank's place at the center.

By August 1952, the center was welcoming its first female student, Lt. Mary Bellas of the U.S. Navy.

Building on experience from World War II and the Korean War, the Special Forces volunteers were trained to infiltrate deep into an enemy's country, then to carry out the ancient military arts of guerrilla or partisan warfare: harassment, ambush, sabotage, demolition and hit-and-run raiding.

The trainees at the center would develop, test and then use the techniques and the growing array of technological gear and gizmos capable of filling the skies, the airways and the screens with a flood of propaganda aimed at all levels of the civilian and military populations.

The center and its school would be the training ground for special operations and the propaganda war, which 50 years later have grown to become cutting-edge weapons in the military's strategic arsenal, and the practitioners of psychological warfare have become the front-line warriors of today.

Implications Of A-Bomb Test Unknown In 1952

Originally published December 12, 2002

For nearly four months in early 1952, men of the 504th Airborne Infantry Regiment (now the 504th Parachute Infantry Regiment) had been eating Texas dust alongside other paratroopers of the 82nd Airborne Division.

Acting as "aggressor" forces in a huge war game, the troopers were packing for a long road trip back home to Fort Bragg.

Then they got one more assignment. Instead of turning east toward North Carolina, they turned west toward the yawning desert of Nevada.

This time, they were to eat atomic dust.

On April 22, 1952, the paratroopers made history by jumping or charging into the monstrous shower of earth and debris cascading down from the explosion of an atomic bomb larger than the device that destroyed the Japanese cities of Hiroshima and Nagasaki in 1945.

The explosion took place at Yucca Flats, the isolated testing ground of the U.S. superbomb program. A "red-tailed" B-50 bomber released the atomic bomb that exploded at 3,500 feet above the great bowl of the flats, raising a deadly plume of the sort now familiar as a "mushroom cloud."

Some 504th men were in foxholes about four miles from the explosion site. But 120 did what paratroopers do. They jumped into the cloud.

The Fayetteville Observer reported the operation in an eyewitness story from a pool reporter who saw the explosion and wrote for The Associated Press:

"For the first time paratroops were also used in the exercise, 120 selected men jumping to aid the ground troops simulating a vicious attack on a enemy strong point by swirling through the murky, dusty and debris-laden atomic blast area."

"The exercise began when a red-tailed B-29 from Kirkland Field in Albuquerque whirred into the area, and then a dramatic play-by-play account was provided by an ABC official booming over a loudspeaker.

"The voice counted down from one to 10, and then the heavens opened in a beautiful but fearful rolling turbulent orange ball of fire with shades of white, blue and purple."

"A deep sigh passed over the witnesses as an unexpected heat wave struck the rocky knoll known as 'news nob' and then the ear-shattering thunder of the explosion reached the observation post.

"The light, like that of 50 suns, bathed the area.

"Then in proud, precise military manner the infantry moved out into the thick of the new-formed dust storm beneath the mushroom's stem.

"They passed by hundreds of test animals imprisoned in cages for radiological study, and around blackened equipment on display.

"And again on schedule the 120 selected paratroopers dropped into the area in support of the ground force and successfully completed the mission."

A risky experiment

The test 50 years ago was part of a long program of nuclear weapons testing that involved tens of thousands of military personnel essentially acting as test subjects in what could have been a deadly experiment.

A half-century later, there is a National Association of Atomic Veterans, which presses for more medical benefits for the men, now mostly in their 70s and 80s, who back then jumped or charged into mushroom clouds.

After the test in 1952, there was no talk of the downside of the future of atomic warfare.

Lt. Gen. Joseph Swing, commanding general of 6th Army, announced that the test was a triumph for the Army, demonstrating that ground troops would have a role in an atomic war.

Said Swing: "It was a complete success and another long stride in the perfection of the use of atomic weapons in conjunction with ground troops."

The fact that 504th paratroopers made their jump into the mushroom cloud may have been because of Swing, who was a famous World War II airborne soldier and commander of the 11th Airborne Division in the Pacific.

Such an advocate of the airborne would have wanted to make sure they got a piece of the action in what he described as a "fantastically smooth operation."

Homecoming

The 504th jumpers got plenty of hometown publicity for their feat.

An Army photo in the Observer depicts the troopers lined up on the railroad platform at Fort Bragg after they returned, dressed in "class A" uniforms with the jaunty garrison caps on their heads.

There have been many descriptions of the day 50 years ago.

Robert Stroup, who was a 29-year-old trooper of Company H of the 504th, was in the ground element.

He told the Association of Atomic Veterans:

"At countdown we were to lay flat in the trenches and cover our eyes with our hands. At count zero, I could see the bones in my hands and I could hear loud crackling noises. A metal loudspeaker behind us told us to stand up and observe.

"The fireball was huge and had all colors of the rainbow with lightning at its base and you had to keep turning to keep from being burned. It hurt the eyes to look it.

"I could see the shock wave coming and I turned my back to it.

"When it hit, it temporarily knocked me out. I do remember it moving me across the trench and back all in a split second. It was like being dropped from 20 feet up onto your back. It was a while before I could breathe. I watched it till it was a mushroom with ice on the top."

John Kessling of Fayetteville described his part in the exercise in a 1994 interview with military editor Henry Cuningham.

Kessling was in the jumping contingent, which exited from C-46 transports onto the valley floor as the cloud rolled down. He was a master sergeant in Company F at the time and retired in 1976 as a sergeant major.

Stroup listed a long litany of medical problems that he thought were related to his experience, but Kessling decided not to pursue the complicated process for additional medical benefits. In fact, he was already 100 percent disabled from later service in Vietnam.

While the world still worries about what might happen in warfare involving what are now called "weapons of mass destruction," we no longer take it quite as calmly as Federal Defense Administrator Millard Caldwell, who assured reporters after the 1952 test that the explosion they had just witnessed "would have wiped out three square miles of New York City at the cost of 500,000 lives."

N. Korean Sub Repeats '60s History

Originally published October 3, 1996

To retired Lt. Gen. William P. Yarborough of Southern Pines, the latest story about a North Korean spy submarine caught off the South Korean coast is deja vu all over again.

It was a hot summer day 31 years ago when Bill Yarborough, then a two-star general, caught the North Koreans red-handed at the same game.

It was 1965, and the decorated former World War II paratrooper and early boss of the Green Berets was the top man for the U.N. Command on the Military Armistice Commission.

For more than a decade, the commission had convened at a nondescript building in the Korean village of Panmunjom, enforcing the peace that had ended the Korean War 12 years earlier.

On one side of a table, members of the U.N. Command team would take their seats. On the other, the North Korean team. The top man on each side would talk in turn.

Caught at last

For months, Yarborough had endured the incessant invective of his North Korean counterpart, a "political general" named Chung Kuk Pak.

Meeting after meeting, Pak accused the United Nations of violating the armistice, either by sending troops across the truce line or, at sea, seizing innocent North Korean fishermen.

Now, on July 27, 1965, Yarborough was about to get his revenge. He had the goods on them.

In a speech laced with well-founded sarcasm, Yarborough formally accused the North Koreans of sending a three-man midget submarine far up the Han River, which flowed from near the truce line into the heart of Seoul, capital of South Korea.

A spying mission, no doubt.

Yarborough referred to the submarine as the "Kim Il Sung" after the North Korean dictator, a name that he personally selected for it.

It was made, he told Pak, of "East German vacuum cleaners, bicycle chains, Japanese batteries, and even an American Evinrude control system."

While Pak sat glumly, Yarborough invited him outside.

As commission members streamed out of the building, a big U.S. Army flatbed truck rolled to a stop beside the building.

Lashed down to the flatbed was a 20-foot long midget submarine, which looked like a fat cigar with a tiny conning tower resembling a turret on a tank.

And then the real touch.

The U.N. Command had commandeered a life-size clothes dummy. Dressed it in a North Korean military uniform.

Stuck it in the conning tower. And in its hand, until the wind blew it off, was a North Korean flag.

Photographers had a field day.

Yarborough reported to Pak that the U.N. Command held one of the three-man crew prisoner.

Two others had been killed by South Korean naval forces.

Pak was nearly tongue-tied. Ignoring the subject of the meeting, he could only blurt to Yarborough: "It is apparent that you are now mentally deranged."

Still trying

For Yarborough, it was a bright moment in an often-frustrating eight-month tour of duty as chief U.N. representative at the Military Armistice Commission.

And as he said this week, it proves that nothing changes along the Korean truce line.

The North Koreans are at it, with bigger submarines but with the same mission. And they still are the world's most inveterate liars at the "peace table."

To understand Yarborough's perspective, you could look at the record of the 193rd formal meeting of the Military Armistice Commission in the wooden building at Panmunjon three months earlier.

The North Korean general spoke first. Reading from a script, he glanced up occasionally to look across the table at the trim, stern-faced American general who was his counterpart for the U.N. Command.

The United Nations, said Pak, had placed itself in "the embarrassing position of criminals."

The United Nations, he said, was guilty of "a shameless and brigandish nature." U.N. forces, he told the American, were "massacring the South Korean people at random."

Eight meetings later, at the 201st session of the Armistice Commission, the same line.

Pak said of the U.N.: "Your recent criminal act of having committed a provocative attack by dispatching an armed group to our territory again exposes the world that the aggressive policy of your headquarters has been running amok."

The United Nations, Pak said, was trying to provoke another war.

Another day, another meeting.

It was June 4, 1965 — 15 years to the month since North Korean forces poured across the 38th parallel, invading South Korea, provoking the Korean War.

Pak opened with yet another attack.

The U.N. Command, he said, was using the Armistice Commission for "malicious propaganda as to slander and calumniate Marxist-Leninism."

Yarborough had all he could take.

For 10 minutes, the soldier-diplomat who was one of the first three company commanders of paratroopers and who had fought on nearly every European battlefield in World War II went after Pak with all the rhetorical fierceness of a tiger chasing a wildebeest.

No longer would he sit silent for "lengthy gas-filled essays delivered at every meeting."

"Why," he demanded of Pak, "should our side at this table believe that you wish to be sincere at this table when your whole evil structure worldwide is the very essence of insincerity?"

Why, indeed, "when you are pathological liars."

Instead of answering carefully worded U.N. requests for information about various alleged armistice violations, Pak had delivered "hackneyed, worn-out, inarticulate and unimaginable rubbish."

The North Korean side, Yarborough said, had made no fewer than 29,442 allegations of "ground violations" by the U.N. forces.

Few with merit

Careful investigations had found only 15 to "have validity" and were acknowledged by the U.N. Command.

Finally, and becoming visibly indignant, Yarborough turned to the fate of 2,254 Americans and other U.N. personnel listing as missing in action in the war that had ended 12 years earlier.

Time after time, year after year, the U.N. Command had pressed for information.

Its requests were ignored.

"Do you need another roster of their names?" Yarborough asked. "I have one here for you here. Shall I read these 2,254 names for you again to jog your memory?"

Pak ignored the question and got personal. "You are a man of base character and low dignity," he said to Yarborough.

And so it went at Panmunjom. A day like any other day.

Except that day when the olive drab flatbed truck rolled in with its unusual cargo.

And it seems to be still going on.

War Has Echoes Of 40 Years Past

Originally published January 4, 2007

At the beginning of a new year, it appears likely that history will require the U.S. Army to deliver more superhuman effort in the years ahead.

The history of the mission in the Middle East will soon be four years old.

It could go on for at least that long again.

I don't hold much with history repeating itself.

But 40 years ago, during the holiday season, Washington was not unlike today.

The Army was preparing to test a huge buildup in Southeast Asia, where after two years of trying, the mission in Vietnam was not succeeding.

In the background, a once highly popular president was slumping in the polls. In the 1966 midterm elections, his party dropped 47 seats in the U.S. House of Representatives.

In the summer just past, the administration of President Lyndon Johnson had projected a six-point mission for the forces in Southeast Asia.

The admirals and generals said the goals were "hopelessly farfetched" for forces currently on the ground.

At the time, there were 116,000 soldiers and 41,000 Marines who rushed to the war zone beginning in the crisis year of 1965 and who staved off defeat after the collapse of the South Vietnamese army in 1966.

The White House responded, and the next six months were marked by a momentous "surge" of U.S. manpower pouring into the highlands, jungles and deltas of Vietnam.

At Christmas in 1966, Gen. William Westmoreland, the U.S. commander, could count on nearly 400,000 soldiers, Marines and airmen for campaigns that he believed would turn the tide against growing Viet Cong and North Vietnamese forces.

A witness in Washington

Along with many others in Washington during that snowy holiday season 40 years ago, I was grimly optimistic.

As a newspaper reporter who was desperately trying to stay concentrated on so-called "domestic" news — civil rights, education, health care, the environment — I was hoping that the burden of Southeast Asia could finally lift.

Like Johnson, I was heartsick that the war was crowding out what seemed to have been such a bright and hopeful era of real progress at home.

But as it is this season, the war crowded all other matters off the agenda. It was already being called "Johnson's War." The fate, and the legacy, of many in and out of government were increasingly tied to the war. It held friend and foe in its mire.

As the current secretary of defense recently said about the Middle East, there are no new ideas about how to proceed.

That was mostly true in Southeast Asia 40 years ago.

But proceed we did.

A U.S. Army that had been relatively unused for a decade entered what historians call "The Year of the Big Battles," days and weeks and months of violent action.

The struggle added new names to military annals: Dak To, Iron Triangle, DMZ, Junction City, Cedar Falls, Task Force Oregon, Bong Son Plains, Klamath Falls and many more.

"Flower children"

By the summer of 1967, the violence in Southeast Asia was creating a wave of peaceful protest at home. The first "flower children" were showing up in Washington streets. Ministers came up from North Carolina to stand silently in anti-war vigils on the Capitol steps. And every evening, Sen. Wayne Morse of Oregon would rise in an empty Senate chamber and deliver yet another anti-war speech for the record. Many times, the only audience was The Associated Press reporter and myself.

Westmoreland came stateside briefly to deliver a progress report to a joint session of Congress. Another 100,000 fighting men joined the war.

And on it went for five more years. Ultimately, more than 4 million Americans went to the war zone. Fifty-eight thousand died in the fighting.

An accurate mournful summary of the Southeast Asia war is from the "Encyclopedia of Military History" by the Dupuys, father and son:

"The U.S. had entered into the longest, oddest and by far the most unpopular war in its history. It was a war without fixed front. The enemy was here, there and everywhere. The enemy had made guerrilla warfare a science. American ground forces held only the ground on which they stood in a war of thousands of savage engagements without a single battle in the conventional modern sense.

"The Vietnam war, a phantasmagoria of brutal combat, political and social entanglements, and increasing frustration would be viewed on television in American homes; military personnel flew to to the area on commercial aircraft, and the military effort was heavily influenced by political considerations in Washington."

History doesn't repeat itself. Instead, it is more like the old song about the times they always are a'changing.

But 40 years after that holiday season so long ago, don't count me as optimistic this time. Only grim.

Khe Sanh Called For Heroic NCOs

Originally published April 9, 1998

Even in the surreal world of the Khe Sanh plateau in late January 1968, it was a strange happening.

Sitting in the team house of the U.S. Special Forces camp at Lang Vei, 1st Sgt. Bill Craig of Special Forces team A-101 looked up as a fully armed North Vietnamese soldier walked in the door.

Craig didn't wait to ask how the visitor got through the barbed wire and past a guard at the camp gate. He grabbed the nearest weapon, a bottle of whiskey, and prepared to engage the intruder.

Pvt. Luong Dinh Du of the 540th Division, North Vietnamese Army, quickly threw up his hands and surrendered to the hardbitten Craig, who at 38 was a veteran of the Korean War and a Green Beret since airborne "graduation" at Fort Bragg in 1960.

Under questioning, Du insisted that NVA forces were preparing to attack Lang Vei. His word fit with other intelligence warnings of an impending offensive aimed at sweeping Specials Forces and Marines off the Khe Sanh plateau, the northernmost bastion of U.S. forces in Vietnam.

Nine miles from the Lang Vei camp, thousands of Marines in bunkers and trenches at Khe Sanh itself were already coming under constant attack by several divisions of NVA regulars.

And a week after Craig confronted Du, beginning at midnight of Feb. 6, 1968, Lang Vei was the scene of a 12-hour melee when tank-led North Vietnamese regulars poured into the camp from three directions.

By the time rescue helicopters arrived on the afternoon of Feb. 7, all but one of the 24 Special Forces soldiers in the camp, and hundreds of native fighters, the Montagnards, were dead, wounded or missing.

Such was the heroism of that fight 30 years ago that 19 defenders would be given the Silver Star Medal, one would receive the Distinguished Service Cross, two the Bronze Star.

And a Wilmington native, Sgt. 1st Class Eugene Ashley, a 36-year-old Green Beret regular who died from a severe chest wound after leading five charges against the attackers, would be posthumously awarded the Medal of Honor.

The fight goes down in miliary history as a classic of small-unit action and a "battle of the sergeants," where senior NCOs showed the leadership that makes them the backbone of the Army.

By 1968, the Khe Sanh plateau was familiar ground to Special Forces. The first SF camp there dated from the summer of 1962.

The Green Beret soldiers were to train mountain tribesmen, the Montagnards known as the "Bru." This was the earliest U.S. effort to build South Vietnamese defenses against Viet Cong rebels and North Vietnamese regulars.

As early as 1964, the camp was rated as a success story. More than 300 Montagnards and their families were linked to the Special Forces team A-101.

Early 1967, however, was a time of trouble for Lang Vei.

In March, a U.S. fighter-bomber mistakenly strafed and bombed the village, killing many.

In May, a North Vietnamese Army company aided by Viet Cong spies infiltrated the camp, inflicted heavy casualties on the Montagnards and killed the Special Forces camp commander and his second in command.

A new commander, Capt. Frank C. Willoughby, arrived in June. He didn't like the location of "old" Lang Vei camp, and received permission to build a new camp a half-mile to the west on a higher flatland straddling the old Indochina highway known as Route 9.

The "new" Lang Vei was a powerful outpost, with buildings, weapons pits and 40-by-25 foot underground operations center built of concrete and steel by a battalion of Seabees.

When the dozen Russian-made North Vietnamese small tanks and infantry attacked at midnight on Feb. 6, the defenders of Lang Vei consisted of SF team A-101 of 5th Special Forces Group; three companies of Montagnards, and a "MIKE Force" of better-trained South Vietnam scouts headed by a half-dozen Green Berets.

Just arrived was Lt. Col. Dan Schungel, the commander of Company C, 5th Special Forces Group, of which A-101 was a unit. And at "old" Long Vei was a recently arrived unit of Laotian irregulars, the "Royal Elephant Battalion," which only a few days earlier fled across the border ahead of North Vietnamese tank columns.

Ashley and two other A-101 soldiers were with them, inoculating new arrivals.

The deadly fight that followed pitted several bands of defenders against the swarming attackers.

The MIKE force post outside the barbed wire put up a big fight before being overrun. Four of its six Green Berets were captured or remain missing in action. Another was killed.

Col. Schungel fought from behind a barricade of concrete-filled oil drums, firing at tanks until he as so badly wounded that he crawled under the concrete camp dispensary building.

Craig and several other A-101 solders fought from a mortar pit with anti-tank weapons.

Willoughby and seven other Green Berets, as well as Montagnards, were trapped in the underground commander bunker, assailed by tanks and battered by North Vietnamese demolition teams.

All eight were wounded, several badly. Willoughby regained consciousness in mid-morning of Feb. 7 and thereafter was in radio contact with outside forces.

Ashley in old Long Vei waited until dawn, then directed fire from artillery and helicopter gunships while organizing a relief force that charged up the plateau in five distinct assaults.

Killed when an artillery blast ripped a huge hole in his chest, Ashley's body was loaded into a jeep by his comrade, Sgt. Richard Allen, and the battered relief force moved out toward Khe Sanh.

While the epic fight raged at the camp, Khe Sanh and higher headquarters were undecided how to proceed. The Marines were ready to write off Lang Vei. They were under siege themselves.

But Col. Daniel Baldwin, the senior Special Forces officer in the northern part of South Vietnam, cut through the indecision and ordered a rescue mission of four big Marine helicopters to head for the besieged camp.

Maj. George Quamo at the 5th Special Forces headquarters at Da Nang commanded what he expected to be a "suicide mission."

But in an hourlong operation, Quamo and a team of veteran Green Berets enlisted men brought out the wounded, as well as a few Montagnard and MIKE force Vietnamese. The battle for Lang Vei was over.

Schungel, Willoughby and Craig recuperated from their wounds and went on to legendary careers in Special Forces.

Within days, Ashley was recommended by his comrades for the Medal of Honor.

In a gruesome twist typical of the confusion that day, the body of the heroic Green Beret was overlooked. It remained in its body bag in the abandoned jeep on Route 9 until found by soldiers of the 1st Air Cavalry Division sweeping the area two months later.

Five years later, Ashley's widow and five children heard the citation "for conspicuous gallantry at the cost of his life above and beyond the call of duty" at a White House ceremony when President Richard M. Nixon presented the country's highest award for heroism in battle.

A new book tells the detailed story of Special Forces heroism at that obscure camp 30 years ago. It is "Night of the Silver Stars: The Battle of Lang Vei" by William R. Phillips. The Naval Institute Press. Annapolis, Md. 186 pages.

* 5th SPECIAL FORCES GROUP

* 1961 — Activated at Fort Bragg

* 1965 — Relocated to Nha Trang, Republic of Vietnam

* 1969 — Awarded Presidential Unit Citation

* 1972 — Relocated to Fort Bragg

* 1987 — Relocated to Fort Campbell, Ky.

Soldier Remembers Lang Vei Aftermath

Originally published April 30, 1998

When retired Sgt. 1st Class Mike Durkot read a Military History story about the 1968 shootout at the Special Forces camp at Lang Vei in Vietnam, it brought back memories.

It also brought a revelation.

Durkot, who now lives on a peaceful "eight acres with a pond" near Rockfish, saw anything but a peaceful sight when his company of the 1st Cavalry Division swept into the deserted Lang Vei camp on April 5, 1968.

Two months earlier, the camp was a bloody battleground when North Vietnamese tanks and infantry overran Special Forces and Montagnard defenders.

Durkot's outfit, Company C of the 12th Cavalry Regiment, was tasked with sweeping the area and recovering the dog tags from the remains of any Special Forces soldiers who didn't make it.

Among the most conspicuous remains in an area strewn with such sad rubbish of war were those of an American sergeant, slumped in an abandoned jeep near the entrance to a underground command bunker where eight Special Forces soldiers held out for a night and a morning during the North Vietnamese attack.

Durkot's company recovered the dog tags.

He didn't know to whom they belonged.

Until 30 years later, in April, when he read the Military History story of the battle at Lang Vei.

The remains were those of Sgt. 1st Class Eugene Ashley, a 38-year-old Wilmington native who would be posthumously awarded the Medal of Honor, the country's highest military honor, "for conspicuous gallantry at the cost of his life above and beyond the call of duty," as a result of his exploits at Lang Vei.

The Green Beret regular died from a chest wound after leading five charges against the attackers.

In a gruesome twist of the confusion that day, the body of the heroic Green Beret never got carried out by either retreating defenders or helicopter-borne rescuers.

Durkot's recollections of the recovery of Ashley's remains dispute those of a recent book about the fight at Lang Vei which provided much of the information for the story here.

The book described the brave sergeant's body being put in a body bag by survivors of the charges that he led against the North Vietnamese attackers. "No, there were no body bags there," Durkot says.

Under fire

Durkot's outfit did more than clean up the Lang Vei camp.

They fought there themselves.

The camp, on a high flatland close to the border with Laos, was a deadly no man's land in 1968, and the 1st Cavalry soldiers came under fire from North Vietnamese tanks arrayed on a high bluff that was the border between Vietnam and Laos.

The retired veteran paints a colorful word picture of their ordeal:

"They couldn't depress their tubes (gun barrels) enough to get at us," he recalls.

But for two days, it was "BANG (the gun firing) and BOOM (the shell landing)" on other troops farther back near the big Marine bastion of Khe Sanh.

Then, when his unit was ordered to fall back, "they mortared us."

"We would time their fire so that we kept one jump ahead of the next round. We would time their fire, and then ran ahead of where they had registered to fire for the next round," he says.

It was a vivid memory for the 68-year-old Durkot, who retired in 1973 after one of the Army's more unusual careers.

Durkot is among a disappearing breed of soldier who was not even a U.S. citizen when he donned the uniform.

He joined in Europe, where he had been a 19-year-old escapee from his native Slovakia, then under Russian control.

As a refugee in Poland, he got a taste of paramilitary life in a camp guard battalion.

He managed to win a five-year hitch in the U.S. Army while in France under a Cold War law that granted citizenship to qualified refugees in return for military service.

Durkot came to the U.S. and went to language school.

He then became the quintessential career soldier, a drill sergeant.

By 1968, the Army was collecting such veteran noncoms to fill ranks of the big units fighting in Southeast Asia.

Durkot saw Fort Bragg shortly before he left to join the 1st Cavalry Division, the famous outfit known as the "First Team."

When he retired, he lived for several years in Mount Olive.

But he and his wife have lived on their tree-covered eight acres near Rockfish for 11 years now. A son lives next door.

If he hadn't settled near Fort Bragg, he may never have learned of his long-ago sad link with one of the Vietnam War's most honored Green Berets, whose portrait is in the special operations hall of heroes at Fort Bragg.

Another note

The Military History report on the 30th anniversary of Lang Vei also brought a note from Col. Jack Donovan, who commands the Joint Readiness Training Center at Camp Polk, La.

Col. Donovan writes:

"Served a second tour in Vietnam in 1992 when I started the POW/MIA search project.

"Went to Lang Vei about a dozen times and participated in the excavation of the command bunker. Had a chance to visit with (Capt.) Frank Willoughby (who subsequently has died) in Alabama upon return from my 15-month tour."

A Reactivated Unit's Historic Distinction

Originally published February 21, 2002

The skies were peaceful and the ceremonies were colorful three weeks ago when the Army reactivated the 2nd Battalion of the 503rd Parachute Infantry Regiment at headquarters of the Southern European Task Force in Vicenza, Italy.

Thirty-five years ago this week, the skies over Vietnam were reverberating with the noise of a fleet of airplanes and helicopters when the same unit made the only airborne combat jump of the Vietnam War.

The 780 jumpers went to battle on Feb. 22, 1967, flying in 13 C-130 transport aircraft, essentially the same "Herk" (officially the Hercules) used for airborne transport today. Another eight aircraft carried heavy equipment.

The parachutists jumped at 1,000 feet, and the entire force was on the ground within 20 minutes. There were no battle casualties, and only 11 minor jump injuries.

Brig. Gen. John R. Deane Jr. jumped with the battalion, with Lt. Col. Robert H. Sigholtz as the task force commander.

The paratroopers went on to fight the land battle of Operation Junction City, which lasted into May 1967.

During seven years in Vietnam, the battalion earned 14 of its 18 battle streamers, plus five of its seven distinguished unit citations, including two Presidential Unit Citations.

The jump 35 years ago was an element in a larger air-assault plan, the first mass use of helicopters in an operation whose mission was to build a cordon of fire around an entire area northwest of the city of Saigon, where the enemy forces of the Viet Cong controlled much of the countryside.

The parachute troops were part of an assault from all four sides of the area.

By using paratroopers, the U.S. command was freeing 66 helicopters that were used to deliver a dozen battalions of infantry and their equipment.

"The Herd"

In Vietnam, the 503rd Parachute Infantry Regiment was part of the famed 173rd Airborne Brigade, the first big U.S. unit to arrive in Southeast Asia and one of the last to leave, nicknamed "the Herd" (short for "thundering herd").

Junction City was the first big demonstration of the concept of "air mobility," with helicopters delivering 9,518 soldiers and 50 tons of cargo a day.

While the 503rd has very little history connected to Fort Bragg, "air mobility" was very much the brainchild of a famous Fort Bragg commander.

The man who wrote the book on helicopters-in-battle was Lt. Gen. John J. "Jack" Tolson, a soft-spoken native of New Bern who was an airborne soldier from the earliest days of World War II. An airborne battalion commander in the Pacific Theater, he became adept at the air delivery of both men and supplies.

Tolson would end his long career in 1971 with the three stars of a lieutenant general, commanding the 18th Airborne Corps at Fort Bragg. He died Dec. 2, 1991, at age 76. The youth center at Fort Bragg was named for him three years ago.

In Vietnam in 1967, Tolson got the opportunity to put his own vision into practice when he took command of the 1st Cavalry Division (Airmobile).

Despite its name from the era of the horse soldier, the division was actually the Army's first all-helicopter infantry division.

Known as the the "First Team," the division became the backbone of many U.S. offensive operations in Southeast Asia from 1965 until it withdrew in 1972.

Tolson is rightly hailed for his leadership in the helicopter army.

But he never lost his admiration for the jumpers of the parachute infantry.

In a history he wrote of the war in Vietnam, Tolson says of the Junction City jumpers that "the practical aspects of making more helicopters available were perhaps colored by emotional and psychological motives of this proud unit, which was anxious to prove the value of the parachute badge; nevertheless the jumpers contributed strongly to the overall attack."

Never got to jump

Tolson also wrote sympathetically of the thousands of airborne soldiers who never got to jump other than in the Junction City operation.

He wrote:

"Every man with jump wings was eager to prove his particular mettle in Vietnam. However, this special talent was not often suited for that enemy, that terrain, and that situation. Nevertheless, I firmly believe that there is a continuing requirement for the airborne capability in the U.S. force structure."

The unit that was reactivated at the 173rd Airborne Brigade home in Italy in January and which conducted the Junction City jump in 1967 has a unique genealogy in the Army's lineage tables.

The original 2nd Battalion of the 503rd went to England, and became the 509th Parachute Infantry Regiment just before the invasion of North Africa in November 1942, when it made the first combat jump of World War II.

But the rest of the regiment went to the Pacific, and another 2nd Battalion was part of the 503rd that conducted the first combat jump of the war there, when in September 1943 the paratroopers of the battalion dropped on the village of Nadzeb on the island of New Guinea.

Among the jumpers was none other than 29-year-old Jack Tolson, who as a lieutenant colonel personally briefed Gen. Douglas MacArthur on the results of the jump.

The 503rd in the Pacific earned its nickname, "The Rock," in February 1945 for its combat jump assault on the Philippine island fortress of Corregidor, the huge rock island at the entrance to Manila Bay that had been captured by the Japanese in 1942.

Now, the "new" 2nd Battalion finds itself once again a part of "the Herd," but serving in an altogether different part of the world.

And Junction City remains one of the rare post-World War II combat jumps of the airborne Army.

CHAPTER SIX
Modern Military

1983 Grenada Mission Showed War's Mix Of Skill, Confusion

Originally published October 28, 1993

By the time the Christmas Barbecue Special rolled down the runway at Salines airport in December 1983, the few battle scars of the invasion of Grenada were already being reclaimed by the tropical vegetation of that Caribbean island.

The special was a C-130 loaded with several jeep loads of — what else? — barbecue and all the fixings.

The idea for a holiday pig-pickin' came from Bobby Suggs, telephone company publicist and president of the Braxton Bragg Chapter of the Association of the U.S. Army.

He wanted to say thanks to the Fort Bragg troops who remained on the island nearly three months after the Oct. 25 invasion.

The Army and the Air Force cooperated. The weather didn't. The big plane and its succulent cargo spent a long morning rain delay at a military airport in Charleston, S.C.

As a result, the feast didn't start until the sun was sinking behind the hill overlooking St. George, the Grenadian capital. The guest list included not only the troops, but several dozen officials of the new Grenadian government. The international party went on into the night.

The serendipity of the Charleston delay was that we got to stay over for the night in Grenada. The Air Force wasn't using Salines after sundown.

The sightseeing tour for Suggs' party, which included the president of his company and several press types, took us on a helicopter flight over of the lush, 21-mile-long island and its offshore dependencies, and a bone-cracking jeep ride over the southern portion.

We saw some buildings pockmarked by small-arms fire. There was the shell of a prison smashed by gunship rounds and some warehouses where examples of Soviet-made weapons were on display.

And painted in large letters on the side of a roadside country store, "Thanks, 82nd Airborne Division."

The really adventurous part of the trip came just at the end, when the press escort, an intrepid colonel (who has since retired), helped everybody load cases of island beer and then directed the jeep driver to take off over a dirt road that seemed to aim directly at Salines.

Instead, the path sort of died out in a cattle pasture several hundred yards short of the end of the runway, which was nearly blocked by huge boulders from the construction work that would a year later turn it into an international airport. Another few minutes and we would have had even another night over.

By December, the American military presence on Grenada was down to a company-sized contingent of MPs and civil affairs soldiers. Headquarters for the command was a beachfront motel where the obligatory perimeter of razor-wire was mostly hidden in the vegetation.

The only notable sightseeing stop in the motel was the public telephone allegedly used in the early hours of the operation by an 82nd Airborne Division soldier to overcome the notorious dearth of inter-service communications that plagued the invasion.

If Grenada was back to its laid-back island ways in December 1983, it was nonetheless a place where a considerable piece of post-World War II history was made for the soldiers of Bragg and the airmen of Pope Air Force Base.

It was a hurry-up call.

Only 96 hours elapsed between the Friday evening when Gen. Ed Trobough, commander of the 82nd, was alerted that his forces could be part of the Caribbean operation and the Tuesday morning when soldiers of the 2nd Battalion of the 325th Parachute Infantry began unloading at Salines.

When the call came, Trobough recalled that, "I went to the world atlas to find out exactly where Grenada was. I was surprised to find out it wasn't in the Middle East."

The operation was billed as a "rescue mission" of American medical students but also wound up toppling a revolutionary government that had seized power.

Before it was over, about 5,000 troops were involved. Other than Marines and Rangers, nearly all were from Fort Bragg.

Two young men from the 82nd were among nine servicemen who died in combat. Capt. Michael D. Ritz and Staff Sgt. Gary Lynn Epps of the 2nd Battalion of the 325th were cut down by the same blast of fire from an ambush just as dawn was breaking on Oct. 26. The soldiers were scouting in preparation for their company to relieve Marine units near the airport.

Eight months later, another Fort Bragg soldier, Pvt. Sean Luketina of the 82nd Signal Battalion, died of wounds suffered when a Navy plane fired on his position.

Luketina was among at least 10 Americans who died of "friendly fire" or from accidents.

Although the major portion of the operation was over within five days, the Pentagon ordered Army troops to keep on coming. The two battalions of the 325th were joined within a week by others from the 508th, 504th and the 505th. In all, the forces included six battalions of infantry, an artillery battalion, and hundreds of service and medical troops from 1st Support Command.

As the veil of official secrecy was lifted, the American public saw on television and read about something new in military operations — scores of women soldiers, including military police and service troops, as well as the traditional Army nurses, serving in a fighting zone.

Even as some units went in, the traffic became two-way.

The 325th units flew back to Fort Bragg on Nov. 12. The tumultuous welcome stood in sharp contrast to the total secrecy of Oct. 25, when they loaded at Pope Air Force Base's Green Ramp. At the beginning, many of the troopers thought they were on their way to a training exercise in a mythical country known as Macapa. It was the high command's way of quelling rumors.

The longest stay of a major unit turned out to be that of the 2nd Battalion of the 505th Regiment, which landed on Oct. 27 and didn't come back until Dec. 12, missing the Barbecue Special by a few days.

By then, the euphoria of the first few days had worn off. There were complaints that U.S. troops were looting, of illegal arms smuggling, and of boredom among soldiers with nothing to do in the rural backcountry of the island.

Over the years, there would be other revelations of more serious problems with the Grenada operation.

There would also be many changes and improvements based on the lessons of Grenada, especially in communications and command.

In the longer sweep of history, Grenada was one of those unexpected missions that always strain the ability of military forces to react quickly and effectively.

In the history of airborne forces, the hurry-up Grenada mission was like the experience of the 82nd Division 40 years earlier in September 1943, when division units boarded their transports for the invasion of Italy even as commanders briefed the pilots at hasty wing-side conferences.

When all was said and done, the Grenada experience of the airborne soldiers and their Air Force transporters was what is so often the usual in military operations: a mixed performance of skill, bravery, confusion and uncertainty. That is always the very nature of war.

Comparing Buildups In Vietnam And Persian Gulf

Originally published January 18, 1991

A guiding tenet of U.S. war strategy in the Persian Gulf is said to be to "hit hard and with everything we've got."

According to some, this tenet is in sharp contrast to the U.S. experience in the last big war, the Vietnam War, wherein, according to one anchorman, the buildup of the American war machine was a "piecemeal strategy."

The word is that today's military and political leaders are not about to repeat that mistake.

It is said that, heeding the "lesson of Vietnam," from the moment of the Iraqi seizure of Kuwait, Gen. Colin Powell, chairman of the Joint Chiefs of Staff, who was briefly a combat commander in Vietnam, recommended the big Persian Gulf buildup. President G.H. Bush, who was a politician during Vietnam, adopted it.

Thus, five months later as the war begins, there are 330,000 American soldiers and Marines on the ground in Saudi Arabia, with another 135,000 on the way, as well as 260 Air Force combat planes and 60 ships at sea.

That is, indeed, "the largest buildup since the Vietnam War."

As this war begins, how does its buildup compare, in time and space, to that in Southeast Asia? And the related question: Was Vietnam's ground buildup really piecemeal?

The first U.S. ground combat forces arrived in Southeast Asia on March 8, 1965, when two Marine battalions came ashore at Danang.

In mid-July 1965, the new American commander, Gen. William Westmoreland, concluded his plan for winning the war and asked President Lyndon Johnson for 44 more battalions to carry out the plan.

Within days, Johnson approved the request.

By October, there were enough forces to fight the first conventional ground action, offensive operations in the Ia Drang Valley.

By December 1965, five months after the request, there were almost 200,000 ground troops in Vietnam.

Given the constraints of logistics and the fact that the bulk of American military power was aimed at the Soviet Union, the Southeast Asia buildup was both massive and swift.

The buildup continued in 1966, with the total force doubling to 400,000. The force grew to a half million in 1967. At the time of the bombing halt in March 1968, the number had grown to almost 550,000.

By late that year, two understrength brigades of the 82nd Airborne Division at Fort Bragg were the only combat units left in the continental United States.

The Vietnam buildup depended on the vast logistical facilities that modern military activities require, such as the port at Camranh Bay.

Construction on that supply and support base began in May 1965, only weeks before General Westmoreland put in his request for ground force. It eventually supported a fourth of the entire force in Vietnam.

By contrast, in the Persian Gulf the United States has been building logistical and even tactical installations in Saudi Arabia for more than a decade, many of them specifically tailored for use by such units as the 82nd Airborne Division.

From another perspective, the total force buildup comparisons between Vietnam and the Persian Gulf tell only part of story.

In the Vietnam era, the United States enlisted 8.7 million volunteers and drafted 2.2 million men, trained them, and sent most of them abroad.

To fill the ranks this time, President Bush has authorized the mobilization of up to 200,000 reservists and Guardsmen. During Vietnam, approximately 60,000 men were eventually culled from the badly-prepared reserve units of that day.

That is the buildup comparison, and you can see that the Vietnam buildup was no more "piecemeal" than the one in the Persian Gulf.

In fact, both Vietnam and the Persian Gulf deployments are examples of one of the oldest and most enduring of U.S. military doctrines. That is, to bring together "everything you've got" before committing to battle.

It is a doctrine as old as the mobilization of the huge Army of the Potomac under General George B. McClellan at the opening of the Civil War.

But then there is the other half of the tenet, which is to "hit hard" with all the force.

In Vietnam, the Army forgot the lesson of that half of the tenet.

Unlike the buildup, the war in Southeast Asia was indeed fought "piecemeal."

At peak strength, the U.S. had the equivalent of nine divisions in Southeast Asia. Yet, at no time did more than three divisions engage in action together and at the same time. Most offensives involved only a single division.

As retired Green Beret Col. Harry Summers, a noted military analyst, has so well concluded, the real "lesson of Vietnam" is that U.S. battle strategy was wrong there.

Based on such concepts as "search and destroy" and the "fire base," the strategy never brought maximum forces to bear in offensive action.

The Vietnam war-fighting commanders forgot the most important part of the tenet. It is the one that U.S. Grant both invented and typified when he slugged south against

General Lee's Confederates in the spring of 1864, using every man he could lay his hands on, and vowing to "fight it out on this line if it takes all summer."

As the first phase of the Persian Gulf War opens, involving the standard massive air operations that were also part of the Southeast Asia strategy, it is too early to say whether or not this war's strategists will indeed be true to the full doctrine of General Grant.

On The Little Bighorn And In The Gulf

Originally published February 13, 1991

As if the footage of bombs tumbling from B-52 bellies and of endless columns of tanks moving through the sand wasn't enough, television this month also gave us the latest version of Gen. George Armstrong Custer's disaster on the Little Bighorn.

It was, of course, one of the few times when U.S. arms failed. Chief Crazy Horse and his huge army of Sioux and Cheyenne braves were able to trap 212 men of the 7th Cavalry and kill them to a man.

This latest version differs from earlier tellings in giving much more credit to the skill, bravery and sheer desperation of the Indians.

But the main cause of the white man's disaster (or the Indians' triumph) remains with Custer himself, with the fatal hubris that throughout his life gripped his heart and led him to insanely foolish actions.

Incredibly brave and skillful though they were, the Plains Indians could not reverse the iron rule that says victory always goes to the big battalions, or at least to those battalions armed with the most sanguinary weapons.

If Custer had waited for John Gibbon's column of infantry and howitzers to come up, the Little Bighorn would not be the monument to Indian bravery, and to white man's foolishness that it is. Six months later, as a matter of fact, General Nelson Miles' howitzers easily brought Crazy Horse to bay.

Which brings us to the contrast — and the similarities — between the Little Bighorn and the Persian Gulf.

To set the stage, it should be said that our enduring fascination with the 7th Cavalry's disaster on the Little Bighorn grows out of its very rarity in the history of a country that — if you look at the record — would rather go to war that eat sweet corn in summer.

Moreover, we are good at it. We seldom lose. Furthermore, we usually only fight when we have an overwhelming abundance of the most advanced weaponry. And finally, we usually are able to use this technological overkill on a foe who is just the opposite — under-armed, overmatched, depending largely on bravery or grit, neither of which weigh very much on the awful scales of war. And then there is the occasional hubris (from the Greek: "overbearing pride, arrogance, presumption") that does get in the way of victory or invite avoidable disaster.

In the Persian Gulf, the latest in the almost endless succession of wars that the United States has provoked or joined, there will be no Little Bighorn.

We have the best-trained armed forces ever. Highly skilled leadership. Five months to get ready before the shooting started. Equipped with an armory of the most dazzlingly deadly weapons ever put in the hands of fighting people. The foe is totally outmatched technologically, doing war as it were with an abacus against an enemy armed with a computer.

Moreover, while there is undoubtedly hubris at work here, it is carefully controlled, tenaciously underplayed. Few successful great captains have been so low-key, even humble, as Gen. Norman Schwarzkopf ("the middle of my name is 'war' ") as he unfolds the latest triumphant news of a three-week air war that has so far cost his side fewer casualties than the traffic on the way to a Carolina beach weekend.

In a word, Crazy Horse and his arrows and carbines had no chance against John Gibbon and Nelson Miles and their howitzers, and neither does Saddam Hussein against Norman Schwarzkopf and his bomb bays jammed with smart bombs.

Except in rare times when hubris got in the way, such as at the Little Bighorn, or 90 years later in Vietnam, that generally has been the story for a country that has engaged in bellicosity throughout most years of its history.

Wait a minute, you say. We're a peace-loving people. We don't fight unless provoked.

Horsefeathers, as a 7th Cavalry horse soldier would say. We like war. We do it more often than most other countries. We initiate as many as we back in to.

Pick a decade. Any decade. A good bet that the United States was either fighting its revolution; slaughtering Indians; chasing pirates; beating up on Mexicans, Filipinos, Spaniards, Nicaraguans, Koreans or Vietnamese; fighting each other in the bloodiest rebellion in history; or engaged in world wars against hopelessly outmatched countries that usually had been our allies and would be again.

You might say that when push comes to shove the United States would rather fight than talk, fight than negotiate, fight than compromise. That is certainly one reading of what is happening in the Persian Gulf.

This bellicosity is unmatched in any other current nation on the globe. Pick one. Germans? Since 1860, they fought three wars. In the same period, let's see, the United States soaked its own earth with the blood of the War of the Rebellion; soaked the west with the blood of Plains Indians for 20 years; declared and fought the Spanish-American War; provoked and fought the Philippine Insurrection, inflicting 116,000 deaths, of whom 100,000 died of famine as a result of U.S. scorced earth tactics; sent Marines to Nicaragua, Cuba, Haiti, the Dominican Republic, and Veracruz, Mexico; mobilized an army of 150,000 troops to chase Pancho Villa's 500-man outlaw band; finally wound up fighting in World War I and World War II, Korea, Vietnam, Panama, and even found somebody to fight on the peaecful Spice Island, Grenada!

Don't let the politicians kid you. Americans do not hate war. They revel in it. A perceptive historian, Goeffrey Perret, has written a book entitled "A Country Made By War" (Random House). He makes a good case for his title.

What Americans do not like about war is to lose at it.

And that is why what happened that June day in 1876 on the Little Bighorn continues to grip our imagination, even in the midst of a real war in the Persian Gulf.

Stormin' Norman Stood On These Shoulders

Originally published March 15, 1991

No one can take anything away from the big captain of the Persian Gulf victory.

Four-star Gen. Norman Schwarzkopf, the "Stormin' Norman" who commanded Desert Shield and Desert Storm, had the command and he executed the mission. But the seeds of victory in the Persian Gulf were planted many years before 1988, the year

Schwarzkopf hung his flag in the headquarters of the U.S. Central Command.

Many were planted in North Carolina, at Fort Bragg and at Camp Lejeune.

The history of success in the Persian Gulf War started a decade ago.

In his victory, Stormin' Norman stood on the shoulders of such soldiers as four-star Gen. Robert Kingston, now retired, who headed the original Rapid Deployment Joint Task Force, predecessor of Central Command, and made the first plans for sending U.S. forces — land, sea and air — to fight in the Middle East.

The RDJTF (later simply RDF) was formed in 1980, its mission largely defined by turmoil in the Middle East, where the United States' major client, the Shah of Iran, had been deposed and U.S. hostages seized in Tehran. Its mission in those days was to respond to a Soviet invasion of Iran.

In 1981, about 4,000 soldiers and airmen designated for the task force got on-the-spot experience in a training exercise in the Egyptian desert known as Bright Star.

In January 1983, the task force was upgraded to Central Command, with headquarters at MacDill Air Force Base, Fla.

Kingston, who had been a junior-officer hero of the Korean War and was head of Special Forces at Fort Bragg, organized the task force and was first head of Central Command.

Major forces on call by Central Command were the paratroopers of the 18th Airborne Corps, the 2nd Marine Division at Camp Lejeune and the Special Forces.

In those days, the tank divisions that Gen. Schwarzkopf was able to send rolling across the Iraqi desert this year were nailed down in Europe as part of the NATO command.

By Christmas of 1981, Gen. Kingston was quoted as saying that he could put an airborne brigade in the Middle East within 48 hours.

By the winter of 1982, 25,000 soldiers, Marines and airmen assigned to the Rapid Deployment Force carried out their first large-scale training together, in the Mojave Desert in California. Later, such training became routine.

Among the Fort Bragg units training in "Gallant Eagle 82" was the 325th Airborne Infantry Regiment.

Nine years later, in August 1990, the 325th turned up as the first U.S. unit on the ground in Saudi Arabia.

In the 100-hour ground campaign against Iraq, the regiment had the distinction of closing the circle of Stormin' Norman's left-wheel drive that sealed off the shattered Iraqi army.

Many of its paratroopers were among the first to come home last week.

During the 1980s, much training and planning at Fort Bragg had a Middle East focus. And during those years, small teams of Bragg-based soldiers, mainly from Special Forces but also from the 82nd, were to be found in Middle East countries.

They trained troops, planned logistical systems, advised allied commanders and generally became the Army's principal experts on what would be required if called on to fight in the region.

Key figures in those years were such commanders of the airborne as generals Jack Mackmull, James J. Lindsay and Leroy Suddath, all now retired, and active-duty generals Carl Stiner, Gary Luck and James H. Johnson Jr.

Lt. Gen. Luck, commanding general of the 18th Airborne Corps, commands all paratroopers in the Persian Gulf.

Maj. Gen. Johnson, three years behind Schwarzkopf at the U.S. Military Academy, commands the 82nd Airborne Division, a unit he has been associated with in one capacity or another since his first assignment 30 years ago.

Stiner now wears the four stars of a full general as commander of Special Operations Command (SOCOM), the overall joint command for Green Berets, Rangers, Navy and Air Force commandos, and other behind-the-scenes forces, a command also based at MacDill.

Ironclad military secrecy precludes any detailed reporting of the accomplishments of these forces in the Persian Gulf War. But they apparently have been so impressive that even Stormin' Norman felt constrained to personally praise them when he gave his famous on-camera briefing describing the 100-hour ground war.

Stiner, who commanded the 82nd and the 18th Corps before taking SOCOM, is a leading military expert on the Middle East.

In the early 1980s, he headed the U.S. mission that trained the army of Saudi Arabia and helped plan logistical facilities that made it possible for allied forces to put half a million troops into that country within seven months.

At the time, Stormin' Norman was in Hawaii.

Lindsay, who preceded Stiner as division, corps and SOCOM commander, participated in or commanded the planning and training of the airborne and Special Forces troops for practically the entire period since Kingston commanded RDJTF.

Mackmull, who commanded the division, the corps Special Forces before retiring in 1984, was a decorated helicopter pilot in Vietnam and former West Point football star. He played a major role in developing Army helicopter forces, which in the Persian Gulf War proved so spectacularly successful logistically and as tank-killers.

Suddath, who was Stormin' Norman Schwarzkopf's roommate at West Point, headed the Army's Special Operations component, formerly known as 1st Special Operations Command, at Fort Bragg as that force beefed up for its worldwide mission.

Another who saw the Middle East first as an airborne planner is three-star Gen. Calvin A.H. Waller, Gen. Schwarzkopf's deputy.

Waller was chief of staff of 18th Airborne Corps and Fort Bragg in 1983-84, and later assistant division commander of the 82nd.

So it happened that when the whistle blew for war in the Persian Gulf, Gen. Norman Schwarzkopf got to play quarterback. But many of the plays and much of the preparation for the mission resulted from the pioneering contributions of these commanders and many others wearing airborne and Special Forces patches.

As the triumphant battle captain of Desert Shield and Desert Storm, Stormin' Norman was standing on these shoulders.

How Will History Judge The Second War In Iraq?

Originally published May 8, 2003

How will the second Iraqi war rate in the long sweep of U.S. military history?

To those in the front echelons, war is always dangerous, unpredictable and nerve-wracking. For the brave young men and women up front, this one was no exception.

But the thing was so one-sided that it is hard to find a parallel.

While the country mourns several dozen young men and women killed in action or accidents and hundreds wounded, the ouster of the Iraqi despot does not ring with the sounds of desperate, sacrificial combat like many of the passages of arms that began on Lexington green in 1775 and carried on through a half-dozen giant conflicts and dozens of small wars.

If a major standard of a war is the quality of the opposition to overcome, even the Native Americans, especially the Native Americans, put up a much more formidable fight than Saddam's supposed minions.

There were few minions.

Maybe we will find "weapons of mass destruction."

But we certainly did not find weapons (or people) of conventional destruction in any great measure.

In fact, it is hard to find in the public war record so far any battles that involved more than a few companies in short, quick, one-sided firefights.

The enemy came at us mostly in guerrilla bands, in wild quick orgies of suicidal assault usually ending with the sky blackened by smoke of blasted Iraqi vehicles and the stench of burned flesh. The vehicles were as likely to be pickup trucks as tanks or mobile artillery.

Many U.S. soldiers were quoted on the "embedded" journalist shows marveling at how much it was like cutting butter with a hot knife, a function of the differences in firepower and training.

If our intelligence was to be believed, or at least if the media briefers were to be believed, Iraq's military, the vaunted Republican Guards, would be a formidable foe, fighting from behind a mysterious "red line" that perhaps would bristle with gas-attack weaponry, or engaging in brutal, one-corner-at-a-time street fighting like the Russians and Germans at Stalingrad in World War II.

Instead, the Iraqi army, if there ever was one, largely melted away.

Soldiers threw off uniforms. Officers melded back into the middle-class population. Tons and tons of small arms, bombs and shells, bullets and grenades, were untouched in their depots and caches. No planes flew to combat.

In history, it will turn out to be the first demonstration in the 21st century of the huge quantum leap in war-fighting technology since the first Iraqi war 13 years ago.

Everything in war is "smart" these days, and the computer chip reigns supreme. For example, a single U.S. soldier wears more smart technology and communications gear than a whole regiment only a few years back.

The fruits of this transformation are expressed in instant communications, precision munitions and tremendous flexibility in command decision-making. It all works with awesome impact, at least when the enemy is as inept and unprepared as Saddam's crew.

Within hours, the overwhelming arsenal of U.S. "smart" technology, delivered by land, air and sea, put Iraqi forces back into the dark ages, so to speak, if they ever were really out of them.

In the sweep of military history, what will be the lessons of the war? Mostly lessons we already knew.

The tank has reigned supreme in the broad terrain of the Middle East almost since the end of World War I. Some of the first "armored cars" were used there during that war 90 years ago.

And the same is true today.

In modern history, the IDF (Israel Defense Force, their euphemism for armed services) demonstrated the tank's supreme role in war after war with its Arab attackers. If the United States is to be the new military presence in the Middle East, the tankers will have to stay.

The other ironclad win-the-war tools of Middle East warfare, sophisticated air supremacy and disruption of command and control, are IDF specialties of long standing, indeed of warfare at least since World War II, and the U.S. performance confirms it, in 21st-century spades of course.

In fact, a major U.S. distinctive in this year's event is also as old as Gen. Ulysses S. Grant in Virginia in 1864.

That is, pile it on, in overwhelming numbers, of men, material and morale. You'll whip them every time.

The difference between Grant's principle of the big battalions (he got it from Napoleon) and today is that today's weaponry allows for a "redundancy" that, unlike in Grant's day, keeps down human casualties, expending rounds of munitions by the millions rather than the bodies of human beings by the thousands.

That is, our side has redundancy. The other side is dead. In big numbers.

Finally, what about "our crowd," our Fort Bragg folks, the airborne and special operations?

Special operations performed its secretive missions in usual secretive fashion, and we will know only gradually how much they contributed. Given the chaos in the Iraqi regime, it must have been considerable, perhaps even decisive.

Special operations is here to stay, even in an era of high tech.

So is psychological operations, waged with all the tools of modern mass communications and public relations. It became a major arm of the war plan, with propaganda aimed not only at the enemy, but at other countries and even at our own population.

And what about the role of the airborne soldier in this chapter of our military history?

In the first Iraqi war, the airborne divisions were misused, relegated to a vehicle march across the far western flank of the allied forces (accompanied by the French).

This time, once again, history will record a misuse.

The paratroopers, arguably the best fighters in the Army, served essentially as supply-line guards. The 101st Airborne Division (Air Assault) mixed its deadly helicopter weaponry in with the other stuff, more of the redundancy.

The 325th Airborne Infantry Regiment and the 101st might have seized Baghdad airport at least two, perhaps four, days earlier than the tanks eventually got there. The whole Baghdad end-game debacle might have been ameliorated at least somewhat.

If the politicians, planners and the big commanders didn't give us many unique lessons in military history, as usual the soldiers themselves performed magnificently.

The U.S. Army of today is the best-trained and motivated, the best led at its fighting echelons, of any fighting force in the nation's history.

The valor, skill and intelligence of the fighting men and women is awesome.

And as the hard part of this war begins, that is, the reconstruction of the country that by their valor they "own," they are showing again the immense personal compassion and good will that are also unique qualities of the men and women of the armed services.

Current Events Are Emerging As History

Originally published February 1, 2007

When does history emerge from current events?

Daily news headlines from the Middle East are current events. But the story of the U.S. Army in the Middle East is stretching into historical years.

The U.S. Civil War was over in four years, start to finish. World War I, the "war to end all wars," lasted a little more than four years. The U.S. Army was involved for a year and a half.

World War II lasted six years. The U.S. was involved for three years. In Vietnam, U.S. forces were involved for about 10 years, depending on how you define involvement.

Afghanistan is rivaling the timelines of these familiar lodestars of military history.

That came home to me this week when Gen. Dan K. McNeill, who grew up in Duplin County, left to become the top military man in Afghanistan. He will be commander of the International Security Assistance Force, the command inspired by NATO and the United Nations.

McNeill went to Afghanistan for the first time in the summer of 2002 as a three-star general.

By then, six months after the first Army Special Forces soldiers went into action, U.S. fighting men and women and Afghan allies had routed the controlling Taliban out of the major population centers and set out on a hunt for Osama bin Laden.

Those early months of the operation in Afghanistan, known as "Enduring Freedom," have become legendary chapters in the history of the U.S. Army.

In a few weeks in late 2001, a handful of men — mostly of the 5th Special Forces Group — linked with Afghan rebels known as the Northern Alliance to press the Taliban into wholesale retreat into the mountains along the Pakistan border. They set the stage for the installation of the government that since has been formally elected and installed in the old capital of Kabul.

That was four years ago.

McNeill stayed for a year and then came out of Afghanistan to become deputy commander, and later commander, of the U.S. Army Forces Command. The command oversees practically all the Army's stateside tactical units that do not belong to U.S. Army Special Operations Command.

When McNeill took the Afghanistan post in 2002, the Army presence there was fewer than 2,000 soldiers, mostly from 5th Special Forces Group, 75th Ranger Regiment and the 10th Mountain Division.

Since then, more than 8,000 men and women of the 82nd Airborne Division have shuttled in and out — and in again — on the Afghanistan battle list. A brigade is on its way there.

When he left in spring 2003, McNeill turned over the command to Lt. Gen. John R. Vines and, at the same time, gave up command of the 18th Airborne Corps and Fort Bragg.

That seems ages ago, and it has been. Three and a half years have gone by. Vines came back to Fort Bragg last year and retired in December.

When Vines took over in Afghanistan in 2003, there were about 8,500 U.S. military personnel there and several thousand from other countries.

The Afghanistan that McNeill returns to this year still has its ancient history and its stark landscape. It also has nightclubs in Kabul, thousands of Westerners working for security companies with contracts from government and private clients, traffic jams and the world's largest opium crop.

It also has a growing Taliban and al-Qaida problem.

U.S. intelligence predicts a resurgent Taliban and its al-Qaida allies will launch major military offensives when spring softens the landscape along the border with Pakistan.

The new Taliban is judged to be a better armed and more professional military force than the old.

Al-Qaida still uses old tactics, such as suicide bombers. In the week before McNeill arrived, a suicide bomber killed 10 workers at the airport at Khost, the main U.S. command center for the war along the border.

But al-Qaida is also technologically sophisticated. It produces full-length documentary propaganda films featuring a 28-year-old American who is the first U.S. citizen to be charged with treason in 50 years.

New experience

For McNeill, the former N.C. State University ROTC student, it is a daunting new experience in military command at the highest international level.

He leads an array of troops that is of historic proportions in the annals of coalition warfare.

The International Security Assistance Force has units from 36 countries, give or take one or two as the list changes. They range from big outfits, such as the British, French, German and Canadian commands, down to platoons and companies from Estonia and the Czech Republic.

The force that McNeill heads counts more than 32,000 soldiers. Of the total security force, the U.S. Army contributes 12,000, while another 14,000 U.S. soldiers are under separate commands dedicated to hunting al-Qaida.

As he arrives, McNeill will greet familiar folks traveling a familiar schedule.

The headquarters of the 82nd Airborne and its 4th Brigade Combat Team are deploying from Fort Bragg for a stint in the combat force.

The role of McNeill's command is a primer on nation-building warfare in the 21st century.

It emphasizes combat and training of military forces, but it also emphasizes building government infrastructure throughout the Texas-sized country by dotting the landscape with dozens of mixed civilian-military groups known as Provincial Reconstruction Teams.

To prepare for this textbook mission in counterinsurgency and joint cooperation, late last year thousands of Fort Bragg soldiers joined with personnel from other services in an intense hands-on training exercise.

In the months ahead, the performance of Dan McNeill and the men and women of the 82nd Airborne will no doubt produce headlines, adding to the calendar of the year's current events.

And in the long view, they will also be making more history in a military enterprise with a timeline rivaling the wars of the generations before them.

U.S., Iran Relations Begin New Chapter

Originally published February 22, 2007

Four years ago this month, the 4th Psychological Operations Group at Fort Bragg advertised for a midlevel civilian expert, salary range $45,000 to $55,000, as a "research specialist on Iran and Central Asia."

The new hire would join an eight-member team of military intelligence experts who would gather information, write reports and present PowerPoint briefings on their area of study.

Just another chapter in the long history of Fort Bragg's links to the Middle East, and especially to Iran.

Because of the long-standing links, it seems I write a variation of this report every year or so, depending on the current temperature level in U.S. relations with Iran. The relations blow hot and cool. Right now, it is heated up again, with evidence that Iranian roadside munitions are being used by some factions in Iraq.

That means the story of Fort Bragg and Iran is in a new chapter because some of those roadside munitions are targeting troops of the 82nd Airborne Division who are the spearpoint of the last-ditch mission attempting to suppress sectarian violence in the streets and slums of Baghdad.

And of course this is playing out on the larger scene as the U.S. administration continues its not-so-subtle effort to scare the Iranians into stopping their program to eventually build a nuclear bomb.

The 82nd soldiers in the Baghdad streets are reading the headlines along with the rest of us about huge aircraft carrier battle groups converging on the Persian Gulf while the sword-rattling in Washington grows ever louder.

Desert One

The climatic chapter in the history of Fort Bragg and Iran is nearly 27 years old.

Today's Army special operations forces have come a long way from that April in 1980 when Operation Eagle Claw came to disaster at a refueling site known as Desert One in the Iranian desert with eight battle deaths — five Air Force men and three Marines.

Their doomed mission was an audacious attempt to rescue Americans held hostage in Tehran, the capital of Iran. Special Forces soldiers were at the heart of the team.

In those days, special operations was a pesky stepchild of the military.

Even then, its training was superb and its techniques were mostly sound.

Nonetheless, there was doubt that such forces would ever be more than a minor and exotic weapon in the military arsenal.

The disaster in the desert changed all that.

In the 27 years since Desert One, the country learned fast that special operations would be a key to warfare in the post-Cold War world.

By 2002, special operations soldiers essentially were winning their own first war when Fort Bragg troops teamed with insurgents to overthrow the Taliban government in Afghanistan.

Twenty-seven years after Desert One, special operations is a muscular big boy in the Army family. Its command center at Fort Bragg swarms with experts on every corner of the globe, including the Middle East and especially Iran.

If the White House decides to go to war with Iran, the work of the experts of the 4th Psychological Operations Group will undergird the war plans.

It won't be the first time that Fort Bragg soldiers would be aimed at Iran.

At one point when the Iranian revolution of the late 1970s was thought to threaten the whole region, at least one war-plan scenario called for the entire 82nd Airborne Division to take part in a full-fledged invasion of Iran, with the paratroopers coming in from the east.

From then until now, Fort Bragg planners and operations officers have accumulated one of the Army's largest stores of knowledge of the Middle East.

Actually the Bragg connection goes back even farther.

I was there on a bright April day in 1962 when the Shah of Iran came winging in to see the troops at Fort Bragg.

Decked out in a resplendent white-and-gold military uniform, the Iranian head of state was guest of President John F. Kennedy, who wore his usual dark suit with no hat.

The two youngish leaders, both wearing fashionable dark sunglasses, observed a full display of everything the airborne could show them.

Then they went down to Camp Lejeune to observe the Marines do their stuff, complete with an amphibious landing on Onslow Beach.

I was a shirttail-dragging reporter who was mainly interested in politics, and as far as I was concerned the military stuff was just so much window-dressing for the intense politicking that Kennedy was laying on the Shah.

Like every U.S. president since Harry Truman, JFK lavished attention on Iran, keenly aware of its oil riches and in those Cold War days of its strategic position on the borders of the Soviet Union.

All that wooing went for naught when the ancient religious impulses of the Iranian people erupted in revolution against the Shah and his Western ways.

For no other reason than for the sake of the troops, let us hope history doesn't come another circle with the U.S. trying again to have its way over Iran with another misbegotten military adventure in the deserts.

Army Is Making Another Fashion Statement

Originally published March 22, 2007

The Army apparently is about to enter yet another era in its long history of uniform changes.

If clothes make the man or the woman, the Army has never been negligent in trying to keep up with the latest fashions.

And, according to the latest news from the fashion front, green is out and blue is in.

New regulations are in the works to make the blue uniform required dress for all occasions except, of course, for most working hours when the new and already controversial Velcro-trimmed Army Combat Uniform is worn.

The plan is to simplify the wardrobe, which now includes, in addition to the ACU, the "Army Green" Class A uniform, which is a jacket, tie and trousers (airborne wears them stuffed in the boots); as well as a blue dress uniform, and also a white uniform and a "mess jacket" blue uniform for social occasions.

Apparently the new wardrobe would include only the dress blue with appropriate neckties (bow tie for formal). For headgear, the "service cap," the pancake-topped visored headgear, would be standard.

Actually, the blue is familiar if you observe television clips of ceremonies at Arlington National Cemetery or at the White House.

The troops of these scenes, mostly soldiers of "The Old Guard" (3rd U.S. Infantry Regiment), wear dress blues as their everyday attire, with white gloves when handling their shoulder weapons.

The new fashion regulations provoke distressed outcries from those who have come to love the Army Green Class A uniform, especially with its opportunities to wear various unit patches on both shoulders (they now show up on the Velcro fields of the ACU).

Despite the grousing, as in every era of the history of Army uniforms, the change will gradually become tradition.

Army Green, as a matter of history, is really not that old, dating from the middle of the 20th century.

World War II, as you recall, was fought in brown or khaki, and World War I in a sort of mixture of brown and khaki (hence the "doughboys").

Often there was a mixture.

A picture of variety

I have a grand panoramic photo of the entire Fort Bragg garrison formed — and in homage — on the occasion of the death of President William G. Harding in 1923.

A striking thing about the soldiers is the variety of dress.

In the band, for instance, some wear light leggings, some dark, some wear cavalry boots. Some have on dark winter outfits, others light-colored summer stuff. Most of the officers and ranks wear the service cap. But a crowd along the back of the formation sport their jaunty campaign hats, the broad-brimmed chateau preferred by such luminaries as Gen. John J. Pershing and that remained a favorite of troops right up to World War II.

Blue has been the most enduring Army color.

Before, during and after the Civil War it was the usual color.

The venerable Fayetteville Independent Light Infantry militia company wore blue until the Civil War but afterward stuck to a sort of gray, reminiscent of the army of the Confederacy.

The story is that when the Spanish-American War came along and the FILI was called to federal service, the unit first stubbornly refused to give up gray for regular blue.

In a defiant gesture, it marched from the train station to the governor's mansion in Raleigh still in gray, but then ceremoniously packed gray away and marched to the regimental bivouac, where it was sworn in U.S. service and then donned the blue.

Washington's preference

None other than George Washington, first commander in chief, preferred blue, and we are familiar with the many portraits of him in the blue frock coat accompanied by the buff of his britches and waistcoat.

Washington wore blue first as a colonel of the Virginia regiment of his colonial home state. In those days before the War of Independence, Washington yearned to wear the red coat of the British army.

But by the time of Lexington and Concord and Bunker Hill, he was joining his ragtag-and-bobtail Patriot soldiers in deriding the British as "lobsters."

Washington was a stickler for uniforms and military paraphernalia.

Even in the darkest moments of the Revolutionary War, his headquarters were issuing detailed regulations about proper dress.

The official uniforming of the Patriot army dates from Oct. 2, 1779, when a general order was published pursuant to a resolution of the Continental Congress "authorizing and directing the Commander-in-Chief, according to circumstances and of supplies of clothing, to fix and prescribe the uniform as well as regard to color and facings as the cut of the fashion of the clothes to be worn by the troops of the respective states and regiments, woolen overalls for winter and linen for summer to be substituted for the breeches."

The order prescribed blue as the color of the coat, but each state and branch of service had a different color for facings.

The Marquis de Lafayette supplied his own officers with a coat costing four guineas, "including the small sword, which he presented to them."

These regulations were all very well. But likely as not, the Continental soldier wore clothing of less exalted description.

Baron Friedrich Wilhelm von Steuben, the noted inspector general of Washington's main army, described the fashion scene at the winter camp at Valley Forge, Pa., where he "saw officers of the grand parade mounting guard in a sort of dressing gown, made of an old blanket or woolen bedcover."

The officers, he said, had coats "of every color and make," and the soldiers were "literally naked."

Steuben's report was seconded by the inspector of the Rhode Island regiment, who "reported that the naked situation of the troops when observed parading for duty, is sufficient to extort the tears of compassion from every human being. There was not two in five who have a shoe or stocking or so much as breeches to render them decent."

Nonetheless, this near-naked crowd went out and eventually whipped the redcoats, proving perhaps that clothes aren't the vital measure of a soldier.